representations
of algebras

PURE AND APPLIED MATHEMATICS

A Program of Monographs, Textbooks, and Lecture Notes

EXECUTIVE EDITORS

Earl J. Taft
Rutgers University
New Brunswick, New Jersey

Zuhair Nashed
University of Delaware
Newark, Delaware

EDITORIAL BOARD

M. S. Baouendi
University of California,
San Diego

Jane Cronin
Rutgers University

Jack K. Hale
Georgia Institute of Technology

S. Kobayashi
University of California,
Berkeley

Marvin Marcus
University of California,
Santa Barbara

W. S. Massey
Yale University

Anil Nerode
Cornell University

Donald Passman
University of Wisconsin,
Madison

Fred S. Roberts
Rutgers University

David L. Russell
Virginia Polytechnic Institute
and State University

Walter Schempp
Universität Siegen

Mark Teply
University of Wisconsin,
Milwaukee

LECTURE NOTES IN PURE AND APPLIED MATHEMATICS

1. N. Jacobson, Exceptional Lie Algebras
2. L.-Å. Lindahl and F. Poulsen, Thin Sets in Harmonic Analysis
3. I. Satake, Classification Theory of Semi-Simple Algebraic Groups
4. F. Hirzebruch et al., Differentiable Manifolds and Quadratic Forms
5. I. Chavel, Riemannian Symmetric Spaces of Rank One
6. R. B. Burckel, Characterization of C(X) Among Its Subalgebras
7. B. R. McDonald et al., Ring Theory
8. Y.-T. Siu, Techniques of Extension on Analytic Objects
9. S. R. Caradus et al., Calkin Algebras and Algebras of Operators on Banach Spaces
10. E. O. Roxin et al., Differential Games and Control Theory
11. M. Orzech and C. Small, The Brauer Group of Commutative Rings
12. S. Thomier, Topology and Its Applications
13. J. M. Lopez and K. A. Ross, Sidon Sets
14. W. W. Comfort and S. Negrepontis, Continuous Pseudometrics
15. K. McKennon and J. M. Robertson, Locally Convex Spaces
16. M. Carmeli and S. Malin, Representations of the Rotation and Lorentz Groups
17. G. B. Seligman, Rational Methods in Lie Algebras
18. D. G. de Figueiredo, Functional Analysis
19. L. Cesari et al., Nonlinear Functional Analysis and Differential Equations
20. J. J. Schäffer, Geometry of Spheres in Normed Spaces
21. K. Yano and M. Kon, Anti-Invariant Submanifolds
22. W. V. Vasconcelos, The Rings of Dimension Two
23. R. E. Chandler, Hausdorff Compactifications
24. S. P. Franklin and B. V. S. Thomas, Topology
25. S. K. Jain, Ring Theory
26. B. R. McDonald and R. A. Morris, Ring Theory II
27. R. B. Mura and A. Rhemtulla, Orderable Groups
28. J. R. Graef, Stability of Dynamical Systems
29. H.-C. Wang, Homogeneous Branch Algebras
30. E. O. Roxin et al., Differential Games and Control Theory II
31. R. D. Porter, Introduction to Fibre Bundles
32. M. Altman, Contractors and Contractor Directions Theory and Applications
33. J. S. Golan, Decomposition and Dimension in Module Categories
34. G. Fairweather, Finite Element Galerkin Methods for Differential Equations
35. J. D. Sally, Numbers of Generators of Ideals in Local Rings
36. S. S. Miller, Complex Analysis
37. R. Gordon, Representation Theory of Algebras
38. M. Goto and F. D. Grosshans, Semisimple Lie Algebras
39. A. I. Arruda et al., Mathematical Logic
40. F. Van Oystaeyen, Ring Theory
41. F. Van Oystaeyen and A. Verschoren, Reflectors and Localization
42. M. Satyanarayana, Positively Ordered Semigroups
43. D. L Russell, Mathematics of Finite-Dimensional Control Systems
44. P.-T. Liu and E. Roxin, Differential Games and Control Theory III
45. A. Geramita and J. Seberry, Orthogonal Designs
46. J. Cigler, V. Losert, and P. Michor, Banach Modules and Functors on Categories of Banach Spaces
47. P.-T. Liu and J. G. Sutinen, Control Theory in Mathematical Economics
48. C. Byrnes, Partial Differential Equations and Geometry
49. G. Klambauer, Problems and Propositions in Analysis
50. J. Knopfmacher, Analytic Arithmetic of Algebraic Function Fields
51. F. Van Oystaeyen, Ring Theory
52. B. Kadem, Binary Time Series
53. J. Barros-Neto and R. A. Artino, Hypoelliptic Boundary-Value Problems
54. R. L. Sternberg et al., Nonlinear Partial Differential Equations in Engineering and Applied Science
55. B. R. McDonald, Ring Theory and Algebra III
56. J. S. Golan, Structure Sheaves Over a Noncommutative Ring
57. T. V. Narayana et al., Combinatorics, Representation Theory and Statistical Methods in Groups
58. T. A. Burton, Modeling and Differential Equations in Biology
59. K. H. Kim and F. W. Roush, Introduction to Mathematical Consensus Theory

60. J. Banas and K. Goebel, Measures of Noncompactness in Banach Spaces
61. O. A. Nielson, Direct Integral Theory
62. J. E. Smith et al., Ordered Groups
63. J. Cronin, Mathematics of Cell Electrophysiology
64. J. W. Brewer, Power Series Over Commutative Rings
65. P. K. Kamthan and M. Gupta, Sequence Spaces and Series
66. T. G. McLaughlin, Regressive Sets and the Theory of Isols
67. T. L. Herdman et al., Integral and Functional Differential Equations
68. R. Draper, Commutative Algebra
69. W. G. McKay and J. Patera, Tables of Dimensions, Indices, and Branching Rules for Representations of Simple Lie Algebras
70. R. L. Devaney and Z. H. Nitecki, Classical Mechanics and Dynamical Systems
71. J. Van Geel, Places and Valuations in Noncommutative Ring Theory
72. C. Faith, Injective Modules and Injective Quotient Rings
73. A. Fiacco, Mathematical Programming with Data Perturbations I
74. P. Schultz et al., Algebraic Structures and Applications
75. L Bican et al., Rings, Modules, and Preradicals
76. D. C. Kay and M. Breen, Convexity and Related Combinatorial Geometry
77. P. Fletcher and W. F. Lindgren, Quasi-Uniform Spaces
78. C.-C. Yang, Factorization Theory of Meromorphic Functions
79. O. Taussky, Ternary Quadratic Forms and Norms
80. S. P. Singh and J. H. Burry, Nonlinear Analysis and Applications
81. K. B. Hannsgen et al., Volterra and Functional Differential Equations
82. N. L. Johnson et al., Finite Geometries
83. G. I. Zapata, Functional Analysis, Holomorphy, and Approximation Theory
84. S. Greco and G. Valla, Commutative Algebra
85. A. V. Fiacco, Mathematical Programming with Data Perturbations II
86. J.-B. Hiriart-Urruty et al., Optimization
87. A. Figa Talamanca and M. A. Picardello, Harmonic Analysis on Free Groups
88. M. Harada, Factor Categories with Applications to Direct Decomposition of Modules
89. V. I. Istrătescu, Strict Convexity and Complex Strict Convexity
90. V. Lakshmikantham, Trends in Theory and Practice of Nonlinear Differential Equations
91. H. L. Manocha and J. B. Srivastava, Algebra and Its Applications
92. D. V. Chudnovsky and G. V. Chudnovsky, Classical and Quantum Models and Arithmetic Problems
93. J. W. Longley, Least Squares Computations Using Orthogonalization Methods
94. L. P. de Alcantara, Mathematical Logic and Formal Systems
95. C. E. Aull, Rings of Continuous Functions
96. R. Chuaqui, Analysis, Geometry, and Probability
97. L. Fuchs and L. Salce, Modules Over Valuation Domains
98. P. Fischer and W. R. Smith, Chaos, Fractals, and Dynamics
99. W. B. Powell and C. Tsinakis, Ordered Algebraic Structures
100. G. M. Rassias and T. M. Rassias, Differential Geometry, Calculus of Variations, and Their Applications
101. R.-E. Hoffmann and K. H. Hofmann, Continuous Lattices and Their Applications
102. J. H. Lightbourne III and S. M. Rankin III, Physical Mathematics and Nonlinear Partial Differential Equations
103. C. A. Baker and L. M. Batten, Finite Geometrics
104. J. W. Brewer et al., Linear Systems Over Commutative Rings
105. C. McCrory and T. Shifrin, Geometry and Topology
106. D. W. Kueke et al., Mathematical Logic and Theoretical Computer Science
107. B.-L. Lin and S. Simons, Nonlinear and Convex Analysis
108. S. J. Lee, Operator Methods for Optimal Control Problems
109. V. Lakshmikantham, Nonlinear Analysis and Applications
110. S. F. McCormick, Multigrid Methods
111. M. C. Tangora, Computers in Algebra
112. D. V. Chudnovsky and G. V. Chudnovsky, Search Theory
113. D. V. Chudnovsky and R. D. Jenks, Computer Algebra
114. M. C. Tangora, Computers in Geometry and Topology
115. P. Nelson et al., Transport Theory, Invariant Imbedding, and Integral Equations
116. P. Clément et al., Semigroup Theory and Applications
117. J. Vinuesa, Orthogonal Polynomials and Their Applications
118. C. M. Dafermos et al., Differential Equations
119. E. O. Roxin, Modern Optimal Control
120. J. C. Díaz, Mathematics for Large Scale Computing

121. P. S. Milojevič, Nonlinear Functional Analysis
122. C. Sadosky, Analysis and Partial Differential Equations
123. R. M. Shortt, General Topology and Applications
124. R. Wong, Asymptotic and Computational Analysis
125. D. V. Chudnovsky and R. D. Jenks, Computers in Mathematics
126. W. D. Wallis et al., Combinatorial Designs and Applications
127. S. Elaydi, Differential Equations
128. G. Chen et al., Distributed Parameter Control Systems
129. W. N. Everitt, Inequalities
130. H. G. Kaper and M. Garbey, Asymptotic Analysis and the Numerical Solution of Partial Differential Equations
131. O. Arino et al., Mathematical Population Dynamics
132. S. Coen, Geometry and Complex Variables
133. J. A. Goldstein et al., Differential Equations with Applications in Biology, Physics, and Engineering
134. S. J. Andima et al., General Topology and Applications
135. P. Clément et al., Semigroup Theory and Evolution Equations
136. K. Jarosz, Function Spaces
137. J. M. Bayod et al., p-adic Functional Analysis
138. G. A. Anastassiou, Approximation Theory
139. R. S. Rees, Graphs, Matrices, and Designs
140. G. Abrams et al., Methods in Module Theory
141. G. L. Mullen and P. J.-S. Shiue, Finite Fields, Coding Theory, and Advances in Communications and Computing
142. M. C. Joshi and A. V. Balakrishnan, Mathematical Theory of Control
143. G. Komatsu and Y. Sakane, Complex Geometry
144. I. J. Bakelman, Geometric Analysis and Nonlinear Partial Differential Equations
145. T. Mabuchi and S. Mukai, Einstein Metrics and Yang–Mills Connections
146. L. Fuchs and R. Göbel, Abelian Groups
147. A. D. Pollington and W. Moran, Number Theory with an Emphasis on the Markoff Spectrum
148. G. Dore et al., Differential Equations in Banach Spaces
149. T. West, Continuum Theory and Dynamical Systems
150. K. D. Bierstedt et al., Functional Analysis
151. K. G. Fischer et al., Computational Algebra
152. K. D. Elworthy et al., Differential Equations, Dynamical Systems, and Control Science
153. P.-J. Cahen, et al., Commutative Ring Theory
154. S. C. Cooper and W. J. Thron, Continued Fractions and Orthogonal Functions
155. P. Clément and G. Lumer, Evolution Equations, Control Theory, and Biomathematics
156. M. Gyllenberg and L. Persson, Analysis, Algebra, and Computers in Mathematical Research
157. W. O. Bray et al., Fourier Analysis
158. J. Bergen and S. Montgomery, Advances in Hopf Algebras
159. A. R. Magid, Rings, Extensions, and Cohomology
160. N. H. Pavel, Optimal Control of Differential Equations
161. M. Ikawa, Spectral and Scattering Theory
162. X. Liu and D. Siegel, Comparison Methods and Stability Theory
163. J.-P. Zolésio, Boundary Control and Variation
164. M. Křížek et al., Finite Element Methods
165. G. Da Prato and L. Tubaro, Control of Partial Differential Equations
166. E. Ballico, Projective Geometry with Applications
167. M. Costabel et al., Boundary Value Problems and Integral Equations in Nonsmooth Domains
168. G. Ferreyra, G. R. Goldstein, and F. Neubrander, Evolution Equations
169. S. Huggett, Twistor Theory
170. H. Cook et al., Continua
171. D. F. Anderson and D. E. Dobbs, Zero-Dimensional Commutative Rings
172. K. Jarosz, Function Spaces
173. V. Ancona et al., Complex Analysis and Geometry
174. E. Casas, Control of Partial Differential Equations and Applications
175. N. Kalton et al., Interaction Between Functional Analysis, Harmonic Analysis, and Probability
176. Z. Deng et al., Differential Equations and Control Theory
177. P. Marcellini et al. Partial Differential Equations and Applications
178. A. Kartsatos, Theory and Applications of Nonlinear Operators of Accretive and Monotone Type
179. M. Maruyama, Moduli of Vector Bundles
180. A. Ursini and P. Aglianò, Logic and Algebra
181. X. H. Cao et al., Rings, Groups, and Algebras
182. D. Arnold and R. M. Rangaswamy, Abelian Groups and Modules
183. S. R. Chakravarthy and A. S. Alfa, Matrix-Analytic Methods in Stochastic Models

184. J. E. Andersen et al., Geometry and Physics
185. P.-J. Cahen et al., Commutative Ring Theory
186. J. A. Goldstein et al., Stochastic Processes and Functional Analysis
187. A. Sorbi, Complexity, Logic, and Recursion Theory
188. G. Da Prato and J.-P. Zolésio, Partial Differential Equation Methods in Control and Shape Analysis
189. D. D. Anderson, Factorization in Integral Domains
190. N. L. Johnson, Mostly Finite Geometries
191. D. Hinton and P. W. Schaefer, Spectral Theory and Computational Methods of Sturm–Liouville Problems
192. W. H. Schikhof et al., p-adic Functional Analysis
193. S. Sertöz, Algebraic Geometry
194. G. Caristi and E. Mitidieri, Reaction Diffusion Systems
195. A. V. Fiacco, Mathematical Programming with Data Perturbations
196. M. Křížek et al., Finite Element Methods: Superconvergence, Post-Processing, and A Posteriori Estimates
197. S. Caenepeel and A. Verschoren, Rings, Hopf Algebras, and Brauer Groups
198. V. Drensky et al., Methods in Ring Theory
199. W. B. Jones and A. Sri Ranga, Orthogonal Functions, Moment Theory, and Continued Fractions
200. P. E. Newstead, Algebraic Geometry
201. D. Dikranjan and L. Salce, Abelian Groups, Module Theory, and Topology
202. Z. Chen et al., Advances in Computational Mathematics
203. X. Caicedo and C. H. Montenegro, Models, Algebras, and Proofs
204. C. Y. Yıldırım and S. A. Stepanov, Number Theory and Its Applications
205. D. E. Dobbs et al., Advances in Commutative Ring Theory
206. F. Van Oystaeyen, Commutative Algebra and Algebraic Geometry
207. J. Kakol et al., p-adic Functional Analysis
208. M. Boulagouaz and J.-P. Tignol, Algebra and Number Theory
209. S. Caenepeel and F. Van Oystaeyen, Hopf Algebras and Quantum Groups
210. F. Van Oystaeyen and M. Saorin, Interactions Between Ring Theory and Representations of Algebras
211. R. Costa et al., Nonassociative Algebra and Its Applications
212. T.-X. He, Wavelet Analysis and Multiresolution Methods
213. H. Hudzik and L. Skrzypczak, Function Spaces: The Fifth Conference
214. J. Kajiwara et al., Finite or Infinite Dimensional Complex Analysis
215. G. Lumer and L. Weis, Evolution Equations and Their Applications in Physical and Life Sciences
216. J. Cagnol et al., Shape Optimization and Optimal Design
217. J. Herzog and G. Restuccia, Geometric and Combinatorial Aspects of Commutative Algebra
218. G. Chen et al., Control of Nonlinear Distributed Parameter Systems
219. F. Ali Mehmeti et al., Partial Differential Equations on Multistructures
220. D. D. Anderson and I. J. Papick, Ideal Theoretic Methods in Commutative Algebra
221. Á. Granja et al., Ring Theory and Algebraic Geometry
222. A. K. Katsaras et al., p-adic Functional Analysis
223. R. Salvi, The Navier-Stokes Equations
224. F. U. Coelho and H. A. Merklen, Representations of Algebras
225. S. Aizicovici and N. H. Pavel, Differential Equations and Control Theory
226. G. Lyubeznik, Local Cohomology and Its Applications

Additional Volumes in Preparation

representations of algebras

proceedings of the conference held in São Paulo

edited by

Flávio Ulhoa Coelho
Héctor A. Merklen
University of São Paulo
São Paulo-SP, Brazil

MARCEL DEKKER, INC.   NEW YORK · BASEL

ISBN: 0-8247-0733-8

This book is printed on acid-free paper.

Headquarters
Marcel Dekker, Inc.
270 Madison Avenue, New York, NY 10016
tel: 212-696-9000; fax: 212-685-4540

Eastern Hemisphere Distribution
Marcel Dekker AG
Hutgasse 4, Postfach 812, CH-4001 Basel, Switzerland
tel: 41-61-261-8482; fax: 41-61-261-8896

World Wide Web
http://www.dekker.com

The publisher offers discounts on this book when ordered in bulk quantities. For more information, write to Special Sales/Professional Marketing at the headquarters address above.

Copyright © 2002 by Marcel Dekker, Inc. All Rights Reserved.

Neither this book nor any part may be reproduced or transmitted in any form or by any means, electronic or mechanical, including photocopying, microfilming, and recording, or by any information storage and retrieval system, without permission in writing from the publisher.

Current printing (last digit):
10 9 8 7 6 5 4 3 2 1

PRINTED IN THE UNITED STATES OF AMERICA

Preface

The Conference on Representations of Algebras–São Paulo (CRASP) was held at the Instituto de Matemática e Estatística of the Universidade de São Paulo. The Scientific Committee consisted of:
- Michael Butler (Liverpool)
- Jose Antonio de la Peña (Mexico)
- Idun Reiten (Trondheim)
- Claus Ringel (Bielefeld)
- Flávio Ulhoa Coelho (São Paulo)

while the local Organizing Committee members were
- Flávio Ulhoa Coelho
- Eduardo N. Marcos
- Maria Izabel R. Martins
- Héctor A. Merklen

Seventy-two researchers from 17 different countries attended this conference. There were 14 invited talks and 32 contributed talks. Many of the contributions are presented in these proceedings.

All papers published here were refereed and consist mainly of original research results. We would like to express our thanks to all the participants of the conference and the contributors to these proceedings. Also, our sincere thanks go to our colleagues who served as referees.

Our thanks are also extended to Joelma Martins Gomes and Sueli Aparecida Paschoal Dian, whose secretarial work was essential to the conference and for the publication of this volume.

The conference was supported by FAPESP, CNPq, CAPES (through its program PROAP), IME-USP, CCInt-USP, CPq-USP, SBM, and CBA. Without their valuable support, this conference would not have been possible.

Flávio Ulhoa Coelho
Héctor A. Merklen

Contents

Preface		*v*
Contributors		*vii*
Invited Talks		*ix*
Participants		*xi*
1.	On the Existence of Left and Right Almost Split Morphisms Lidia Angeleri-Hügel	1
2.	Actions of Hopf Algebras on Quantum Polynomials Vyacheslav A. Artamonov	11
3.	Strongly Simply Connected Derived Tubular Algebras Ibrahim Assem	21
4.	H^1 and Presentations of Finite Dimensional Algebras Michael J. Bardzell and Eduardo N. Marcos	31
5.	Tame Tilted Algebras with Almost Regular Connecting Components Grzegorz Bobiński	39
6.	Reflexive Modules Are Not Closed under Submodules Gabriella D'Este	53
7.	Fibre Sum Functors and the Bimodule Ext Peter Dräxler	65
8.	Smooth Automorphism Group Schemes Daniel R. Farkas, Christof Geiss, and Eduardo N. Marcos	71
9.	A Combinatorial Characterization of Hereditary Categories Containing Simple Objects Dieter Happel and Idun Reiten	91
10.	Symmetric Quasi-Schurian Algebras Octavio Mendoza Hernández	99

11.	On Lattices at the Ends of Connected Components of the Auslander-Reiten Quiver *Alfredo Jones*	117
12.	Factorisations of Morphisms for Wild Hereditary Algebras *Otto Kerner*	121
13.	A Note on Concealed-Canonical Artin Algebras *Dirk Kussin and Zygmunt Pogorzaly*	129
14.	Koszul Algebras and the Gorenstein Condition *Roberto Martínez-Villa*	135
15.	Some Remarks about the "Double Extension" Algebra of a Finite Poset *Teresita Noriega*	157
16.	Coil Algebras that Are Derived-Tame *Jose Antonio de la Peña and Bertha Tomé*	165
17.	One-Point Extensions of Quasitilted Algebras by Modules on Stable Tubes *Jose Antonio de la Peña and Sonia E. Trepode*	177
18.	Combinatorial Partial Tilting Complexes for the Brauer Star Algebras *Mary Schaps and Evelyne Zakay-Illouz*	187
19.	Almost Split Sequences in Categories of Representations of Quivers II *Sverre O. Smalø*	209
20.	Cotilting Objects and Dualities *Robert Wisbauer*	215
21.	Coherent Components of Auslander-Reiten Quivers whose DTr-Orbits Are Finite *Hailou Yao*	235
22.	Twisted Hopf Algebras *Pu Zhang and Li-Bin Li*	269

Contributors

Lidia Angeleri-Hügel   Universität Munchen, Munich, Germany

Vyacheslav A. Artamonov   Moscow State University, Moscow, Russia

Ibrahim Assem   Université de Sherbrooke, Sherbrooke, Quebec, Canada

Michael J. Bardzell   Salisbury State University, Salisbury, Maryland

Grzegorz Bobiński   Nicholas Copernicus University, Toruń, Poland

Gabriella d'Este   Università di Milano, Milan, Italy

Peter Dräxler   Universität Bielefeld, Bielefeld, Germany

Daniel R. Farkas   Virginia Polytechnic Institute and State University, Blacksburg, Virginia

Christof Geiss   Instituto de Matemáticas, UNAM, Ciudad Universitaria, México D.F., Mexico

Dieter Happel   Technische Universität Chemnitz, Chemnitz, Germany

Octavio Mendoza Hernández   Universidad Nacional del Sur, Bahía Blanca, Argentina

Alfredo Jones   Centro de Matemática, Montevideo, Uruguay

Otto Kerner   Heinrich-Heine-Universität, Düsseldorf, Germany

Dirk Kussin   Universität Paderborn, Paderborn, Germany

Li-Bin Li   University of Science and Technology of China, Hefei, P. R. China

Eduardo N. Marcos   Universidade de São Paulo, São Paulo, SP, Brazil

Roberto Martinez-Villa   Universidad Nacional Autónoma de México, México D.F., Mexico

Teresita Noriega   Universidad de la Habana, Havana, Cuba

Jose Antonio de la Peña   UNAM, Ciudad Universitaria, México D.F., Mexico

Zygmunt Pogorzaly   Nicholas Copernicus University, Toruń, Poland

Idun Reiten   Norwegian University of Science and Technology, Trondheim, Norway

Mary Schaps   Bar-Ilan University, Ramat-Gan, Israel

Sverre O. Smalø   NTNU, Trondheim, Norway

Bertha Tomé   UNAM, Ciudad Universitaria, México D.F., Mexico

Sonia E. Trepode   Universidad de Mar del Plata, Mar del Plata, Argentina

Robert Wisbauer   University of Düsseldorf, Düsseldorf, Germany

Hailou Yao   Beijing Polytechnic University, Beijing, P. R. China

Evelyne Zakay-Illouz   Jordan Valley College, Jordan Valley, Israel

Pu Zhang   University of Science and Technology of China, Hefei, P. R. China

Invited Talks

M. BAROT, *A characterization of positive unit forms.*

R. BAUTISTA, *Representations over rational functions.*

C. CIBILS, *Noncommutative tensor products of sets.*

Y. DROZD, *Noncommutative nodes and their derived categories.*

C. GEISS, *Derived clannish algebras.*

E. GREEN, *Some results on Hochschild cohomology and related topics.*

H. KRAUSE, *Morphisms determined by objects and Brown representability.*

H. LENZING, *Two-orbits of tubular type.*

H. MERKLEN, *Standardly stratified algebras and the characteristic module.*

M. I. PLATZECK, *Module of finite projective dimension over standardly stratified algebras.*

J. SCHRÖER, *Module varieties over some canonical algebras.*

A. SKOWROŃSKI, *Selfinjective algebras of quasitilted type.*

S. SMALØ, *Almost split sequences in categories of finite dimensional representations of quivers.*

P. ZHANG, χ-*Hopf algebras, Hall algebras and Lusztig-Green-Ringel classes.*

Participants

ALVARES, EDSON RIBEIRO Departamento de Matemática-IME, Universidade de São Paulo, Rua do Matão, 1010, CEP 05508-900, São Paulo, Brazil.
e-mail: era@ime.usp.br

ANGELERI-HÜGEL, LIDIA Matematisches Institut der Universität Theresienstr 39, D-80333, München, Germany.
e-mail: angeleri@rz.mathematik.uni-muenchen.de

AQUINO, REGINA MARIA DE Departamento de Matemática-IME, Universidade de São Paulo, Rua do Matão, 1010, CEP 05508-900, São Paulo, Brazil.
e-mail: aquino@ime.usp.br

ASSEM, IBRAHIM Departement de Mathematiques et Informatique, Université de Sherbrooke, Sherbrooke, Quebec, J1K 2R1, Canada.
e-mail: ibrahim.assem@dmi.usherb.ca

BAROT, MICHAEL Instituto de Matematicas, UNAM, Ciudad Universitaria, Mexico, D.F., C. P. 04510, Mexico.
e-mail: barot@matem.unam.mx

BAUTISTA, RAYMUNDO Instituto de Matematicas, UNAM, Ciudad Universitaria, Mexico, D.F., C. P. 04510, Mexico.
e-mail: raymundo@matem.unam.mx

BEKKERT, VIKTOR Faculty of Mechanics and Mathematics Kyiv Taras Shevchenko, University Vladimirskaya Str, 64 252033 Kyiv, Ukraine.
e-mail: biblio@akcecc.kiev.ua

BOBIŃSKI, GRZEGORZ Faculty of Mathematics and Informatics, Nicholas Copernicus University, Ul. Chopina 12/18, 87-100, Toruń, Poland.
e-mail: gregbob@mat.uni.torun.pl

BOVDI, VICTOR Departamento de Matemática-IME, Universidade de São Paulo, Rua do Matão, 1010, CEP 05508-900, São Paulo, Brazil.
e-mail: vbovdi@math.klte.hu

BRAGA, CLÉZIO APARECIDO Departamento de Matemática-IME, Universidade de São Paulo, Rua do Matão, 1010, CEP 05508-900, São Paulo, Brazil.
e-mail: clezio@ime.usp.br

BRENNER, SHEILA Department of Mathematical Sciences, University of Liverpool, Liverpool, L69 3BX, UK.
e-mail: sbrenner@liv.ac.uk

BUAN, ASLAK BAKKE Institutt for Matematiske Fag, NTNU, Lade, N-7491, Trondheim, Norway.
e-mail: aslakb@math.ntnu.no

BUSTOS, CRISTIAN PATRICIO NOVOA Universidade Católica de Goiás, Rua 227-A Número 72 apto 1204, setor leste universitario, 74610-096, Goiania, Brazil.
e-mail: cristian@ime.usp.br

BUTLER, MICHAEL C. R. Department of Mathematical Sciences, University of Liverpool, Liverpool L69 3BX, UK.
e-mail: mcrb@liv.ac.uk

CIBILS, CLAUDE Departement de Mathematiques, Université de Montpellier 2, F-34980, Montpellier, Cedex 5, France.
e-mail: cibils@math.univ-montp2.fr

COELHO, FLÁVIO ULHOA Departamento de Matemática-IME, Universidade de São Paulo, Rua do Matão, 1010, CEP 05508-900, São Paulo, Brazil.
e-mail: fucoelho@ime.usp.br

DMYTRENKO, VASYL Department of Mathematical Sciences, University of Delaware, 19711 Newark, DE, USA.
e-mail: dmytrenk@math.udel.edu

DROZD, YURY Dept. of Mathematics, Kieve Taras Shevchenko University, Volodimirska 64, 252033 Kiev, Ukraine.
e-mail: yuriy@drozd.org

ESCUDER, CECILIA TOSAR Departamento de Matemática-IME, Universidade de São Paulo, Rua do Matão, 1010, CEP 05508-900, São Paulo, Brazil.
e-mail: mtosar@varela.reu.edu.uy

FACCHINI, ALBERTO Dipartimento di Matematica, Universitá di Udine I-33100

Participants

e Informatica, Via Delle Scienze, 206, Italy.
e-mail: facchini@dimi.uniud.it

FARKAS, DANIEL Math Department, Virginia Tech, Blacksburg, VA 24061 - 0123, USA.
e-mail: farkas@math.vt.edu

FERNANDEZ, ELSA A. Universidad de la Patagonia, Auda Roca 1890, (9120) Puerto Madryn, Chubut, Argentina.
e-mail: unonetti@cenpat.edu.ar

FERRAZ, RAUL Departamento de Matemática-IME, Universidade de São Paulo, Rua do Matão, 1010, CEP 05508-900, São Paulo, Brazil.
e-mail: raul@ime.usp.br

FERREIRA, VITOR DE OLIVEIRA Departamento de Matemática-IME, Universidade de São Paulo, Rua do Matão, 1010, CEP 05508-900, São Paulo, Brazil.
e-mail: vofer@ime.usp.br

FERRERO, MIGUEL Instituto de Matemática, UFRGS, Porto Alegre, 90420-160, Brazil.
e-mail: ferrero@mat.ufrgs.br

GASTAMINZA, SUSANA Departamento de Matemática, Universidad Nacional Del Sur, Av Alem 1253 8000, Bahia Blanca, Argentina.

GEISS, CHRISTOF Instituto de Matematicas, UNAM, Ciudad Universitaria, Mexico, D.F., C. P. 04510, Mexico.
e-mail: christof@matem.unam.mx

GREEN, EDWARD L. Dept of Math., Virginia Tech., 24061-0123, Blacksburg, USA.
e-mail: green@math.vt.edu

HOUARI, MOHAMMED EL Departamento de Matemática-IME, Universidade de São Paulo, Rua do Matão, 1010, CEP 05508-900, São Paulo, Brazil.
e-mail: elhouari@ime.usp.br

HUARD, FRANÇOIS Bishop's University, Lennoxville, Quebec, Canada.
e-mail: Fhuard@ubishops.ca

IKEMOTO, LUCIA Departamento de Matemática-IME, Universidade de São Paulo, Rua do Matão, 1010, CEP 05508-900, São Paulo, Brazil.
e-mail: ikemoto@ime.usp.br

JENSEN, BERNT TORE Institutt for Matematiske Fag, NTNU, Lade, N-7491, Trondheim, Norway.
e-mail: berntj@math.ntnu.no

JONES, ALFREDO Centro de Matemática, Facultad de Ciencias, Iguá 4225, Montevideo, Uruguay.
e-mail: ajones@cmat.edu.uy

KERNER, OTTO Mathematisches Institut, Heinrich-Heine-Universität, D40225, Dusseldorf,Germany.
e-mail: kerner@mx.cs.uni-dusseldon.de

KRAUSE, HENNING Department of Mathematics, University of Bielefeld, D-33501, Bielefeld, Germany.
e-mail: henning@mathematik.uni-bielefeld.de

LANZILOTTA, MARCELO AMÉRICO Departamento de Matemática-IME, Universidade de São Paulo, Rua do Matão, 1010, CEP 05508-900, São Paulo, Brazil.
e-mail: marclan@cmat.edu.uy

LENZING, HELMUT Fachbereich Mathematik 33095, Informatik, Universität Paderborn, Germany.
e-mail: helmut@uni-paderborn.de

LOCATELI, ANA CLÁUDIA Departamento de Matemática, Universidade Federal do Espírito Santo, Vitória, Brazil.
e-mail: locatelli@cce.ufes.br

LOPES, ANA TERESA TAVARES Av. Caxingui 95, ap 62, Butantã, CEP 05579-000, São Paulo, SP, Brazil.
e-mail: analopes@ime.usp.br

MADSEN, DAG Institutt for Matematiske Fag, NTNU, Lade, N-7491, Trondheim, Norway.
e-mail: dagma@math.ntnu.no

MALICKI, PIOTR Faculty of Mathematics and Informatics, Nicholas Copernicus University, Ul. Chopina 12/18, 87-100, Toruń, Poland.
e-mail:pmalicki@mat.uni.torun.pl

MARCOS, EDUARDO DO NASCIMENTO Departamento de Matemática-IME, Universidade de São Paulo, Rua do Matão, 1010, CEP 05508-900, São Paulo, Brazil.
e-mail: enmarcos@ime.usp.br

MARTINS, MARIA IZABEL R. Departamento de Matemática-IME, Universidade

Participants

de São Paulo, Rua do Matão, 1010, CEP 05508-900, São Paulo, Brazil.
e-mail: bel@ime.usp.br

MENDOZA, OCTAVIO HERNÁNDEZ Humboldt 2870, 8000, Bahia Blanca, prov. Buenos Aires, Argentina.
e-mail: omendoza@uniba.edu.ar

MERKLEN, HÉCTOR ALFREDO Departamento de Matemática-IME, Universidade de São Paulo, Rua do Matão, 1010, CEP 05508-900, São Paulo, Brazil.
e-mail: merklen@ime.usp.br

MICHELENA, SANDRA Departamento de Matemática, Universidad Nacional Del Sur, Av Alem 1253 8000, Bahia Blanca, Argentina.
e-mail: michele@criba.edu.ar

NORIEGA, TERESITA Facultad de Matematica y Computación, Universidad de La Habana, San Lazaro y L La, Habana 4, Cuba.
e-mail: noriega@matcom.uh.cv

OLIVEIRA, ALEGRIA GLADYS CHALOM DE Departamento de Matemática-IME, Universidade de São Paulo, Rua do Matão, 1010, CEP 05508-900, São Paulo, Brazil.
e-mail: agchalom@ime.usp.br

PEÑA, MARIA INÉS Dpto. de Matemática, Fac.Cs.Ex., Universidad Nacional de Mar del Plata, Funes 3350, 7600, Argentina.
e-mail: mapena@mdp.edu.ar

PLATZECK, MARIA INES Departamento de Matemática, Universidad Nacional Del Sur, Av Alem 1253 8000, Bahia Blanca, Argentina.
e-mail: impiovan@criba.edu.ar

PRATTI, NILDA ISABEL Dpto. de Matemática, Fac.Cs.Ex., Universidad Nacional de Mar del Plata, Funes 3350, 7600, Argentina.
e-mail: nilprat@mdp.edu.ar

REDONDO, MARIA JULIA Departamento de Matemática, Universidad Nacional Del Sur, Av Alem 1253 8000, Bahia Blanca, Argentina.
e-mail: mredondo@criba.edu.ar

REITEN, IDUN Institutt for Matematiske Fag, NTNU, Lade, N-7491, Trondheim, Norway.
e-mail: idunr@math.ntnu.no

RODRIGUES, VIRGINIA SILVA Rua Amambaí, no. 107, Monte Castelo, Juiz de

Fora, MG, CEP 36081-060, Brazil.

RODRIGUES, WALTER MARTINS Departamento de Matemática-IME, Universidade de São Paulo, Rua do Matão, 1010, CEP 05508-900, São Paulo, Brazil.
e-mail: walterm@ime.usp.br

SÁENZ, EDITH CORINA Facultad de Guanajuato, Apdo. Postal 402, Guanajuata, Gto. México.
e-mail: corina@fractal.cimat.mx

SALAZAR, HERNAN ALONSO GERALDO Departamento de Matemática-IME, Universidade de São Paulo, Rua do Matão, 1010, CEP 05508-900, São Paulo, Brazil.
e-mail: hernang@ime.usp.br

SALORIO, MARIA JOSÉ SOUTO Faculdade Informática, Universidade La Coruña, Campus de Elvina, 15017, Coruña, Spain.
e-mail: mariaj@udc.es

SAORIN, MANUEL Departamento de Matematicas, Universidad de Murcia, aptdo. 4021, 30100, Espinardo, MU, Spain.
e-mail: msaorin@fcu.um.es

SAVIOLI, ANGELA MARTA PEREIRA DAS DORES Departamento de Matemática-IME, Universidade de São Paulo, Rua do Matão, 1010, CEP 05508-900, São Paulo, Brazil.
e-mail: marta@ime.usp.br

SCHRÖER, JAN Department of Mathematics, University of Bielefeld, D-33501, Bielefeld Germany.
e-mail: jschroe@mathematik.uni-bielefeld.de

SKOWROŃSKI, ANDRZEJ Faculty of Mathematics and Informatics, Nicholas Copernicus University, Ul. Chopina 12/18, 87-100, Toruń, Poland.
e-mail: skowron@mat.uni.torun.pl

SLUNGAARD, INGER HEIDI Institutt for Matematiske Fag, NTNU, Lade, N-7491, Trondheim, Norway.
e-mail: ingersl@math.ntnu.no

SMALØ, SVERRE O. Institutt for Matematiske Fag, NTNU, Lade, N-7491, Trondheim, Norway.
e-mail: sverresm@math.ntnu.no

SOLBERG, OYVIND Institutt for Matematiske Fag, NTNU, Lade, N-7491, Trondheim, Norway.

Participants

e-mail: oyvinso@math.ntnu.no

TREPODE, SONIA ELISABET Dpto. de Matemática, Fac.Cs.Ex., Universidad Nacional de Mar del Plata, Funes 3350, 7600, Argentina.
e-mail: strepode@mdp.edu.ar

TOMÉ, BERTHA Departamento de Matematicas, Facultad de Ciencias, UNAM, Mexico.
e-mail: bta@fciencias.unam.mx

VARGAS, ROSANA RETSOS SIGNORELLI Departamento de Matemática-IME, Universidade de São Paulo, Rua do Matão, 1010, CEP 05508-900, São Paulo, Brazil.
e-mail: rosana@ime.usp.br

ZACHARIA, DAN Dept. Mathematics, Syracuse NY 13244, USA.
e-mail:zacharia@syr.edu

ZHANG, PU Department of Mathematics, University of Science and Technology of China, Hefei 230026, P. R. China.
e-mail: pzhang@ustc.edu.cn

ZUAZUA, RITA E. Instituto de Matematicas, UNAM, Ciudad Universitaria, Mexico, D.F., C. P. 04510, Mexico.
e-mail: zuazua@matem.unam.mx

ZWARA, GRZEGORZ Faculty of Mathematics and Informatics, Nicholas Copernicus University, Ul. Chopina 12/18, 87-100, Toruń, Poland.
e-mail: gzwara@mat.uni.torun.pl

On the existence of left and right almost split morphisms

LIDIA ANGELERI-HÜGEL Mathematisches Institut der Universität, Theresienstraße 39, D-80333 München, e-mail: angeleri@rz.mathematik.uni-muenchen.de

ABSTRACT We discuss the existence of left and right almost split morphisms for a skeletally small category $\mathcal{M}$ of modules over an arbitrary ring R. To this end, we associate to $\mathcal{M}$ a certain R-module M and investigate finiteness conditions on M viewed as a module over its endomorphism ring.

INTRODUCTION

Left and right almost split morphisms are usually studied for categories of finitely generated modules over artin algebras. In this small note, we consider left and right almost split morphisms for a skeletally small subcategory $\mathcal{M}$ of the category of all (right) modules over an arbitrary ring. If the modules M_i, $i \in I$, are representatives of the isomorphism classes of $\mathcal{M}$, then the existence of left and right almost split morphisms can be interpreted in terms of finiteness conditions for the modules $M = \coprod_{i \in I} M_i$ and $N = \prod_{i \in I} M_i$ viewed as modules over their endomorphism rings S and T, respectively. An important role in this context is played by the radicals $\mathrm{r}(M,C)_S$ and $_T\mathrm{r}(A,N)_S$ where A and C are R-modules. In case that $\mathcal{M}$ is a finite subcategory consisting of modules with local endomorphism ring, we can restrict ourselves to the Jacobson radical $J(S)$ of S, and we obtain for instance that $\mathcal{M}$ has left (respectively, right) almost split morphisms if and only if $J(S)$ is a finitely generated left (respectively, right) S-module.

We also study generalized right almost split morphisms, a concept introduced by Brune in [3]. For a right artinian ring R, the existence of such maps means that R is right pure-semisimple, see [11]. This is a consequence of Brune's work on the so-called Kulikov property, relying on a functorial approach. We now obtain a direct, module-theoretical proof of this result and also a dual characterization of pure-semisimple rings in terms of generalized left almost split morphisms.

The paper is divided into three sections. The first section is devoted to some preliminaries. In Section 2, we relate finiteness conditions for $_SM$ and $_TN$ to the notion of a finitely (co)generated family of homomorphisms which was introduced by Auslander in [1]. These results are then applied in Section 3 to the study of left and right almost split morphisms.

The author acknowledges a HSPIII-grant of the University of Munich.

1 PRELIMINARIES

Let us first introduce some notation. If R is a ring, we write $J(R)$ for the Jacobson radical of R, and denote by **Mod** R the category of all and by **mod** R the category of all finitely presented right R-modules.

Throughout the paper, we fix a skeletally small subcategory $\mathcal{M}$ of Mod R which is closed under isomorphic images, and let $\{M_i \mid i \in I\}$ be a complete irredundant set of representatives of the isomorphism classes of $\mathcal{M}$. Further, we set $M = \coprod_{i \in I} M_i$ with $S = \mathrm{End}_R M$, and $N = \prod_{i \in I} M_i$ with $T = \mathrm{End}_R N$. Moreover, we denote by **add** $\mathcal{M}$ the class consisting of all modules isomorphic to direct summands of finite direct sums of modules of $\mathcal{M}$.

Recall that for two modules X_R, Y_R the radical $\mathbf{r}(X,Y)$ denotes the collection of all homomorphisms $f: X \to Y$ such that there is no isomorphism of the form $Z \to X \xrightarrow{f} Y \to Z$ where Z is a module with local endomorphism ring. Then $r(X,Y)$ is an $\mathrm{End}_R Y - \mathrm{End}_R X$ - subbimodule of $\mathrm{Hom}_R(X,Y)$. Let us collect some easy properties of this bimodule.

LEMMA 1.1. *Let Y_R be a module with endomorphism ring $E = \mathrm{End}_R Y$, and let X_R be an indecomposable module.*

(1) $J(E) \subset r(Y,Y)$, with equality if $Y = \coprod_{i=1}^n Y_i$ and all Y_i have local endomorphism ring.

(2) $\mathrm{Hom}_R(X,Y)/r(X,Y)$ is either zero or a simple left E-module.

(3) $r(X,Y)$ is a noetherian left E-module if and only if $\mathrm{Hom}_R(X,Y)$ is a noetherian left E-module.

Proof. The first assertion in (1) is well-known. For the second assertion, assume that Y has a decomposition as stated. This means that E is semiperfect, that is, $\mathrm{id}_Y = \sum_{i=1}^n e_i$ for local idempotents $e_1, \ldots, e_n \in E$. Then every $f \in r(X,Y)$ has the form $f = \sum_{i=1}^n f e_i$ where $E f e_i$ is properly contained in $E e_i$ and therefore lies in $J(E) e_i$, which shows $f \in J(E)$.

(2) Assume that there is a nonzero element $\bar{f} \in \mathrm{Hom}_R(X,Y)/r(X,Y)$. Since X is indecomposable, f is a split monomorphism and therefore generates the left E-module $\mathrm{Hom}_R(X,Y)$. This yields the claim.

(3) Apply statement (2) on the exact sequence of E-modules

$$0 \longrightarrow {}_E\, r(X,Y) \longrightarrow {}_E \mathrm{Hom}_R(X,Y) \longrightarrow {}_E \mathrm{Hom}_R(X,Y)/r(X,Y) \longrightarrow 0.$$

$\square$

2 FAMILIES OF HOMOMORPHISMS

Following Auslander [1, §1], we say that a family of homomorphisms $(a_k : A \to X_k)_{k \in K}$ is **finitely cogenerated** if there is a finite subset $K_0 \subset K$ such that the product map $a : A \to \coprod_{k \in K_0} X_k$ induced by the a_k with $k \in K_0$ has the property that all a_k, $k \in K$, factor through a. We start out with two propositions describing when families of homomorphisms in $\bigcup_{i \in I} r(A, M_i)$ are finitely cogenerated.

PROPOSITION 2.1. *The following statements are equivalent for a module A.*

(1) The family of all homomorphisms in $\bigcup_{i \in I} r(A, M_i)$ is finitely cogenerated.

(2) The left T-module $r(A, N)$ is generated by finitely many maps whose images are contained in a finite subproduct $\prod_{i \in I_0} M_i$ of N.

(3) There is a map $a \in r(A, X)$ such that $X \in \mathrm{add}\mathcal{M}$ and all maps $h \in r(A, Y)$ where $Y \in \mathrm{add}\mathcal{M}$ factor through a.

If A is finitely generated, the following statement is further equivalent.

(4) $r(A, M)$ is a finitely generated left S-module.

Proof. $(1) \Rightarrow (3)$: By assumption there are indices $i_1, \ldots, i_r \in I$ and maps $a_k \in r(A, M_{i_k})$, $1 \leq k \leq r$, such that all maps in $\bigcup_{i \in I} r(A, M_i)$ factor through the product map $a : A \to X = \coprod_{k=1}^r M_{i_k} \in \mathrm{add}\mathcal{M}$ induced by the a_k, and of course $a \in r(A, X)$.

$(3) \Rightarrow (2)$: By the universal property of products, every map $f \in r(A, N)$ factors through a. Moreover, there is a split monomorphism $\iota : X \to \bigoplus_{k=1}^n X_k$ for some $X_1, \ldots, X_n \in \mathcal{M}$, and for any $1 \leq k \leq n$ there is $i_k \in I$ such that $X_k \cong M_{i_k}$, giving rise to a split monomorphism $\alpha_k : X_k \to N$. Then the maps

$$a_k : A \xrightarrow{a} X \xrightarrow{\iota} \bigoplus_{k=1}^n X_k \xrightarrow{pr_k} X_k \xrightarrow{\alpha_k} N,$$

$1 \leq k \leq n$, form a generating set of $r(A, N)$ over T with the required property.

$(2) \Rightarrow (1)$: Let a_k, $1 \leq k \leq n$, be a generating set of the left T-module $r(A, N)$ such that there is a finite subset $I_0 \subset I$ with $\mathrm{Im}\, a_k \subset \prod_{i \in I_0} M_i$ for all $1 \leq k \leq n$. Denote by $p_i : N \to M_i$, $i \in I$, the canonical projections. Then it is easy to check that all maps in $\bigcup_{i \in I} r(A, M_i)$ factor through the product map induced by the $a_{k_i} = p_i a_k \in r(A, M_i)$, $i \in I_0$, $1 \leq k \leq n$.

Finally, if A is finitely generated, each map in $r(A, M)$ has its image in a finite subcoproduct of M, hence in a direct summand of N. With this in mind, it is easy to see that (4) and (2) are equivalent. □

PROPOSITION 2.2. *The following statements are equivalent for a module A.*

(1) Every family of homomorphisms in $\bigcup_{i \in I} r(A, M_i)$ is finitely cogenerated.

(2) $r(A, N)$ is a noetherian left T-module.

If A is finitely generated, then (1) and (2) are further equivalent to

(3) $r(A, M)$ is a noetherian left S-module.

Proof. Denote by $p_i : N \to M_i$ and $e_i : M_i \to N$, $i \in I$, the canonical projections and injections, respectively.

(1)$\Rightarrow$(2): For a submodule $_T U \subset r(A, N)$ we consider the family of homomorphisms $\{p_i f \mid f \in U, i \in I\}$ in $\bigcup_{i \in I} r(A, M_i)$. By assumption there are indices $i_1, \ldots, i_n \in I$ and maps $f_k \in U$, $1 \leq k \leq n$, such that the product map $a : A \to \coprod_{k=1}^n M_{i_k}$ induced by the $p_{i_k} f_k$ has the property that all other maps of the form $p_i f$ with $i \in I$ and $f \in U$ factor through a. Then also all $f \in U$ factor through a, and so the f_k, $1 \leq k \leq n$, form a generating set of $_T U$.

(2)$\Rightarrow$(1): Let now $(a_k : A \to M_{i_k})_{k \in K}$ be a family in $\bigcup_{i \in I} r(A, M_i)$, and consider the T-submodule $U = \sum_{k \in K} T f_k$ of $r(A, N)$ given by the maps $f_k : A \xrightarrow{a_k} M_{i_k} \xrightarrow{e_{i_k}} N$ in $r(A, N)$. By assumption $_T U = \sum_{k \in K_0} T f_k$ for some finite subset $K_0 \subset K$. This implies that all f_k, and therefore also all a_k, factor through the product map $a : A \to \coprod_{k \in K_0} M_{i_k}$ induced by the a_k with $k \in K_0$.
If A is finitely generated, the equivalence of (1) and (3) is proven with similar arguments, taking into account the fact that each map in $r(A, M)$ has its image in a finite subcoproduct of M. $\square$

Recall that a module is said to be **endonoetherian** if it is noetherian as a module over its endomorphism ring. As observed by Huisgen-Zimmermann [7], the module N is endonoetherian if and only if for all finitely generated modules A_R every family of homomorphisms in $\bigcup_{i \in I} \text{Hom}_R(A, M_i)$ is finitely cogenerated. Over a semilocal ring, we can now restrict ourselves to families of homomorphisms in $\bigcup_{i \in I} r(A, M_i)$ with A indecomposable.

PROPOSITION 2.3. *Assume that R is semilocal. Then the following statements are equivalent.*

(1) N is endonoetherian.

(2) M is endonoetherian.

(3) For all finitely generated (or equivalently, for all finitely presented) indecomposable modules A_R every family of homomorphisms in $\bigcup_{i \in I} r(A, M_i)$ is finitely cogenerated.

Proof. (2)⇔ (3): M is endonoetherian if and only if $\text{Hom}_R(A, M)$ is a noetherian left S-module for every finitely generated (or equivalently, for every finitely presented) module A_R. Since R is semilocal, every finitely generated module has a finite decomposition in indecomposables [6, 1.14]. So, it suffices to consider finitely generated (or finitely presented) indecomposable modules A, and the statement follows from 2.2 and 1.1. The equivalence (1)⇔ (3) is proven by the same arguments. □

We now consider the dual situation. A family of homomorphisms $(b_k : X_k \to C)_{k \in K}$ is said to be **finitely generated** [1, §1] if there is a finite subset $K_0 \subset K$ such that the coproduct map $b : \coprod_{k \in K_0} X_k \to C$ induced by the b_k with $k \in K_0$ has the property that all b_k, $k \in K$, factor through b. Let us describe when families of homomorphisms in $\bigcup_{i \in I} r(M_i, C)$ are finitely generated. The arguments are dual to those employed above, and we will therefore omit the proofs.

PROPOSITION 2.4. *The following statements are equivalent for a module C.*

(1) The family of all homomorphisms in $\bigcup_{i \in I} r(M_i, C)$ is finitely generated.

(2) The right S-module $r(M, C)$ is generated by finitely many maps whose kernels contain a cofinite subcoproduct $\coprod_{i \in I \setminus I_0} M_i$ of M.

(3) There is a map $b \in r(X, C)$ such that $X \in \text{add}\mathcal{M}$ and all maps $h \in r(Y, C)$ where $Y \in \text{add}\mathcal{M}$ factor through b.

PROPOSITION 2.5. *The following statements are equivalent for a module C.*

(1) Every family of homomorphisms in $\bigcup_{i \in I} r(M_i, C)$ is finitely generated.

(2) $r(M, C)$ is a noetherian right S-module.

We now give a dual version of Proposition 2.3. Recall that R is said to be a **right Morita ring** if it is right artinian and the minimal injective cogenerator of $\text{Mod}\, R$ is finitely generated.

PROPOSITION 2.6. *Assume that R is semilocal. Further, let W_R be a minimal injective cogenerator of $\text{Mod}R$, and $M_S^* = \text{Hom}_R(M, W)_S$. The following statements are equivalent.*

(1) M_S^* is noetherian.

(2) For all finitely cogenerated indecomposable modules C_R every family of homomorphisms in $\bigcup_{i \in I} r(M_i, C)$ is finitely generated.

If R is a right Morita ring, then following statement is further equivalent.

(3) For all finitely generated indecomposable modules C_R every family of homomorphisms in $\bigcup_{i \in I} r(M_i, C)$ is finitely generated.

Proof. Since R is semilocal, there are only finitely many simple right R-modules $S_1, \cdots, S_n$ up to isomorphism. So, $W \cong \coprod_{i=1}^{n} C_i$ where $C_i = E(S_i)$ is an injective envelope of S_i.

(1)$\Rightarrow$(2): If C is finitely cogenerated, then $\mathrm{Hom}_R(M,C)_S$ is finitely M^*-cogenerated, and the claim follows from Proposition 2.5.

(2)$\Rightarrow$(1): By Proposition 2.5 we know that $\mathrm{r}(M,C_i)_S$ is noetherian for all $1 \le i \le n$, hence by the dual version of statement (3) in Lemma 1.1 also $\mathrm{Hom}_R(M,C_i)_S$ is noetherian for all $1 \le i \le n$, and M_S^* is noetherian.

Assume now that R is a right Morita ring. Since R is right artinian, all finitely generated modules are finitely cogenerated, which yields (2)$\Rightarrow$(3). Moreover, since all C_i are finitely generated, we deduce (3)$\Rightarrow$(1) as above from 2.5 and the dual version of statement (3) in Lemma 1.1. $\square$

3 LEFT AND RIGHT ALMOST SPLIT MORPHISMS

Assume now that C is a module in $\mathcal{M}$ with local endomorphism ring. Recall that a homomorphism $b: X \to C$ with $X \in \mathrm{add}\mathcal{M}$ is said to be **right almost split in add $\mathcal{M}$** if b is not a split epimorphism and any homomorphism $h: Y \to C$ where $Y \in \mathrm{add}\mathcal{M}$ and h is not a split epimorphism factors through b. We have seen in Proposition 2.4 that C has a right almost split morphism in add$\mathcal{M}$ if and only if the family of all homomorphisms in $\bigcup_{i \in I} \mathrm{r}(M_i, C)$ is finitely generated. Inspired by Brune's work [3], we will further say that C has a **generalized right almost split morphism in add $\mathcal{M}$** if every family of homomorphisms in $\bigcup_{i \in I} \mathrm{r}(M_i, C)$ is finitely generated. The notions of a **left almost split morphism** and a **generalized left almost split morphism in add $\mathcal{M}$** are defined dually. Finally, if $\mathcal{M}$ consists of modules with local endomorphism ring, we say that $\mathcal{M}$ **has (generalized) right, respectively left, almost split morphisms** if every module belonging to $\mathcal{M}$ has a (generalized) right, respectively left, almost split morphism in add$\mathcal{M}$.

The results in Section 2 can now be interpreted as characterizations for the existence of (generalized) left and right almost split morphisms. In particular, we have the following two Corollaries. Observe that statement (3) in Corollary 3.1 has independently been obtained by Dung [4, 2.3], [5, 3.11].

COROLLARY 3.1. *Assume that all M_i have local endomorphism ring.*

(1) $\mathcal{M}$ has generalized right almost split morphisms if and only if $\mathrm{r}(M, M_i)$ is a noetherian right S-module for all $i \in I$.

(2) $\mathcal{M}$ has generalized left almost split morphisms if and only if $\mathrm{r}(M_i, N)$ is a noetherian left T-module for all $i \in I$.

(3) If $\mathcal{M}$ consists of finitely generated modules, then $\mathcal{M}$ has (generalized) left almost split morphisms if and only if $\mathrm{r}(M_i, M)$ is a finitely generated (noetherian) left S-module for all $i \in I$.

COROLLARY 3.2. *Assume that $\mathcal{M}$ is a finite subcategory consisting of modules with local endomorphism ring.*

(1) $\mathcal{M}$ has left (respectively, right) almost split morphisms if and only if $J(S)$ is a finitely generated left (respectively, right) S-module.

(2) $\mathcal{M}$ has generalized left (respectively, right) almost split morphisms if and only if $J(S)$ is a noetherian left (respectively, right) S-module.

(3) Assume that all M_i have finite length. Then $\mathcal{M}$ has left (respectively, right) almost split morphisms if and only if S is left (respectively, right) artinian if and only if $\mathcal{M}$ has generalized left (respectively, right) almost split morphisms.

Proof. Note first that we have $M = N$ and $S = T$. Then (1) and (2) follow from 2.1 and 2.2, or 2.4 and 2.5, respectively, and the fact observed in Lemma 1.1 that $J(S) = \mathrm{r}(M, M) \simeq \bigoplus_{i=1}^{n} \mathrm{r}(M_i, M) \simeq \bigoplus_{i=1}^{n} \mathrm{r}(M, M_i)$.

(3) If all M_i have finite length, then so does M, and S is therefore semiprimary. Hence S is left artinian if and only if $_S J(S)$ is finitely generated if and only if $_S J(S)$ is noetherian, and the symmetric statement holds on the right side. This proves the claim. □

Note that there are categories with left and right almost split morphisms where the module M is not endonoetherian.

EXAMPLE 3.3. Let R be a hereditary artin algebra, and denote by $\mathcal{P} = \{P_j \mid j \in J\}$, respectively $\mathcal{I} = \{I_k \mid k \in K\}$, a complete irredundant set of representatives of the isomorphism classes of the indecomposable preprojective right modules, respectively of the indecomposable preinjective left modules. Set $P_R = \coprod_{j \in J} P_j$, and $_R I = \coprod_{k \in K} I_k$. Then add $\mathcal{P}$ is closed under submodules, hence it is covariantly finite in mod R by [2, 4.7], and it has right and left almost split morphisms. Dually, add $\mathcal{I}$ is contravariantly finite in R mod and has right and left almost split morphisms. However, P_R is endonoetherian if and only if R is of tame representation type, and $_R I$ is endonoetherian if and only if R is of finite representation type.

To prove the latter statements, we argue as in [12, Corollary 12]. Observe first that for the usual artin algebra duality $D : \mathrm{mod}\, R \to R\,\mathrm{mod}$ we have $D(P_R) \cong \prod_{j \in J} D(P_j) \cong \prod_{k \in K} I_k$. Moreover, it follows from [12, Proposition 3] that P_R satisfies the ascending (descending) chain condition for finite matrix subgroups if and only if the module $D(P_R)$ satisfies the descending (ascending) chain condition for finite matrix subgroups, which is further equivalent to the fact that $_R I$ satisfies the descending (ascending) chain condition for finite matrix subgroups. But we know from [12, Observation 8] that for the pure-projective modules P_R and $_R I$ the ascending chain condition for finite matrix subgroups is equivalent to endonoetherianness. So, P_R, respectively $_R I$, is endonoetherian if and only if $_R I$, respectively P_R, is Σ-pure-injective. Now, it was shown in [9, 4.6] that $_R I$ is Σ-pure-injective if and only if the algebra is tame. Furthermore, if P_R is Σ-pure-injective, then we know from [12, Proposition 4] that the class of homomorphisms between modules in

$\mathcal{P}$ is noetherian, and this implies by [1, 5.6] that R is of finite representation type. Conversely, if R is of finite representation type, then every module is endofinite [12, Theorem 6] and hence endonoetherian. So, the proof is complete.

We close this chapter with an application to pure-semisimple rings. Namely, we can now rediscover a characterization of these rings in terms of generalized right almost split morphisms given by Brune in [3] and later improved by Zimmermann in [11]. Moreover, we give a dual characterization in terms of generalized left almost split morphisms. Recall that a ring R is said to be **right pure-semisimple** if every right R-module is a direct sum of finitely presented modules.

THEOREM 3.4. (1) (cp. [3, §3, Corollary 1] and [11, p. 372]) A right artinian ring R is right pure-semisimple if and only if every finitely presented indecomposable right R-module has a generalized right almost split morphism in $\mathrm{mod}\,R$.

(2) A semilocal ring R is left pure-semisimple if and only if every finitely presented indecomposable right R-module has a generalized left almost split morphism in $\mathrm{mod}\,R$.

Proof. (1) Let $\{M_i \mid i \in I\}$ be a complete irredundant set of representatives of the isomorphism classes of the finitely presented right R-modules, and set $M = \coprod_{i \in I} M_i$ with $S = \mathrm{End}_R M$. Further, let W_R be a minimal injective cogenerator of $\mathrm{Mod}\,R$, and $E = \mathrm{End}_R W$. As observed by Zimmermann in [11, p. 372], we can assume that R is a right Morita ring. In fact, by [11, Theorem 4], this property follows from the existence of right almost split morphisms in $\mathrm{mod}\,R$ for each simple non-projective module C_R, and is therefore satisfied whenever every finitely presented indecomposable right R-module has a generalized right almost split morphism in $\mathrm{mod}\,R$ or when R is right pure-semisimple. So, the module W_R induces a Morita-duality $\mathrm{mod}\,R \to E\,\mathrm{mod}$, and we have that $M^* = {}_E\mathrm{Hom}_R(M,W)_S$ is an E-S-bimodule with $E \cong \mathrm{Hom}_R(\mathrm{Hom}_R(M,W) \otimes_S M, W) \cong \mathrm{End}_S M^*$ and $S \cong (\mathrm{End}_E M^*)^{op}$. Moreover, the ${}_E\mathrm{Hom}_R(M_i,W)$ form a complete irredundant set of representatives of the isomorphism classes of the finitely presented left E-modules. Further, we know from [8, Theorem 7] or [10] that R is right pure-semisimple if and only if so is E, which means by [12, Theorem 6] that every pure-projective left E-module is endonoetherian, or in other words, that the left E-module $\coprod_{i \in I} \mathrm{Hom}_R(M_i,W)$ is endonoetherian. But by Proposition 2.3 the latter is equivalent to $\prod_{i \in I} \mathrm{Hom}_R(M_i,W) \cong {}_E M^*$ being endonoetherian. So, we see that R is right pure-semisimple if and only if M_S^* is noetherian, and by Proposition 2.6 the proof is complete.

(2) Take M as in (1). By 2.3 we know that every finitely presented indecomposable module A_R has a generalized left almost split morphism if and only if M is endonoetherian. But the latter means that every pure-projective module is endonoetherian, which by [12, Theorem 6] is equivalent to R being left pure-semisimple. □

REFERENCES

[1] M. AUSLANDER: A functorial approach to representation theory, Lecture Notes in Math. 944 (1982), 105-179.

[2] M. AUSLANDER, S. O. SMALØ: Preprojective modules over artin algebras, J. Algebra 66 (1980), 61-122.

[3] H. BRUNE, On a theorem of Kulikov for artinian rings, Comm. Alg. 10 (1982), 433-448.

[4] N. V. DUNG, Preinjective modules and finite representation type of artinian rings, to appear in Comm. Algebra.

[5] N. V. DUNG, Strong preinjective partitions and almost split morphisms, preprint.

[6] A. FACCHINI, Module Theory. Endomorphism rings and direct sum decompositions in some classes of modules. Progress in Math. 167, Birkhäuser (1998).

[7] B. HUISGEN-ZIMMERMANN, unpublished.

[8] H. L. HULLINGER: Stable equivalence and rings whose modules are a direct sum of finitely generated modules, J. of Pure and Appl. Algebra 16 (1980) 265-273.

[9] H. LENZING: Homological transfer from finitely presented to infinite modules, Lecture Notes in Math. 1006 (1983), 734-761.

[10] D. SIMSON: Pure semisimple categories and rings of finite representation type (Corrigendum), J. Algebra 67 (1986) 254-256.

[11] W. ZIMMERMANN: Auslander-Reiten sequences over artinian rings, J. Algebra 119 (1988), 366-392.

[12] B. ZIMMERMANN-HUISGEN, W. ZIMMERMANN: On the sparsity of representations of rings of pure global dimension zero, Trans. Amer. Math. Soc. 320 (1990), 695-711.

Actions of Hopf algebras on quantum polynomials[1]

VYACHESLAV A. ARTAMONOV Department of Algebra, Faculty of Mechanics and Mathematics, Moscow State University, Moscow, 119899, RUSSIA, email: artamon@mech.math.msu.su

To the memory of my mother

ABSTRACT In this paper we classify actions of quantum groups on quantum affine spaces in terms of their coordinate algebras of affine spaces and quantum groups. In other words we classify coactions of commutative Hopf algebras on quantum polynomials.

Let k be a field with a fixed matrix $Q = (q_{ij}) \in \mathrm{Mat}(n, k)$ whose entries $q_{ij} \in k^*$, $i, j = 1, \ldots, n$ are called *multiparameters*. It is assumed that multiparameters satisfy the relations $q_{ii} = q_{ij}q_{ji} = 1$ for all i, j. Let r be an integer $0 \leq r \leq n$.

DEFINITION 1. Denote by

$$\Lambda = k_Q[X_1^{\pm 1}, \ldots, X_r^{\pm 1}, X_{r+1}, \ldots, X_n] \tag{1}$$

the associative k-algebra with a unit element generated by elements

$$X_1, X_1^{-1}, \ldots, X_r, X_r^{-1}, X_{r+1}, \ldots, X_n$$

subject to defining relations

$$\begin{aligned} X_i X_i^{-1} &= X_i^{-1} = 1, & i &= 1, \ldots, r; \\ X_i X_j &= q_{ij} X_j X_i, & 1 &\leq i, j \leq n. \end{aligned} \tag{2}$$

The algebra (2) is an algebra of *quantum polynomials*.

---

[1] Research partially supported by Grants RFFI 99-01-00382, 00-15-96128, Program "Universities of Russia"-5527

It is pretty clear that every element of Λ has a unique representation as a linear combination of monomials

$$X_1^{i_1} \cdots X_r^{i_r} X_{r+1}^{i_r} \cdots X_n^{i_r}, \quad i_j \in \mathbb{Z}, \quad i_{r+1}, \ldots, i_n \geq 0.$$

DEFINITION 2. The algebra Λ is a *general* algebra of quantum polynomials if all multiparameters $q_{ij}, 1 \leq i < j \leq n$, are independent in the multiplicative group k^* of the field k. It means that each equality $\prod_{1 \leq i > j \leq n} q_{ij}^{m_{ij}} = 1$ for some integers m_{ij} implies $m_{ij} = 0$ for each pair of indices $i < j$.

A survey of results on quantum polynomials can be found in [A1]. In the paper [A2] we study Morita-equivalent general quantum polynomials and prove Zarisky-type theorem for these algebras. In the paper [AW] the authors study automorphisms of general algebras of quantum polynomials and their invariants. The paper [A3] is concerned with automorphisms of division rings of fractions of general algebras of quantum polynomials (quantum rational functions). In these papers we study actions of group algebras kG on the algebra Λ (and its division ring of fractions). A group algebra is a cocommutative cosemisimple Hopf algebra and its dual $(kG)^*$ is a direct sum of $|G|$ copies of the field k provided G is a finite group. In the present paper we generalize the situation and consider coactions of a commutative semisimple Hopf algebra on a general algebra of quantum polynomials. Throughout the paper we shall assume that Λ is a general algebra of quantum polynomials and $0 \leq r \leq n$.

For further consideration we need [AW, Theorem 2.1]. For reader's convenience we present a modified proof of this theorem.

THEOREM 1. Suppose that α is an endomorphism of the algebra Λ and there exist at least three variables which do not belong to the kernel of α. Then there exists an integer $\epsilon = \pm 1$ such that, for any $w = 1, \ldots, n$

$$\alpha(X_w) = \alpha_w X_w^\epsilon, \text{ where } \alpha_w \in k. \tag{3}$$

If $r < n$ then $\epsilon = 1$.

Proof. We shall consider a lexicographic order on the set of multi-indices $\mathbb{Z}^n$. Denote by $a_i, i = 1, \ldots, n$, the smallest (the leading) term of $\alpha(X_i)$ provided $\alpha(X_i) \neq 0$.

Suppose that $\alpha(X_i), \alpha(X_j) \neq 0$. Observe that the smallest (the leading) term of a product of non zero polynomials in Λ is equal to the product of the smallest (the leading) terms of factors. Thus (2) implies

$$a_i a_j = q_{ij} a_j a_i. \tag{4}$$

Suppose that

$$a_i = \beta X_1^{l_1} \cdots X_n^{l_n}, \quad a_j = \gamma X_1^{t_1} \cdots X_n^{t_n}, \text{ where } \beta, \gamma \in k^*.$$

Now (2) (see [MP]) implies

$$\beta\gamma \left(\prod_{r>s} q_{rs}^{l_r t_s} \right) X_1^{l_1+t_1} \cdots X_n^{l_n+t_n} = \beta\gamma q_{ij} \left(\prod_{r>s} q_{rs}^{t_r l_s} \right) X_1^{l_1+t_1} \cdots X_n^{l_n+t_n}.$$

Hence
$$\left(\prod_{r>s} q_{rs}^{l_r t_s}\right) = q_{ij} \left(\prod_{r>s} q_{rs}^{t_r l_s}\right). \tag{5}$$

Suppose that $i > j$. If $r > s$ then Definition 2 and (5) imply
$$l_r t_s = \delta_{ri}\delta_{sj} + t_r l_s. \tag{6}$$

Consider the matrix
$$\begin{pmatrix} l_1 & \cdots & l_j & \cdots & l_i & \cdots & l_n \\ t_1 & \cdots & t_j & \cdots & t_i & \cdots & t_n \end{pmatrix}, \quad n \geq 3.$$

Let for example $l_p \neq 0$ and $p \neq j, i$. For every index $q \neq p$ By (6) we have
$$\begin{vmatrix} l_p & l_q \\ t_p & t_q \end{vmatrix},$$
and therefore $t_q = t_p l_q l_p^{-1}$. In particular
$$t_i = t_p l_i l_p^{-1}, \quad t_j = t_p l_j l_p^{-1},$$
that is
$$l_i t_j - l_j t_i = l_i t_p l_j l_p^{-1} - l_j t_p l_i l_p^{-1} = 0,$$
which contradicts (6). Thus $l_p = 0$ for all $p \neq i, j$. Similarly one can prove that $t_p = 0$ if $p \neq i, j$.

Hence
$$a_i = \beta X_i^{l_i} X_j^{l_j}, \quad a_j = \gamma X_i^{t_i} X_j^{t_j}, \quad \text{where } l_i t_j - l_j t_i = 1.$$

By the assumption there exists a third variable X_u such that $\alpha(X_u) \neq 0$. The preceding argument applied to the pairs of indices $(i,u), (j,u)$ shows that
$$a_u = \delta X_u^{r_u} X_i^{r_i} = \lambda X_u^{d_u} X_j^{d_j}, \quad \text{where } \delta, \lambda \in k^*.$$

Finally $r_i = d_j = 0$, that is $a_u = \delta X_u^{r_u}$. Similarly
$$a_i = \beta X_i^{l_i}, \quad a_j = \gamma X_j^{t_j}, \quad l_i t_j = 1,$$
and therefore $l_i = t_j = \epsilon = \pm 1$.

We have now to show that the least and the leading terms of $\alpha(X_i)$ coincide and therefore they are equal to a_i. If this is not the case, then $r = n$ and the least term of $\alpha(X_i)$ has the form
$$\gamma_i' X_i^{-1}, \quad \gamma_i' \in k^*, \quad i = 1, \ldots, n.$$

Then for and index $i = 1, \ldots, n$
$$\alpha(X_i) = \gamma_i' X_i^{-1} + \sum_s \gamma_i''(s) X_1^{m_{i1}(s)} \cdots X_n^{m_{in}(s)}, \tag{7}$$

where $\gamma_i', \gamma_i''(s) \in k^*$ and the sum is taken over all multi-indices
$$(m_{i1}(s), \ldots, m_{in}(s)) \in \mathbb{Z}^n$$
such that
$$(0, \ldots, 0, \overset{i}{-1}, 0, \ldots, 0) < (m_{i1}(s), \ldots, m_{in}(s)) \leq (0, \ldots, 0, \overset{i}{1}, 0, \ldots, 0).$$
Thus,
$$m_{i1}(s) = \cdots = m_{i,i-1}(s) = 0, \quad m_{ii}(s) = -1, 0, 1;$$
$$m_{i,i+1}(s) \begin{cases} > 0, & i < n, \\ \leq 0, & i < n, \end{cases} \quad \begin{array}{l} \text{if } m_{ii}(s) = -1, \\ \text{if } m_{ii}(s) = 1. \end{array} \tag{8}$$

For any index $i = 1, \ldots, n$ pick the least monomial $\gamma_i'' X_i^{m_{ii}} \cdots X_n^{m_{in}}$ in (7). Then $\alpha(X_i)\alpha(X_j) = q_{ij}\alpha(X_j)\alpha(X_i), i < j$, implies

$$\gamma_i' X_i^{-1} \gamma_j' X_j^{-1} + \gamma_i' X_i^{-1} \gamma_j'' X_j^{m_{jj}} \cdots X_n^{m_{jn}} + \tag{9}$$
$$\gamma_i'' X_i^{m_{ii}} \cdots X_n^{m_{in}} \gamma_j' \gamma_j' X_j^{-1} + \cdots =$$
$$q_{ij} \gamma_j' X_j^{-1} \gamma_i' X_i^{-1} + q_{ij} \gamma_j' X_j^{-1} \gamma_i'' X_i^{m_{ii}} \cdots X_n^{m_{in}} + \tag{10}$$
$$q_{ij} \gamma_j'' X_j^{m_{jj}} \cdots X_n^{m_{jn}} \gamma_i'' \gamma_i' X_i^{-1} + \cdots, \tag{11}$$

where $+ \cdots$ is the sum of monomials $\delta X_1^{l_1} \cdots X_n^{l_n}$, $\delta \in k^*$, such that

$$(l_1, \ldots, l_n) > \min((0, \ldots, 0, m_{ii}, \ldots, m_{i,j-1}, m_{ij} - 1, m_{i,j+1}, \ldots, m_{in}),$$
$$(0, \ldots, 0, \overset{i}{-1}, 0, \ldots, 0, m_{jj}, \ldots, m_{jn})).$$

The equalities $\gamma_i' X_i^{-1} \gamma_j' X_j^{-1} = q_{ij} \gamma_j' X_j^{-1} \gamma_i' X_i^{-1}$ and (9) imply

$$\gamma_i' X_i^{-1} \gamma_j'' X_j^{m_{jj}} \cdots X_n^{m_{jn}} + \gamma_i'' X_i^{m_{ii}} \cdots X_n^{m_{in}} \gamma_j' \gamma_j' X_j^{-1} + \cdots$$
$$q_{ij} \gamma_j' X_j^{-1} \gamma_i'' X_i^{m_{ii}} \cdots X_n^{m_{in}} + q_{ij} \gamma_j'' X_j^{m_{jj}} \cdots X_n^{m_{jn}} \gamma_i'' \gamma_i' X_i^{-1} + \cdots. \tag{12}$$

Suppose first that

$$(0, \ldots, 0, m_{ii}, \ldots, m_{i,j-1}, m_{ij} - 1, m_{i,j+1}, \ldots, m_{in}) \leq$$
$$(0, \ldots, 0, \overset{i}{-1}, 0, \ldots, 0, m_{jj}, \ldots, m_{jn}).$$

Then $m_{ii} = -1$ and $i + 1 \leq j - 1$ implies then $m_{i,i+1} = 0$, which contradicts (8). Hence $i \leq j - 2$ implies

$$(0, \ldots, 0, m_{ii}, \ldots, m_{i,j-1}, m_{ij} - 1, m_{i,j+1}, \ldots, m_{in}) >$$
$$(0, \ldots, 0, \overset{i}{-1}, 0, \ldots, 0, m_{jj}, \ldots, m_{jn}),$$

and therefore in (12) we obtain

$$\gamma_i' X_i^{-1} \gamma_j'' X_j^{m_{jj}} \cdots X_n^{m_{jn}} = q_{ij} \gamma_j'' X_j^{m_{jj}} \cdots X_n^{m_{jn}} \gamma_i'' \gamma_i' X_i^{-1}.$$

Then as in (5) we obtain
$$1 = q_{ji}^{-1} \prod_{r \geq j > i} q_{ri}^{-m_{jr}}.$$

Thus,
$$m_{jr} = \begin{cases} -1, & r = j, \\ 0, & r > j, \end{cases}$$

a contradiction with (8).

We have shown that if $1 \leq i \leq j - 2 < j \leq n$ then either $\alpha(X_i)$, or $\alpha(X_j)$ is of the form (3) with $\epsilon = -1$ provided $w = i, j$. Let (3) holds for some $w = 1, \ldots, n$. The preceding considerations show that the leading term of each $\alpha(X_r) \neq 0$ is of the form $\gamma_r X_r^{-1}$, i. e. (3) holds for any index. $\square$

THEOREM 2. *Let H be a commutative semisimple Hopf k-algebra and Λ an H-comodule algebra with a structure morphism $\rho : \Lambda \to H \otimes_k \Lambda$. Assume that either condition is satisfied:*

1) $3 \leq r < n$;

2) $r \leq 2$ and the algebra H is finite dimensional.

Then there exist elements $\alpha_1, \ldots, \alpha_n \in H$ such that, for any $i = 1 \ldots, n$

$$\rho(X_i) = \alpha_i \otimes X_i, \text{ where } \Delta(\alpha_i) = \alpha_i \otimes \alpha_i, \quad \varepsilon(\alpha_i) = 1, \quad S(\alpha_i) = \alpha_i^{-1}.$$

Proof. By the assumption ρ is a k-algebra homomorphism and H is a subdirect sum of fields $K_j, j \in J$. Let $\pi_j : H \to K_j$ be a corresponding projection. Then $\pi_j \rho : \Lambda \to K_j \otimes_k \Lambda$ is a k-algebra homomorphism inducing an embedding of the field k into the field K_j. Thus $\pi_j \rho$ induces an endomorphism of the general quantum polynomials algebra $K_j \otimes_k \Lambda$ over the field K_j. By the assumption $\pi_j \rho(X_1), \pi_j \rho(X_2), \pi_j \rho(X_3) \neq 0$. According to Theorem 1 there exists an integer $\epsilon_j = \pm 1$ such that for each index $j = 1, \ldots, n$

$$\pi_j \rho(X_i) = \alpha_i \otimes X_i^{\epsilon_j}, \quad \alpha_i \in K_j.$$

Moreover $r < n$ implies $\epsilon_j = 1$.

Suppose now that $r \leq 2$ and the algebra H has finite dimension. By [M, Theorem 2.3.1] there exists a finite group G and a field extension F/k such that $F \otimes_k H \simeq (FG)^*$ where $(FG)^*$ is the Hopf dual of the group algebra FG. In addition $F \otimes_k \Lambda$ is a FG-module algebra [M, Lemma 1.6.4]. It means that the group G acts as an automorphism group of F-algebra $F \otimes_k : \Lambda$.

Let $g \in G$. According to Theorem 1 there exist elements $\theta_1, \ldots, \theta_n \in F$ such that

$$g(X_1) = \theta_1 X_1, \ldots, g(X_n) = \theta_n X_n. \tag{13}$$

Then (13) holds for every $g \in FG$. By [M, Lemma 1.6.4], for each $i = 1, \ldots, n$,

$$\rho(X_i) = \alpha_i \otimes X_i \in (F \otimes_k H \otimes_k \Lambda) \cap (H \otimes_k \Lambda) \implies \alpha_i \in H, \tag{14}$$

and $\alpha_i(g) = \theta_i$. It follows $\rho(X_i) = \alpha_i \otimes X_i$ that for each index i.

In both cases considered in Theorem we have proved that
$$\rho(X_i) = \alpha_i \otimes X_i, \quad i = 1\ldots, n.$$
Now
$$\alpha_i \otimes \alpha_i \otimes X_i = (1 \otimes \rho)\rho(X_i) = (\Delta \otimes 1)\rho(X_i) = \Delta(\alpha_i) \otimes X_i$$
and $X_i = \varepsilon(\alpha_i)X_i$. It follows that
$$\Delta(\alpha_i) = \alpha_i \otimes \alpha_i, \quad \varepsilon(\alpha_i) = 1, \quad S(\alpha_i) = \alpha_i^{-1}.$$
$\square$

THEOREM 3. *Let H be a commutative Artinian semisimple Hopf k-algebra and Λ an H-comodule algebra with a structure morphism $\rho : \Lambda \to H \otimes_k \Lambda$. Suppose that $n = r \geq 3$. Then $H = A \oplus B$ is a direct sum of ideals A, B such that $B \subseteq \ker \varepsilon$ and for any $i = 1, \ldots, n$ there exist invertible elements $\lambda_i \in A, \mu_i \in B$ for which*
$$\rho(X_i) = \lambda_i \otimes X_i + \mu_i \otimes X_i^{-1}. \tag{15}$$
Moreover
$$\Delta(\lambda_i) = \lambda_i \otimes \lambda_i + \mu_i \otimes \mu_i^{-1} \in (A \otimes A) \oplus (B \otimes B) \subset H \otimes H,$$
$$\Delta(\mu_i) = \lambda_i \otimes \mu_i + \mu_i \otimes \lambda_i^{-1} \in (A \otimes B) \oplus (B \otimes A) \subset H \otimes H,$$
$$\varepsilon(\lambda_i) = 1, \quad \varepsilon(\mu_i) = 0, \quad S(\lambda_i) = \lambda_i^{-1}, \quad S(\mu_i) = \mu_i. \tag{16}$$

Proof. It follows from the assumption that
$$H = K_1 \oplus \cdots \oplus K_m, \tag{17}$$
where $K_1, \ldots, K_m$ are field extensions of k. Observe that the counit $\varepsilon : H \to k$ is an algebra homomorphism. So we can always assume that $K_1 = k$, $\ker \varepsilon = K_2 \oplus \cdots \oplus K_m$. From the definition of ρ it follows that $(\varepsilon \otimes 1)\rho(X_i) = X_i$ for any $i = 1, \ldots, n$. Hence $\epsilon_1 = 1$.

Suppose now that $\epsilon_j = 1$ for $j = 1, \ldots, p$ and $\epsilon_j = -1$ if $j = p+1, \ldots, m$. Let $p < m$. Set
$$A = K_1 \oplus \cdots \oplus K_p, \quad B = K_{p+1} \oplus \cdots \oplus K_m.$$
Then $B \subseteq \ker \varepsilon$. Moreover, for any index $i = 1, \ldots, n$,
$$\rho(X_i) = (\lambda_i \otimes X_i) \oplus (\mu_i \otimes X_i^{-1}) \in (A \otimes \Lambda) \oplus (B \otimes \Lambda),$$
where $\lambda_i \in A, \mu_i \in B$. Since the elements $X_1, \ldots, X_n$ are invertible in Λ, the elements λ_i, μ_i are invertible in A and B, respectively. Moreover in $H \otimes H \otimes \Lambda$ we have
$$(\Delta \otimes 1)\rho(X_i) = (1 \otimes \rho)\rho(X_i) =$$
$$[(\lambda_i \otimes \lambda_i) \oplus (\mu_i \otimes \mu_i^{-1})] \otimes X_i + [(\lambda_i \otimes \mu_i) \oplus (\mu_i \otimes \lambda_i^{-1})] \otimes X_i^{-1},$$
and therefore
$$\Delta(\lambda_i) = (\lambda_i \otimes \lambda_i) \oplus (\mu_i \otimes \mu_i^{-1}) \in (A \otimes A) \oplus (B \otimes B); \tag{18}$$
$$\Delta(\mu_i) = (\lambda_i \otimes \mu_i) \oplus (\mu_i \otimes \lambda_i^{-1}) \in (A \otimes B) \oplus (B \otimes A). \tag{19}$$

Since $B \subseteq \ker \varepsilon$, (18) implies $\lambda_i = \lambda_i \varepsilon(\lambda_i)$. Therefore $\varepsilon(\lambda_i) = 1$ for each index $i = 1, \ldots, n$, because λ_i is invertible in A. Moreover (18) and (19) imply

$$1 = \varepsilon(\lambda_i) = S(\lambda_i)\lambda_i + S(\mu_i)\mu_i^{-1}; \tag{20}$$
$$0 = \varepsilon(\mu_i) = S(\lambda_i)\mu_i + S(\mu_i)\lambda_i^{-1}. \tag{21}$$

But $H\lambda_i = A, H\mu_i = B$. Thus (20), (21) imply

$$S(\lambda_i)\lambda_i = 1_A, \quad S(\mu_i)\mu_i^{-1} = 1_B, \quad S(\lambda_i)\mu_i = S(\mu_i)\lambda_i^{-1} = 0.$$

Since the elements λ_i, μ_i are invertible in A and B, respectively, we deduce that $S(\lambda_i) \in A$ and $S(\mu_i) \in B$. Thus we have

$$S(\lambda_i) = \lambda_i^{-1} \in A, \quad S(\mu_i) = \mu_i, \quad \varepsilon(\lambda_i) = 1, \quad \varepsilon(\mu_i) = 0.$$

$\square$

COROLLARY 1. *Let $H, \Lambda, \rho, A, B, \lambda_i, \mu_i$ be from Theorem 3 and H has finite dimension. Then there exists a positive integer d such that*

$$\lambda_1^d = \cdots = \lambda_n^d = 1_A, \quad \mu_1^d = \cdots = \mu_n^d = 1_B. \tag{22}$$

Proof. Since $\lambda_i \mu_i = 0$ it follows form (16) that for any $m \geq 0$

$$\Delta(\lambda_i^m) = [(\lambda_i \otimes \lambda_i) + (\mu_i \otimes \mu_i)]^m = (\lambda_i^m \otimes \lambda_i^m) + (\mu_i^m \otimes \mu_i^{-m}). \tag{23}$$

Denote by $\pi : H \to A$ the natural projection which is a k-algebra homomorphism. Since the multiplication map $m : H \otimes H \to H$ is also an algebra homomorphism the mapping $\zeta = \pi m \Delta : A \to A$ is again a k-algebra homomorphism and by (23) $\zeta(\lambda_i^m) = \lambda_i^{2m}$. Take a minimal polynomial $f(T) \in k[T]$ for an element $\lambda_i \in A$. Applying the algebra homomorphism ζ we see that $f(\lambda^{2^s}) = 0$, for any $s \geq 1$. But A is a direct sum of some fields K_j from (17). So projections of invertible elements $\lambda^{2^s}, s \geq 0$, have finitely many images in each K_j. Clearly there exists positive integers $m > m'$ such that $\lambda^{2^m} = \lambda^{2^{m'}}$. Put $d = 2^m - 2^{m'}$.

By (23)

$$\Delta(1_A) = \Delta(\lambda_i^d) = \Delta(\lambda_i)^d =$$
$$(\lambda_i^d \otimes \lambda_i^d) + (\mu_i^d \otimes \mu_i^d) = (1_A \otimes 1_B) + (\mu_i^d \otimes \mu_i^d). \tag{24}$$

But $\Delta(1_A)$ is an idempotent of $H \otimes H$ lying in $(A \otimes A) \oplus (B \otimes B)$. Hence (24) shows that an invertible element $\mu_i^d \otimes \mu_i^d \in B$ is an idempotent. An easy exercise shows that $\mu_i^d \otimes \mu_i^d = 1_B \otimes 1_B$ and therefore $\mu_i^d = 1_B$. $\square$

THEOREM 4. *Let $H, \Lambda, \rho, \alpha_1, \ldots, \alpha_n$ be as in Theorem 2. The subalgebra of coinvariants $\Lambda_H = \Lambda^{coH}$ is generated by all monomials $X_1^{m_1} \cdots X_n^{m_n}$ such that*

$$\alpha_1^{m_1} \cdots \alpha_n^{m_n} = 1. \tag{25}$$

In particular, Λ is a finitely generated left (right) Λ_H-module if and only if

$$\alpha_1, \ldots, \alpha_n$$

are roots of 1.

Proof. Recall that Λ_H consists of all elements $g \in \Lambda_H$ such that $\rho(g) = 1 \otimes g$. Let

$$g = \sum \nu_{m_1,\ldots,m_n} X_1^{m_1} \cdots X_n^{m_n}, \quad \nu_{m_1,\ldots,m_n} \in k.$$

By Theorem 2

$$\rho(g) = \sum \alpha_1^{m_1} \cdots \alpha_n^{m_n} \otimes \nu_{m_1,\ldots,m_n} X_1^{m_1} \cdots X_n^{m_n} \in H \otimes \Lambda.$$

Thus $g \in \Lambda_H$ if and only if (25) holds.

If $\alpha_1, \ldots, \alpha_n$ are roots of degree d of 1, then $X_1^d, \ldots, X_n^d \in \Lambda_H$ and therefore Λ is a finitely generated left (right) Λ_H-module.

Conversely if Λ is a finitely generated left (right) Λ_H-module then according to [AW, Theorem 3.17] there exists an integer d such that $X_i^d \in \Lambda_H$ for every $i = 1, \ldots, n$. Then $\alpha_i^d = 1$. □

COROLLARY 2. *Let $H, \Lambda, \rho, \alpha_1, \ldots, \alpha_n$ be as in Theorem 2, and the algebra H has finite dimension. Then Λ is a finitely generated left (right) Λ_H-module.*

Proof. Let G be the set of group-like elements in H. Then G is a group and $\alpha_1, \ldots, \alpha_n \in G$. The group algebra kG is a subalgebra in H. Hence G is finite and $\alpha_1, \ldots, \alpha_n$ have finite orders. □

Note that Corollary 2 follows also from [M, Lemma 1.7.2, Theorem 4.2.1].

THEOREM 5. *Let $H, \rho, A, B, \lambda_i, \nu_i$ be from Theorem 3. Suppose that H has a finite dimension. Then Λ is a finitely generated left (right) module over the subalgebra of coinvariants Λ_H.*

Proof. By Corollary 1 there exists a positive integer d such that (22) holds. Put $f_i = X_i^d + X_i^{-d}$. Since $\lambda_i \mu_i = 0$ we have

$$\Delta(f_i) = \Delta(X_i)^d + \Delta(X_i)^{-d} =$$
$$[\lambda_i \otimes X_i + \mu_i \otimes X - i^{-1}]^d + [\lambda_i^{-1} \otimes X_i^{-1} + \mu_i^{-1} \otimes X_i]^d =$$
$$(\lambda_i^d + \mu_i^{-d}) \otimes X - i^d + (\lambda_i^{-d} + \mu_i^d) \otimes X_i^{-d} =$$
$$(1_A + 1_B) \otimes X_i^d + (1_A + 1_B) \otimes X_i^{-d} = 1 \otimes f_i.$$

Thus $f_i \in \Lambda_H$. Denote by Γ the subalgebra in Λ generated by the elements $f_1, \ldots, f_n$. Then $\Gamma \subseteq \Lambda_H$ and Λ is a finitely generated Γ-module [AW]. □

Consider now the dual space $H^* = \hom_k(H, k)$. Then H^* is an algebra with respect to convolution multiplication $l * l'$ where, for any $h \in H$

$$(l * l')(h) = \sum_h l(h_{(1)}) l'(h_{(2)}),$$

and

$$\Delta(h) = \sum_h h_{(1)} \otimes h_{(2)} \in H \otimes H.$$

Moreover Λ is a left H^*-module algebra with respect to the action

$$l \circ X_i = l(\alpha_i) X_i, \quad i = 1, \ldots, n. \tag{26}$$

If H has finite dimension over k then $(H \otimes H)^* = H^* \otimes H^*$ and H^* is a Hopf algebra with respect to comultiplication, counit and antipode defined as follows

$$(\Delta l)(h_1 \otimes h_2) = l(h_1 h_2), \quad (\varepsilon l)(h) = l(1), \quad (Sl)(h) = l(Sh).$$

COROLLARY 3. *Let $H, \Lambda, \rho, \alpha_1, \ldots, \alpha_n$ be as in Theorem 2. Then the convolution algebra H^* is commutative.*

Proof. For any monomial $f = X_1^{m_1} \cdots X_n^{m_n}$ we have

$$\rho(f) = \alpha_1^{m_1} \cdots \alpha_n^{m_n} \otimes f. \tag{27}$$

Therefore if $\lambda \in H^*$, then

$$\lambda \circ f = \lambda(\alpha_1^{m_1} \cdots \alpha_n^{m_n}) f.$$

Suppose now that $\lambda, \mu \in H^*$. Then

$$(\lambda * \mu)(\alpha_1^{m_1} \cdots \alpha_n^{m_n}) = (\lambda \otimes \mu)(\Delta(\alpha_1^{m_1} \cdots \alpha_n^{m_n})) =$$
$$(\lambda \otimes \mu)(\alpha_1^{m_1} \cdots \alpha_n^{m_n} \otimes \alpha_1^{m_1} \cdots \alpha_n^{m_n}) = \lambda(\alpha_1^{m_1} \cdots \alpha_n^{m_n}) \mu(\alpha_1^{m_1} \cdots \alpha_n^{m_n}) =$$
$$(\mu * \lambda)(\alpha_1^{m_1} \cdots \alpha_n^{m_n}).$$

It means that $(\lambda * \mu) \circ f = (\mu * \lambda) \circ f$. $\square$

We have already mentioned above that if H is a finite dimensional commutative semisimple Hopf algebra over an algebraically closed field k then there exists a finite group G such that $H \simeq (kG)^*$, [M, Theorem 2.3.1]. Moreover by [M, Lemma 1.7.2] the subalgebra of invariants Λ^{H^*} coincides with subalgebra of coinvariants Λ_H. Recall also that a subalgebra in Λ generated by $X_1, \ldots, X_n$ is an H-comodule subalgebra of Λ.

Let be given an arbitrary $\mathbb{Z}^n$-graded algebra A

$$A = \oplus A_l, \quad l \in \mathbb{Z}^n. \tag{28}$$

Then $A_l A_{l'} \subseteq A_{l+l'}$ for any $l, l' \in \mathbb{Z}^n$. Take a free multiplicative Abelian group G with free generators $g_1, \ldots, g_n$. By [M, Example 4.1.7, p. 41] a grading (28) in A is equivalent to an existence on A of a structure of a kG-comodule algebra with a structure map $\tau : A \to kG \otimes A$, where

$$\tau(a) = g_1^{l_1} \cdots g_n^{l_n} \otimes a, \text{ if } a \in A_{(l_1, \ldots, l_n)}.$$

THEOREM 6. *Let H be a commutative bialgebra with group-like elements*

$$\alpha_1, \ldots, \alpha_n$$

and A from (28). A linear map $\rho : A \to H \otimes A$ such that

$$\rho(a) = \alpha_1^{l_1} \cdots \alpha_n^{l_n} \otimes a, \text{ where } a \in A_{(l_1, \ldots, l_n)},$$

defines an H-comodule structure on A.

Proof. Let $\lambda : kG \to H$ be an algebra homomorphism such that $\lambda(g_i) = \alpha_i$ for $i = 1, \ldots, n$. Then

$$(\lambda \otimes \lambda)\Delta(g_i) = \lambda(g_i \otimes g_i) = \alpha_i \otimes \alpha_i = \Delta(\alpha_i) = \Delta(\lambda(g_i)),$$
$$S(\lambda(g_i)) = \alpha_i^{-1} = \lambda(S(\alpha_i), \quad \varepsilon(\lambda(g_i)) = \varepsilon(\alpha_i) = 1 = \varepsilon(g_i).$$

Thus λ is a Hopf algebra homomorphism. For every element $a \in A_{(l_1,\ldots,l_n)}$ we have

$$(\lambda \otimes 1)\tau(a) = (\lambda \otimes 1)(g_1^{l_1} \cdots g_n^{l_n} \otimes a) = \alpha_1^{l_1} \cdots \alpha_n^{l_n} \otimes a = \rho(a).$$

It follows that $(\lambda \otimes 1)\tau = \rho$, i. e. ρ introduces a structure of a H-comodule algebra on A. □

In noncommutative geometry [D] an quantum polynomial algebra Λ with $r = 0$ is considered together with quantum Grassman algebra Γ. The algebra Γ is generated by elements $\xi_1, \ldots, \xi_n$ subject to defining relations

$$\xi_i^2 = 0, \quad \xi_i \xi_j = p_{ij} \xi_j \xi_i \tag{29}$$

where $p_{ij} \in k^*$, and $p_{ii} = p_{ij} p_{ji} = 1$ for any $i, j = 1, \ldots, n$. The algebra of functions on matrices of size n is introduced in the book [D] as a universal bialgebra $M_{P,Q}(n)$ coacting on the pair of algebras Λ, Γ. In Theorem 6 we have actually considered universal commutative Hopf algebras coacting on Λ, Γ.

REFERENCES

[A1] Artamonov V. A., *Quantum Serre's conjecture*, Uspehi mat. nauk. **53** (1998), N 4., p. 3-76.

[A2] Artamonov V. A., *General quantum polynomials: irreducible modules and Morita-equivalence*, Izv. RAN, ser. math. **63** (1999), N 5. P. 3-36.

[A3] Artamonov V. A., *Division ring of quantum rational functions*, Uspehi mat. nauk. **54** (1999), N 5, p. 154-155.

[AW] Artamonov V. A., Wisbauer R., *Homological properties of quantum polynomials*, Algebras and representation theory (to appear).

[D] Demidov E. E., *Quantum group*, Moscow: Factorial, 1998, 127P.

[MP] McConnell J.C., Pettit J.J., *Crossed products and multiplicative analogues of Weyl algebras*, J. London Math. Soc. **38** (1) (1988), 47-55.

[M] Montgomery S., *Hopf algebras and their actions on rings*, Regional Conference Series in Mathematics, v. **82** – American Math. Soc.: Providence R. I., 1993.

Strongly simply connected derived tubular algebras

IBRAHIM ASSEM Mathématiques et Informatique, Université de Sherbrooke, Sherbrooke, Québec, J1K 2R1, Canada, e-mail: ibrahim.assem@dmi.usherb.ca

ABSTRACT In this note, we show that a derived tubular algebra is strongly simply connected if and only if it contains no full convex subcategory which is hereditary of type $\tilde{\mathbb{A}}_m$, and give several other characterisations of the strong simple connectedness of (derived) tubular algebras.

INTRODUCTION

The objective of this note is to give a simple (and fairly visual) characterisation of the strong simple connectedness of a derived tubular algebra. A finite dimensional algebra A over an algebraically closed field is called derived tubular if there exists a tubular algebra B and an equivalence of triangulated categories between the derived categories $D^b(\text{mod} A) \cong D^b(\text{mod} B)$ of bounded complexes of finitely generated right A- and B-modules, respectively. It is shown in [8] that a derived tubular algebra is simply connected. We here give criteria for such an algebra to be strongly simply connected (in the sense of [17]).

We say that an algebra is *strongly $\tilde{\mathbb{A}}$-free* if it contains no full convex subcategory which is hereditary of type $\tilde{\mathbb{A}}_m$, for any $m \geq 1$. Skowroński has asked in [17], Problem 2, whether it is true that a simply connected algebra is strongly simply connected if and only if it is strongly $\tilde{\mathbb{A}}$-free. The answer to this question is known to be positive if the algebra is iterated tilted of euclidean type [2], or tame tilted [5]. Also, it was shown that there exists a close connection between the strong simple connectedness of an algebra, and the shape of the orbit graphs of the directed components of its Auslander-Reiten quiver [11,4,13]. The main result of this note states that a derived tubular algebra A is strongly simply connected if and only if it is strongly $\tilde{\mathbb{A}}$-free. Further, we give other criteria (using, among others, orbit graphs of directed components), showing that it suffices to consider two particular full subcategories of A. In the case where A is tubular, we (predictably) obtain a much stronger characterisation, using different techniques. The results of this note are applied in [1] to yield a complete characterisation of the (strongly) simply connected tame quasi-tilted and semiregular iterated tubular algebras.

1 STRONGLY SIMPLY CONNECTED TUBULAR ALGEBRAS.

1.1. Throughout this paper, k denotes an algebraically closed field. By algebra is meant a basic and connected finite dimensional k-algebra and by module a finitely generated right module. Such an algebra A can be written as a bound quiver algebra $A \cong kQ_A/I$, where the pair (Q_A, I) is called a *presentation* of A, and may equivalently be considered as a locally bounded k-category, whose object set is denoted by A_0, see [11]. An algebra A is called *triangular* if Q_A has no oriented cycle. A full subcategory C of A is called *convex* if, for any path $a_0 \to a_1 \to \cdots \to a_t$, with $a_0, a_t \in C_0$, we have $a_i \in C_0$ for all i. We use freely properties of the module category mod$A$, the Auslander-Reiten quiver $\Gamma(\text{mod}A)$, and the Auslander-Reiten translations $\tau = \text{DTr}$ and $\tau^{-1} = \text{TrD}$, as can be found, for instance, in [16]. A component Γ of $\Gamma(\text{mod}A)$ is called *directed* if, for all indecomposable modules M in Γ, there exists no sequence $M = M_0 \xrightarrow{f_1} M_1 \xrightarrow{f_2} \cdots \xrightarrow{f_t} M_t = M$ of non-zero non-isomorphisms between indecomposable A-modules. Given a component Γ of $\Gamma(\text{mod}A)$, its *orbit graph* $\mathcal{O}(\Gamma)$ is defined as follows: the points of $\mathcal{O}(\Gamma)$ are the τ-orbits M^τ of the modules M in Γ, there exists an edge $M^\tau \to N^\tau$ if there exist $m, n \in \mathbb{Z}$ and an irreducible morphism $\tau^m M \to \tau^n N$ or $\tau^n N \to \tau^m M$, and the number of such edges equals $\dim_k \text{Irr}(\tau^m M, \tau^n N)$ or $\dim_k \text{Irr}(\tau^n N, \tau^m M)$, respectively (here, $\text{Irr}(X, Y)$ denotes the space of irreducible morphisms from X to Y).

For tubular algebras, we refer the reader to [16]. We need the following facts. Let A be a tubular algebra, then A contains exactly two tame concealed full convex subcategories, denoted by $C^{(0)}$ and $C^{(\infty)}$, and every object of A belongs to $C^{(0)}$ or to $C^{(\infty)}$. Also, A is an extension of $C^{(0)}$, and a coextension of $C^{(\infty)}$, by truncated branches (in the terminology of [9]), and the tubular type of A is one of $(2,2,2,2)$, $(2,3,6)$, $(2,4,4)$, $(3,3,3)$. Further, $\Gamma(\text{mod}A)$ has a postprojective and a preinjective components, which coincide respectively with the postprojective component of $\Gamma(\text{mod}C^{(0)})$ and the preinjective component of $\Gamma(\text{mod}C^{(\infty)})$.

Finally, for simply connected and strongly simply connected algebras, we refer the reader to [17, 6, 3].

1.2. We denote by $B[M]$ the one-point extension of an algebra B by a B-module M.

LEMMA. Assume that B is a representation-finite triangular algebra, and that $A = B[M]$ is simply connected. Then B is simply connected.

Proof. Assume that this is not the case, and let (Q_B, I') be a presentation of B such that the fundamental group $\pi_1(Q_B, I')$ is not trivial. There exists a presentation (Q_A, I) of A such that $I \cap kQ_B = I'$. Let G be an arbitrary abelian group. By [6] (2.4), there exists an exact sequence of abelian groups

$$\text{Hom}(\pi_1(Q_A, I), G) \longrightarrow \text{Hom}(\pi_1(Q_B, I'), G) \longrightarrow G^m$$

(where m is defined as in [6] (2.4)). Since A is simply connected, the first term of this sequence vanishes. Since B is a triangular representation-finite algebra, it is standard, hence the group $\pi_1(Q_B, I')$ is free [15](3.9)(4.3), therefore Hom $(\pi_1(Q_B, I'), G) \neq 0$. We infer that $m \neq 0$, and hence B contains a full subcategory of the form

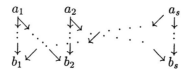

where each of the shown paths is non-zero in B. The full subcategory of B generated by the points $a_1, \ldots, a_s, b_1, \ldots, b_s$ is clearly hereditary of type $\tilde{\mathbb{A}}$, a contradiction to the representation-finiteness of B. □

1.3. LEMMA. *Let A be a tubular algebra, and B be a proper full convex subcategory of A. Then:*

(a) If B is representation-infinite, then B is tilted of euclidean type.

(b) If B is representation-finite, then B is tilted of Dynkin type or of euclidean type $\neq \tilde{\mathbb{A}}$.

Proof. (a) If B is representation-infinite, then it contains a tame concealed full convex subcategory. Now A contains exactly two tame concealed full convex subcategories $C^{(0)}$ and $C^{(\infty)}$. Therefore, B contains one of them (but not both, because, by hypothesis, $B \neq A$). We may assume, by duality, that B contains $C^{(0)}$. Now B is a truncated branch extension of $C^{(0)}$ and A is obtained from B by a sequence of one-point (tubular) extensions. In particular, B is not a tubular algebra. Therefore the tubular type of B is domestic. By [16] (4.9), B is tilted of euclidean type.

(b) Assume that B is representation-finite. Since A is tubular, then it is quasi-tilted, and hence so is B, by [14] (1.15). Since B is representation-finite, it is actually tilted [14] (3.6). On the other hand, the Euler quadratic form $q_B = q_{A|B}$ of B is positive semi-definite, because q_A is. Therefore, B is tilted of Dynkin or of euclidean type. By [10], there remains to show that B is simply connected. Since B is convex in A, there exists a sequence $A = A_0 \supset A_1 \supset \cdots \supset A_t = B$ of full convex subcategories with each A_i either a one-point extension or a one-point coextension of A_{i+1}. Let j be the largest index such that A_j is representation-infinite (thus, A_{j+1} is representation-finite). Note that, since A is tubular, then $t \geq 2$ and $j \geq 1$. Assume that A_{j+1} is not simply connected. By (1.2) and its dual, A_j is not simply connected. Now A_j is a representation-infinite full convex subcategory of A, and it is proper (because $j \geq 1$). By (a) above, A_j is tilted of euclidean type. Since A_j is not simply connected, it is of type $\tilde{\mathbb{A}}$, by [8]. It is thus a truncated branch extension, or coextension, of a hereditary algebra of type $\tilde{\mathbb{A}}$, by [7]. Since A_{j+1} is representation-finite, the unique point of A_j which is not in A_{j+1} cannot lie in the branches, hence it lies on the unique cycle in the bound quiver of A_j. Since this point is either a source or a sink in A_j, then A_{j+1} is tilted of type $\mathbb{A}$. But then A_{j+1} is simply connected, a contradiction. This shows that A_{j+1} is simply connected. Since B is a full convex subcategory of A_{j+1}, then B is simply connected [12] (2.8). □

1.4. COROLLARY. *Let A be a tubular algebra which is not strongly simply connected. Then $C^{(0)}$ or $C^{(\infty)}$ is hereditary of type $\tilde{\mathbb{A}}$.*

Proof. Let C be a full convex subcategory of A which is not simply connected. Since A itself is simply connected, $C \neq A$. By (1.3), C is tilted of Dynkin or euclidean type. Since it is not simply connected, then it is of type $\tilde{\mathbb{A}}$, by [8]. Applying (1.3) again, C is representation-infinite. Therefore C contains as (unique tame concealed) full convex subcategory C' a hereditary algebra of type $\tilde{\mathbb{A}}$. Now A has only two tame concealed full convex subcategories, namely $C^{(0)}$ and $C^{(\infty)}$. Therefore $C' = C^{(0)}$ or $C' = C^{(\infty)}$. □

1.5. LEMMA. *Let A be a truncated branch extension of a tame concealed algebra C. If A satisfies the separation condition, then C is not hereditary of type $\tilde{\mathbb{A}}$.*

Proof. This follows from the fact that, for each $x \in C_0$, the indecomposable projective modules $e_x C$ and $e_x A$ (where e_x denotes the primitive idempotent corresponding to x) coincide when considered as A-modules. □

1.6. We are now able to state and prove the main result of this section.

THEOREM. *Let A be a tubular algebra. The following conditions are equivalent:*
 (a) A is strongly simply connected.
 (b) The orbit graph of each of the directed components of $\Gamma(\bmod A)$ is a tree.
 (c) A is strongly $\tilde{\mathbb{A}}$-free.
 (d) $C^{(0)}$ and $C^{(\infty)}$ are not hereditary of type $\tilde{\mathbb{A}}$.
 (e) A and A^{op} satisfy the separation condition.

Proof. (a) implies (b). The directed components of $\Gamma(\bmod A)$ are the postprojective and preinjective components which are, respectively, the postprojective and the preinjective components of a tilted algebra. The result then follows from [2] (1.3) or [13] (4.2).

(b) implies (c). The postprojective component of $\Gamma(\bmod A)$ is the same as that of $\Gamma(\bmod C^{(0)})$ and its preinjective component is that of $\Gamma(\bmod C^{(\infty)})$. Thus, neither $C^{(0)}$ nor $C^{(\infty)}$ is hereditary of type $\tilde{\mathbb{A}}$. However, if A contains a full subcategory C which is hereditary of type $\tilde{\mathbb{A}}$, then C must coincide with either $C^{(0)}$ or $C^{(\infty)}$, a contradiction.

(c) implies (d). This is trivial.
(d) implies (a). This follows from (1.4).
(a) implies (e). This is trivial.
(e) implies (d). This follows from (1.5) and its dual. □

1.7. EXAMPLES. (a) Let A be given by the quiver

bound by $\alpha\beta = \gamma\delta$, $\varepsilon\gamma = \lambda\mu$, $\nu\mu\delta = 0$. Then A is tubular of type $(3,3,3)$ and strongly simply connected. The orbit graph of the postprojective component of $\Gamma(\bmod A)$ is

and that of its preinjective component is

(b) There exist tubular algebras satisfying the separation condition, which are not strongly simply connected (thus, one cannot improve condition (e) of the theorem). Let A be given by the quiver

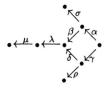

bound by $\alpha\sigma = 0$, $\gamma\rho = 0$ and $\alpha\beta\lambda = \gamma\delta\lambda$. Thus A is tubular of type $(3,3,3)$, satisfies the separation condition, but is not strongly simply connected. Here, A^{op} does not satisfy the separation condition: $C^{(\infty)}$ is hereditary of type $\tilde{\mathbb{A}}$.

(c) The statement of the theorem does not hold for derived tubular algebras. Let A be given by the quiver

bound by $\lambda\alpha = \mu\gamma$, $\lambda\beta = \mu\delta$, $\alpha\nu = c.\beta\eta$, $\gamma\nu = c.\delta\eta$ (for some $c \in k \setminus \{0,1\}$) and $\lambda\alpha\nu = 0$. Then A is derived tubular: indeed, reflecting at the unique sink yields a tubular algebra of type $(2,2,2,2)$. Let A_- (or A_+) denote the full convex subcategory of A generated by all points except the unique source (or sink, respectively). Then the postprojective (or preinjective) component of $\Gamma(\text{mod}A)$ coincides with that of $\Gamma(\text{mod}A_-)$ (or $\Gamma(\text{mod}A_+)$, respectively). Moreover, A_- (or A_+) is tilted of type $\tilde{\mathbb{D}}$ and has a complete slice in the postprojective (or preinjective, respectively) component of its Auslander-Reiten quiver. Thus the orbit graph of each of the postprojective and the preinjective component of $\Gamma(\text{mod}A)$ is

On the other hand, both A and A^{op} satisfy the separation condition. But A is not strongly simply connected, because it is not strongly $\tilde{\mathbb{A}}$-free. Finally, notice that A is a quasi-tilted algebra, so the statement of the theorem does not apply either to quasi-tilted algebras (see, however, [1]).

2 STRONGLY SIMPLY CONNECTED DERIVED TUBULAR ALGEBRAS.

2.1. By [8], a derived tubular algebra is always simply connected. Hence, if it is representation-finite, it is strongly simply connected [12] (2.8). We are thus only interested in the representation-infinite case.

If an algebra A is representation-infinite, then, by [9] (2.5), it is derived tubular if and only if it is isomorphic to a branch enlargement of a tame concealed algebra (in the sense of [9] (2.2)) and its tubular type equals that of the corresponding tubular algebra. Consequently, there exist a source a_+ and a sink a_- (lying in the branches of A) such that each of the full convex subcategories A_+ and A_-, generated respectively by the objects of $A_0 \setminus \{a_-\}$ and $A_0 \setminus \{a_+\}$, is iterated tilted of euclidean type. Notice that the points a_+ and a_- are usually not unique. A pair (a_+, a_-) as above will be called a *tubular pair* of A.

2.2. LEMMA. *Let A be a derived tubular algebra which is not strongly simply connected, and (a_+, a_-) be a tubular pair of A. Then one of the algebras A_+ and A_- is not strongly simply connected.*

Proof. Let B denote the tame concealed full convex subcategory of A. Since A is not strongly simply connected, then, by [3] (1.3), its bound quiver contains an irreducible cycle which is not a contour, or an irreducible contour which is not naturally contractible. Let C denote this cycle. If C lies entirely inside A_+ or A_-, we are done. If not, then both a_+ and a_- lie on C. Now, the cycle C cannot lie entirely inside any individual branch and each walk between two branches passes through B. Since a_+ lies on C, we deduce that a_+ is the root of an extension branch, thus is an extension point of B. Moreover, a_+, being a source of A, is also a source of C. Similarly, a_- is the root of a coextension branch and thus is a coextension point of B. Moreover, a_- is a sink of C.

Now, a_+, being the root of a branch, is a separating point, hence, by [6] (2.2), if $\alpha : a_+ \to a$ and $\beta : a_+ \to b$ are the two arrows on C starting at a_+, there exists a minimal relation $\lambda_1 \alpha v + \lambda_2 \beta w + \sum_{j \geq 3} \lambda_j u_j$ where $\lambda_i \in k$ (for all i) and v, w, u_j (for all j) are paths from a_+ to $c \in B_0$ (say). Let a', b' denote the last points of v, w, respectively, which lie on C, and c' be the first common point of v and w. Observe that, since a', b', c' are predecessors of $c \in B_0$, and proper successors of a_+, then $a', b', c' \in B_0$. Denote by v', w' the subpaths of v, w, respectively, from a' to c' and from b' to c', and by u', u'' the subwalks of C from a' to a_- and from b' to a_-, respectively. Then

$$C' = u'u''^{-1}w'v'^{-1} : a' \;\text{—}\; \overset{u'}{\cdots} \;\text{—}\; a_- \;\text{—}\; \overset{u''}{\cdots}\; \text{—}\; b' \longrightarrow \overset{w'}{\cdots} \longrightarrow c' \longleftarrow \overset{v'}{\cdots} \longleftarrow a'$$

is a closed walk entirely contained inside A_-. We claim that C' is an irreducible cycle which is not a contour.

Indeed, we notice first that $c' \neq a_-$, because $c' \in B_0 \subseteq (A_+)_0$, while $a_- \notin (A_+)_0$. There is no path from a_- to c', because a_- is a sink of A, and no path from c' to a_-, because the existence of such a path would contradict the irreducibility of C. On the other hand, since $u'u''^{-1}$ is a subwalk of C, it has no self-intersections. By definition, $v'w'^{-1}$ has no self-intersections. Finally, there is no common point between $v'w'^{-1}$ and $u'u''^{-1}$, because the existence of such a path would contradict the irreducibility of C. Next, C' is clearly irreducible, because C is, and, finally, C' is not a contour, because it has at least two different sinks, namely c' and a_-.

We have established the existence of an irreducible cycle C' which is not a contour, and lies entirely inside A_-. Hence, again by [3] (1.3), A_- is not strongly simply connected. □

2.3. The above lemma reduces the study of the strong simple connectedness of a derived tubular algebra A to that of the two iterated tilted algebras of euclidean type A_+ and A_-, which was characterised in [2] (3.3). We are able to state and prove the main result of this section.

THEOREM. Let A be a representation-infinite derived tubular algebra, and (a_+, a_-) be a tubular pair of A. The following conditions are equivalent:

(a) A is strongly simply connected.

(b) A is strongly $\tilde{\mathbb{A}}$-free.

(c) A_+ and A_- are strongly $\tilde{\mathbb{A}}$-free.

(d) A_+ and A_- are strongly simply connected.

(e) The orbit graph of each directed component of $\Gamma(\mathrm{mod} A_+)$ and $\Gamma(\mathrm{mod} A_-)$ is a tree.

Proof. Clearly, (a) implies (b), which implies (c). It follows from (2.2) that (c) implies (a). Thus, the first three conditions are equivalent. Since A_+ and A_- are iterated tilted algebras of euclidean type, the equivalence of (c), (d) and (e) follows directly from [2] (3.3). □

2.4. The above theorem yields further criteria for the strong simple connectedness of a tubular algebra. Let indeed A be a tubular algebra, then clearly $a_+ \in C_0^{(\infty)}$ and $a_- \in C_0^{(0)}$, so that A_+ and A_- are tilted algebras of euclidean type. We then have the following corollary.

COROLLARY. Let A be a tubular algebra, and (a_+, a_-) be a tubular pair of A. The following conditions are equivalent:

(a) A is strongly simply connected.

(b) A_+ and A_- are strongly $\tilde{\mathbb{A}}$-free.

(c) A_+ and A_- are strongly simply connected.

(d) The orbit graph of each directed component of $\Gamma(\mathrm{mod} A_+)$ and $\Gamma(\mathrm{mod} A_-)$ is a tree.

(e) The orbit graph of the preinjective component of $\Gamma(\mathrm{mod} A_+)$ and the orbit graph of the postprojective component of $\Gamma(\mathrm{mod} A_-)$ are trees.

Proof. The equivalence of (a), (b), (c) and (d) follows from (2.3). It is clear that (d) implies (e). In order to show that (e) implies (c), we note that, since $a_+ \in C_0^{(\infty)}$, then A_- is a domestic truncated branch extension of $C^{(0)}$, hence there is a complete slice in the preinjective component of $\Gamma(\mathrm{mod} A_-)$. Similarly, there is a complete slice in the postprojective component of $\Gamma(\mathrm{mod} A_+)$. The result now follows at once from [2] (2.3) and its dual. □

2.5. EXAMPLES. (a) Let A be given by the quiver

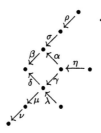

bound by $\alpha\beta = \gamma\delta, \rho\sigma = 0, \sigma\beta = 0, \eta\alpha = 0$ and $\gamma\mu\nu = 0$. Then A is derived tubular of type $(2,3,6)$, but is not tubular. Clearly, A is strongly simply connected. We have here two possible choices for a_+ (and only one for a_-).

(b) It does not suffice to have A_+ (or A_-) strongly simply connected for A to be strongly simply connected, as is shown by the example (1.8)(b) above.

ACKNOWLEDGEMENTS.

The author gratefully acknowledges partial support from the NSERC of Canada. He is also grateful to Flávio Coelho and Sonia Trepode for useful discussions.

REFERENCES

[1] I. Assem, F. U. Coelho and S. Trepode: Simply connected tame quasi-tilted algebras, to appear.

[2] I. Assem and S. Liu: Strongly simply connected tilted algebras, Ann. Sci. Math. Québec 21 (1997), No. 1, 13–22.

[3] I. Assem and S. Liu: Strongly simply connected algebras, J. Algebra 207 (1998), 449–477.

[4] I. Assem, S. Liu and J. A. de la Peña: The strong simple connectedness of a tame tilted algebra, to appear in Comm. Algebra.

[5] I. Assem, E. N. Marcos and J. A. de la Peña: The simple connectedness of a tame tilted algebra, to appear.

[6] I. Assem and J. A. de la Peña: The fundamental groups of a triangular algebra, Comm. Algebra 24(1) 187–208 (1996).

[7] I. Assem and A. Skowroński: Iterated tilted algebras of type $\tilde{\mathbb{A}}_n$, Math. Z. 195 (1987) 269–290.

[8] I. Assem and A. Skowroński: On some classes of simply connected algebras, Proc. London Math. Soc. (3)56 (1988) 417–450.

[9] I. Assem and A. Skowroński: Algebras with cycle-finite derived categories, Math. Ann. 280 (1988) 441–463.

[10] I. Assem and A. Skowroński: Quadratic forms and iterated tilted algebras, J. Algebra 128 (1990) 55–85.

[11] K. Bongartz and P. Gabriel: Covering spaces in representation theory, Invent. Math. 65 (1981) 331–378.

[12] O. Bretscher and P. Gabriel: The standard form of a representation-finite algebra, Bull. Soc. Math. France 111 (1983) 21–40.

[13] S. Gastaminza, J. A. de la Peña, M. I. Platzeck, M. J. Redondo and S. Trepode: Finite dimensional algebras with vanishing Hochschild cohomology, J. Algebra 212 (1999) 1–16.

[14] D. Happel, I. Reiten and S. O. Smalø: Tilting in abelian categories and quasi-tilted algebras, Memoirs Amer. Math. Soc., No.575, Vol.120 (1996).

[15] R. Martínez-Villa and J. A. de la Peña: The universal cover of a quiver with relations, J. Pure Applied Algebra 30 (1983) 277–292.

[16] C. M. Ringel: Tame algebras and integral quadratic forms, Lecture Notes in Math., 1099 (1984), Springer-Verlag, Berlin-Heidelberg-New York.

[17] A. Skowroński: Simply connected algebras and Hochschild cohomologies, Can. Math. Soc. Conf. Proc. Vol.14 (1993) 431–447.

H^1 and Presentations of Finite Dimensional Algebras

MICHAEL J. BARDZELL Salisbury State University, Salisbury, Maryland
email: mjbardzell@ssu.edu

EDUARDO N. MARCOS [1] Universidade De São Paulo, São Paulo, Brazil
email: enmarcos@ime.usp.br

ABSTRACT In this paper we study the first Hochschild cohomology group $H^1(\Lambda)$ of certain finite dimensional algebras and how it relates to presentations of Λ. In particular, we consider this relationship for monomial, directed, and a generalization of Schurian algebras. The relationship between presentations and the fundamental group $\pi_1(\Lambda)$ is also studied.

1 INTRODUCTION

The purpose of this paper is to study $H^1(\Lambda)$ for a finite dimensional algebra $\Lambda = k\Gamma/I$, where Γ is a finite quiver, k is a field, and I is an admissible ideal. We will focus primarily on the relationship between $H^1(\Lambda)$ and presentation properties of certain classes of algebras. In [ARS] Open Problem 5 asks for an invariant characterization of monomial algebras. Such a characterization is provided in [BG]. The approach of that paper is based on $H^1(\Lambda)$ and group gradings (coverings). The general characterization, however, is not algorithmic. It depends on the existence of a certain grading. An algorithmic solution is provided for a certain class of algebras. An algebra Λ is said to be *constricted* if $\dim_k o(a)\Lambda t(a) = 1$ and $\dim_k v\Lambda v = 1$ for each $a \in \Gamma_1$ and $v \in \Gamma_0$. The result on constricted algebras is that Λ is a monomial algebra if and only if $\dim_k H^1(\Lambda) = 1 - |\Gamma_0| + |\Gamma_1| = \chi(\Gamma)$, the reduced Euler characteristic of Γ. For this class of algebras it is shown that if Λ has a monomial presentation, then every presentation is monomial. Constricted algebras include schurian, narrow, and incidence algebras (see [Ha]). In the first section of this paper we show that the assumption that $\dim_k v\Lambda v = 1$ for each $v \in \Gamma_0$ can be dropped. Thus we obtain an algorithmic solution for Problem 5 for a more general

---

[1] The second author gratefully acknowledges financial support in the form of a research scholarship from CNPq - Brasil

class of algebras than the constricted algebras in [BG]. We also give a necessary and algorithmic condition for any algebra to be monomial. A different approach to the monomial characterization problem can be found in [GS].

The algebras satisfying $\dim_k o(a)\Lambda t(a) = 1$ for each $a \in \Gamma_1$ also have nice properties regarding the fundamental group π_1. In section 3 we will see that the fundamental group of an algebra in this class does not depend on the presentation. The final presentation problem we consider is the former conjecture that $H^1(\Lambda) = 0$ implies Γ has no oriented cycles. A counterexample for the general case has recently been given in [BL]. We consider some algebras where the conjecture does hold even for undirected cycles. Our approach is based on the combinatorics of the quiver and relations of an algebra Λ.

Throughout this paper Γ_0 will denote the vertex set, Γ_1 will denote the arrow set, and R will denote a generating set for I. We use $[u,v]$ to denote the set of all paths starting at u and ending at v. We can compute $H^1(\Lambda)$ via the complex $0 \longrightarrow P_0^* \xrightarrow{\phi_1^*} P_1^* \xrightarrow{\phi_2^*} P_2^*$, i.e. $\dim_k H^1(\Lambda) = \dim_k \ker \phi_2^* - \dim_k \operatorname{im} \phi_1^*$. Here $P_0^* = \coprod_{v \in \Gamma_0} v\Lambda v$, $P_1^* = \coprod_{a \in \Gamma_0} o(a)\Lambda t(a)$, and $P_2^* = \coprod_{r \in R} o(r)\Lambda t(r)$. These terms and maps can be found by applying $\operatorname{Hom}_{\Lambda^e}(\ ,\Lambda)$ to the first three terms of the projective resolution $P_2 \longrightarrow P_1 \longrightarrow P_0 \longrightarrow \Lambda \longrightarrow 0$ as discussed in [BM]. Throughout this paper we also use the notation $\bar{v} \in P_0^*$ to denote the vertex v in the vth component and $\bar{a} \in P_1^*$ denote the arrow a in the ath component.

2 DERIVATIONS, $H^1(\Lambda)$, AND MONOMIAL ALGEBRAS

In this section we provide a necessary homological condition for an algebra to be a monomial algebra (Theorem 2.1). This is an algorithmic solution for one direction of Open Problem 5 from [ARS]. We also provide necessary and sufficient conditions for algebras satisfying $\dim_k o(a)\Lambda t(a) = 1$ for each $a \in \Gamma_1$ to be monomial (Theorem 2.4).

THEOREM 2.1. *Let $\Lambda = k\Gamma/I$ be a monomial algebra. Then $\dim_k H^1(\Lambda) \geq \chi(\Gamma)$.*

Proof. Let $\{p_1, ... p_n\}$ be a minimal set of generating paths for I. As we stated in the introduction, the first cohomology group can be computed from the complex $0 \longrightarrow \coprod_{v \in \Gamma_0} v\Lambda v \xrightarrow{\phi_1^*} \coprod_{a \in \Gamma_1} o(a)\Lambda t(a) \xrightarrow{\phi_2^*} \coprod_{i=1}^n o(p_i)\Lambda t(p_i)$ using the maps described in [BM]. To simplify notation, write $\coprod_{v \in \Gamma_0} v\Lambda v = A \oplus B$. Here $A = \coprod_{v \in \Gamma_0} kv$, $B = \coprod_{v \in \Gamma_0} \widehat{v\Lambda v}$, and $\widehat{v\Lambda v}$ is the k span of all paths starting and ending at v, excluding v. Similarly, write $\coprod_{a \in \Gamma_1} o(a)\Lambda t(a) = C \oplus D$ where $C = \coprod_{a \in \Gamma_1} ka$ and $D = \coprod_{a \in \Gamma_1} \widehat{o(a)\Lambda t(a)}$. Note that the first boundary map can be decomposed as $\phi_1^* = f \oplus g$. Also, $f(A) \subseteq C$ and $f(B) \subseteq D$ since ϕ_1^* is multiplication by arrows, i.e. ϕ_1^* raises degrees of P_0^* elements by 1 or else sends them to 0. Similarly, write $\phi_2^* = h \oplus l$. Then the complex becomes $0 \longrightarrow A \oplus B \xrightarrow{f \oplus g} C \oplus D \xrightarrow{h \oplus l} \coprod_{i=1}^n o(p_i)\Lambda t(p_i)$. Now, $\dim_k \operatorname{im} f = |\Gamma_0| - 1$ (see the constricted case in [BG]). Also, $C = \coprod_{a \in \Gamma_1} ka \subseteq \ker \phi_2^*$. So $\dim_k \ker h - \dim_k \operatorname{im} f = 1 - |\Gamma_0| + |\Gamma_1| = \chi(\Gamma)$. Using the fact that $\operatorname{im} g \subseteq \ker l$, the result follows. $\square$

The following result was proved in [BM] using weight functions. However, it follows immediately from Theorem 2.1.

COROLLARY 2.2. *Let $\Lambda = k\Gamma/I$ be a monomial algebra. Then $H^1(\Lambda) = 0$ if and only if Γ is a tree.*

An alternate proof of Theorem 2.1 can be constructed using the fact that there is a monomorphism $\text{Hom}(\pi_1(\Gamma), k) \longrightarrow H^1(\Lambda)$ for any presentation (see [AP], [FGM], [PS]) and that for monomial algebras π_1 is the free group on $\chi(\Gamma)$ letters.

Throughout the rest of this section we will assume that $\dim_k o(a)\Lambda t(a) = 1$ for each arrow a. Note that this implies the quiver has no loops and no parallel arrows. In addition, if a path $p \ne a$ in Γ is parallel to an arrow a then $p \in I$. Let $\text{Der}_{\Lambda_0}(\Lambda)$ denote the set of all derivations on Λ that fix Λ_0. That is, $\text{Der}_{\Lambda_0}(\Lambda)$ = $\{\delta \in \text{Der}(\Lambda) : \delta(v) = v$ for all $v \in \Lambda_0\}$. Before our next result we need the following definition.

DEFINITION 2.1. *([Ad], [FGM]) Let (Γ, I) be a bounded quiver. A relation $\sum_{j \in J} \lambda_j \gamma_j \in I$ is called minimal if, for every proper subset L of J, we have $\sum_{l \in L} \lambda_l \gamma_l \notin I$.*

PROPOSITION 2.3. *Let Λ be an algebra satisfying the aforementioned hypotheses. Then $\dim_k \text{Der}_{\Lambda_0}(\Lambda) \le |\Gamma_1|$. In addition, Λ is a monomial algebra if and only if $\dim_k \text{Der}_{\Lambda_0}(\Lambda) = |\Gamma_1|$.*

Proof. Since $\dim_k o(a)\Lambda t(a) = 1$ for each arrow a, given $d \in \text{Der}_{\Lambda_0}(\Lambda)$ and $\alpha \in \Gamma_1$, $d(\alpha) = \lambda_\alpha \alpha$ for some $\lambda_\alpha \in k$. From this it follows that $\text{Der}_{\Lambda_0}(\Lambda) \subseteq k^{\Gamma_1}$ and $\dim_k \text{Der}_{\Lambda_0}(\Lambda) \le |\Gamma_1|$. It is also easy to see that if I is generated by monomials then all the elements of k^{Γ_1} are derivations.

Now assume that Λ is not monomial and $\dim_k \text{Der}_{\Lambda_0}(\Lambda) = |\Gamma_1|$, i.e. $\text{Der}_{\Lambda_0}(\Lambda) = k^{\Gamma_1}$. Let $\beta = \sum_{i=1}^{n} \lambda_i \mu_i \in [u,v]$, where $n \ge 2$, be a minimal non-monomial relation. Since $\dim_k o(a)\Lambda t(a) = 1$ for each arrow a, we know there are no double arrows in the quiver. Let $\mu_1 = a\alpha_1\omega_1$ and $\mu_2 = a\alpha_2\omega_2$ with $\alpha_1 \ne \alpha_2$. Let d be the derivation corresponding to the map δ_{α_1}. That is, $d(\alpha) = \delta_{\alpha\alpha_i}(\alpha_i)$ for any arrow α (here δ is the Kroneker delta). Then $\delta_{\alpha_1}(\alpha_2) = 0$. Moreover, $\delta_{\alpha_1}(\beta) = \sum_{\alpha_1 \in \text{supp}\mu_i} \lambda_i \mu_i$ is a relation, contradicting the fact that β is minimal. $\square$

The following is a generalization of Theorem 4.1 from [BG].

THEOREM 2.4. *Let $\Lambda = k\Gamma/I$ be a connected algebra such that $\dim_k o(a)\Lambda t(a) = 1$ for each arrow a. The following are equivalent:*

i) $\dim_k \text{Der}_{\Lambda_0}(\Lambda) = |\Gamma_1|$
ii) $\dim_k H^1(\Lambda) = \chi(\Gamma)$
iii) I is monomial.

Proof. From the previous Proposition we have $i \Longleftrightarrow iii$. To establish $ii \Longleftrightarrow iii$, first assume that $I = <p_1, ... p_n>$ is monomial. Then we can compute $H^1(\Lambda)$ from the complex $0 \longrightarrow A \oplus B \xrightarrow{f \oplus g} C \xrightarrow{\phi_2^*} \coprod_{i=1}^{n} o(p_i)\Lambda t(p_i)$. As before $\dim_k \text{im} f = |\Gamma_0| - 1$.

Note that g is the zero map and $C = \coprod_{a \in \Gamma_1} ka = \ker \phi_2^*$. So $\dim_k \ker \phi_2^* - \dim_k \operatorname{im} \phi_1^* = 1 - |\Gamma_0| + |\Gamma_1| = \chi(\Gamma)$.

To show the other direction, assume that I is not a monomial algebra. Let $r = \sum_{i=1}^{n} \lambda_i p_i$ be a non-monomial generator. Following the argument in the proof of Theorem 4.1 from [BG], construct an arrow a that divides p_1 but not all the other paths $p_2, ..., p_n$. Then $\bar{a} \notin \ker \phi_2^*$ and it follows that $\dim_k \ker \phi_2^* < |\Gamma_1|$. Since $\dim_k \operatorname{im} \phi_1^* = \dim_{k\operatorname{Im}} f = |\Gamma_0| - 1$, we have $\dim_k H^1(\Lambda) < \chi(\Gamma)$. □

Note that algebras satisfying the hypotheses of Theorem 2.4 can be broken into two classes based on $H^1(\Lambda)$. All the monomial algebras satisfy $\dim_k H^1(\Lambda) = \chi(\Gamma)$ and all the non-monomial algebras satisfy $\dim_k H^1(\Lambda) < \chi(\Gamma)$. From the proof of Theorem 2.4 we see that if there exists a non-monomial presentation of Λ, then Λ is a non-monomial algebra. This gives us the following result on presentations of this type of algebra.

COROLLARY 2.5. *If Λ is a monomial algebra satisfying the hypotheses of Theorem 2.4, then every presentation of Λ is monomial.*

3 π_1 AND PRESENTATIONS

In this section we will examine the first homotopy group π_1 and presentations of certain algebras. Let us first recall the definition of π_1 and some related terminology. Assume the quiver Γ is connected.

DEFINITION 3.1. *For an arrow $a : u \longrightarrow v$, denote by a^{-1} the formal inverse. A walk in Γ from u to v is a formal composition $a_1^{\epsilon_1} \cdots a_t^{\epsilon_t}$ where $a_i \in \Gamma_1$ and $\epsilon_i \in \{+1, -1\}$. Denote by e_u the stationary path at u.*

DEFINITION 3.2. *Define a homotopy relation $\sim$ on (Γ, I) to be the smallest equivalence relation on the set of all walks in Γ satisfying the following:*

i) For each arrow $a : u \longrightarrow v$ in Γ we have $aa^{-1} \sim e_u$ and $a^{-1}a \sim e_v$.

ii) For each minimal relation $\sum_{j \in J} \lambda_j \gamma_j \in I$ we have $\gamma_i \sim \gamma_j$ for all $i, j \in J$.

iii) If p, q, w, and w' are walks and $p \sim q$, then $wpw' \sim wqw'$ whenever these products are defined.

DEFINITION 3.3. *Fix a base vertex $u \in \Gamma$. Then the group $\pi_1(\Gamma, I)$ of all homotopy classes of closed walks which start and end at u is called the first homotopy group of (Γ, I). See [Ad] for more details on π_1.*

In [FGM] it is proved that one can take *any* set of minimal relations generating I to define the homotopy group. Note that π_1 need not be an invariant of the algebra. That is, it is possible for two different presentations of the same algebra to produce two different homotopy groups. A triangular algebra is called simply connected if all presentations of the algebra give the trivial homotopy group. It can be difficult to determine if a given algebra is simply connected since one has to check the vanishing of the homotopy groups on *every* presentation. So it is important to describe some classes of algebras where the homotopy groups do not depend on a given presentation. We will show that this is the case for algebras that satisfy $\dim_k o(a)\Lambda t(a) = 1$ for each arrow $a \in \Gamma_1$. This includes schurian algebras and therefore triangular algebras of finite representation type. Before we get to this result we first need some technical lemmas.

Throughout this section let $\bar{\alpha}$ denote $\alpha + I$ for any $\alpha \in k\Gamma$.

LEMMA 3.1. *Let $\Lambda = k\Gamma/I$ and choose any complete set of primitive idempotents $\{\bar{e}_1, ...\bar{e}_n\}$. Then there is an invertible element μ such that $\bar{e}_i = \mu^{-1}\bar{v}_{\phi(i)}\mu$ for some $\phi(i) \in S_n$.*

Proof. After reordering we can assume by Krull Schmidt that there is an isomorphism $\Lambda\bar{v}_i \xrightarrow{\phi_i} \Lambda\bar{e}_i$ which takes $\bar{v}_i$ to $\bar{e}_i$. So we get a left module automorphism $\Lambda = \amalg \Lambda\bar{v}_i \xrightarrow{\phi = \amalg \phi_i} \amalg \Lambda\bar{e}_i$ such that $\phi(\bar{v}_i) = \bar{e}_i$. If we let $\mu = \phi(1)$ then μ is invertible since $\Lambda = \Lambda\mu$. Since $\phi(\lambda) = \lambda\mu$ for all $\lambda \in \Lambda$ and ϕ is an epimorphism, $\bar{e}_i = \bar{v}_i\mu$. So $\Lambda\bar{e}_i = \Lambda\bar{v}_i\mu$. Also, $1 = \sum \mu^{-1}\bar{v}_i\mu = \sum \bar{e}_i$ and $\mu^{-1}\bar{v}_i\mu \in \Lambda\bar{e}_i$. Thus, $\bar{e}_i = \mu^{-1}\bar{v}_i\mu$. □

LEMMA 3.2. *Let $\Lambda = k\Gamma/I$ have vertex set $\{\bar{v}_1, ..., \bar{v}_n\}$. Let $\{\mu^{-1}\bar{v}_i\mu\}$ be any other complete set of primitive idempotents. Then there is an isomorphism $\psi : k\Gamma \longrightarrow \Lambda$ such that $\psi(v_i) = \mu^{-1}\bar{v}_i\mu$ and $\ker \psi = I$.*

Proof. Let $\pi : k\Gamma \longrightarrow k\Gamma/I$ be the natural projection. Define $\psi : k\Gamma \longrightarrow \Lambda$ by $\psi(\gamma) = \mu^{-1}\gamma\mu$. Then $\psi(v_i) = \mu^{-1}\bar{v}_i\mu$ and it is clear that $\gamma \in \ker \pi$ if and only if $\gamma \in \ker \psi$. □

The former Lemma has a nice interpretation. Let $\Lambda_1 = k\Gamma/I$ and $\{e_1, ..., e_n\}$ be a complete set of primitive orthogonal idempotents. We always can assume that $\{e_1, ...e_n\}$ is the vertex set without changing the ideal I.

COROLLARY 3.3. *Let $\Lambda = k\Gamma/I$ and assume $\dim_k o(a)\Lambda t(a) \leq 1$ for all $a \in \Gamma_1$. If $\delta : k\Gamma \longrightarrow \Lambda$ is any epimorphism, then there is another epimorphism $\psi : k\Gamma \longrightarrow \Lambda$ with $\ker \psi = \ker \delta$ and, for each $a \in \Gamma_1$, $\psi(a) = \alpha_a \bar{a}$ for some $\alpha_a \in k$.*

Proof. We can change δ by ψ and assume $\psi(v) = \bar{v}$ for each $v \in \Gamma_0$. Since $\dim_k o(a)\Lambda t(a) \leq 1$ for all $a \in \Gamma_1$, $\psi(a) = \alpha_a \bar{a}$. □

COROLLARY 3.4. *Let $\Lambda = k\Gamma/I$ and assume $\dim_k o(a)\Lambda t(a) \leq 1$ for all $a \in \Gamma_1$. Let $\psi : k\Gamma \longrightarrow \Lambda$ be any epimorphism with $\ker \psi = I'$. Furthermore, let $\mu = \sum_{i=1}^{n} \alpha_i \eta_i$, $\alpha_i \in k\backslash\{0\}$, be a minimal relation. Then there is a minimal relation $\tilde{\mu} \in \ker \psi$ with $\tilde{\mu} = \sum_{i=1}^{n} \beta_i \eta_i$, $\beta_i \in k\backslash\{0\}$.*

Proof. By Corollary 3.3 there is an isomorphism $k\Gamma/I \cong k\Gamma/\ker \psi$ which takes the class of any nonzero arrow a to a nonzero multiple of $\bar{a}$. So $0 = \sum_{i=1}^{n} \alpha_i \bar{\eta}_i$ goes to $0 = \sum_{i=1}^{n} \beta_i \bar{\eta}_i$ where $\psi(\alpha_i \bar{\eta}_i) = \beta_i \bar{\eta}_i$. Then $\sum_{i=1}^{n} \beta_i \eta_i \in \ker \psi$ and, if it is not minimal, then we can assume there $1 \leq j < n$ such that $\sum_{i=1}^{j} \beta_i \bar{\eta}_i = 0$. This means $\sum_{i=1}^{j} \alpha_i \bar{\eta}_i = 0$ and μ is not minimal. $\square$

THEOREM 3.5. *Let (Γ, I) and (Γ, I') be two presentations of the same finite dimensional algebra Λ such that $\dim_k o(a)\Lambda t(a) \leq 1$ for all $a \in \Gamma_1$. Then $\pi_1(\Gamma, I) = \pi_1(\Gamma, I')$.*

Note that this corollary tells us that π_1 of a schurian algebra Λ does not depend on the presentation of Λ. In fact we can even comment on binomial algebras. Recall that an ideal I is called binomial if it has a generating set consisting solely of monomials and $k-$ linear combinations of two monomials. We have the following.

COROLLARY 3.6. *Let $\Lambda \cong k\Gamma/I_1 \cong k\Gamma/I_2$ be two presentations of a schurian algebra. Then*

i) I_1 *is binomial*
ii) $\pi_1(\Gamma, I_1) = \pi_1(\Gamma, I_2)$.

Proof. Let $u, v \in \Gamma_0$ and $\bar{\mu}_1, \bar{\mu}_2$ two classes of nonzero paths in $k\Gamma/I_1$ from u to v. Then $\bar{\mu}_1 = \lambda \bar{\mu}_2$ for some $\lambda \in k$ since Λ is schurian. Hence I_1 is binomial. Part 2 follows from the previous corollary since schurian algebras satisfy $\dim_k o(a)\Lambda t(a) \leq 1$ for all $a \in \Gamma$. $\square$

4 $H^1(\Lambda)$ AND CYCLES

In this section we consider the vanishing of $H^1(\Lambda)$ and its relationship to whether or not the underlying quiver of an algebra has cycles. It was conjectured for awhile that $H^1(\Lambda) = 0$ implies that Γ has no oriented cycles. In [Ha] it is shown that the conjecture is true when I is homogeneous and char $k = 0$ and also when Λ is 3-nilpotent. In [BM], it is also shown true for monomial algebras (see also Corollary 2.2 of this paper). A counterexample to the general case has recently been given in [BL]. In this section we will construct some other cases where the conjecture still holds. In fact, we will be able to say something about quivers with undirected cycles also. Our approach is based on using the boundary maps from [BM] and considering the combinatorics of the quiver. Throughout this section assume that the characteristic of k is zero. As in section 2 we write $A = \amalg_{\Gamma_0} kv$, $B = \amalg_{\Gamma_0} \widehat{v \Lambda v}$, $C = \amalg_{\Gamma_1} ka$, $D = \amalg_{\Gamma_1} \widehat{o(a)\Lambda t(a)}$, and $\phi_1^* = f \oplus g$. Now suppose that $\Lambda = k\Gamma/I$ where I is generated by the reduced gröbner basis (see [FFG]) $R = \{p_1, ..., p_m, r_1, ..., r_s\}$. Here $p_1, ..., p_m$ are paths and $r_1, ..., r_s$ are non-monomial generators. Since we know the conjecture is true for monomial algebras, we will assume that Λ is not a monomial algebra and that R contains at least one generator that is not a path. Of course, it is possible for there to be no paths in R. Then $P_2^* = \amalg_{r \in R} o(r)\Lambda t(r) = E \oplus F$, where $E = \amalg_{i=1}^{m} o(p_i)\Lambda t(p_i)$ and $F = \amalg_{i=1}^{s} o(r_i)\Lambda t(r_i)$.

So we may compute $H^1(\Lambda)$ from the complex $0 \longrightarrow A \oplus B \xrightarrow{f \oplus g} C \oplus D \xrightarrow{\phi_2^*} E \oplus F$. Note that ϕ_2^* can send some elements of C (D) into E and other elements into F.

PROPOSITION 4.1. *Suppose that* $\Lambda = k\Gamma/I$ *where* Γ *contains the cycle (not necessarily directed)* $v_1 \xrightarrow{a_1} v_2 \xrightarrow{a_2} \cdots v_e \xrightarrow{a_e} v_1$. *Let* $\alpha = \left(\bigoplus_{i=1}^{e} \delta_i \bar{a}_i \right) \oplus \Omega \in P_1^*$ *where* $\delta_i \in \{0,1\}$ *for* $i = 1,...e$ *and* $\Omega \in D$. *If* $\delta_i = 1$ *for at least one i, then* $\alpha \notin \mathrm{im}\, \phi_1^*$.

Proof. First note that $\phi_1^* = f \oplus g$ maps elements of B to zero or linear combinations of radical squared elements. So we only need to consider $\phi_1^*(A) = f(A)$. Let $\{a_{e+1},...,a_{\Gamma_1}\}$ denote $\Gamma_1 \setminus \{a_1,...a_e\}$. Then we have the following:

$$\phi_1^*(\bar{v}_1) = (-1)^{i_1} \bar{a}_1 \oplus (-1)^{i_e} \bar{a}_e \oplus \Delta_1$$
$$\phi_1^*(\bar{v}_2) = (-1)^{i_1+1} \bar{a}_1 \oplus (-1)^{i_2} \bar{a}_2 \oplus \Delta_2$$
$$\vdots$$
$$\phi_1^*(\bar{v}_e) = (-1)^{i_{e-1}} \bar{a}_{e-1} \oplus (-1)^{i_e+1} \bar{a}_e \oplus \Delta_e$$

where $\Delta_1,...,\Delta_e \in C$ are contained in the $a_{e+1},...,a_{\Gamma_1}$ components and $i_1,...i_e \in \{0,1\}$. The i_j are determined by the orientation of the arrows. If the arrow a ends at v_j then $\bar{a}$ will have positive coefficient in $\phi_1^*(\bar{v}_j)$. If the arrow a begins at v_j then $\bar{a}$ will have negative coefficient in $\phi_1^*(\bar{v}_j)$. Now let $\sigma = k_1 \bar{v}_1 \oplus \cdots \oplus k_e \bar{v}_e$. Suppose $\phi_1^*(\sigma) = \alpha$. Then we obtain the following system of equations:

$$\begin{aligned} -k_1 + k_2 &= \delta_1 \\ -k_2 + k_3 &= \delta_2 \\ &\vdots \\ -k_{e-1} + k_e &= \delta_{e-1} \\ k_1 \qquad\qquad - k_e &= \delta_e \end{aligned}$$

Adding we obtain $0 = \sum_{i=1}^{e} \delta_i$. Since char $k = 0$ and we have assumed that $\delta_i = 1$ for at least one i, we have contradiction. □

Note that this proposition gives us a recipe for showing that certain undirected algebras have nonzero first Hochschild cohomology group. We simply look for elements of the form α that are in ker ϕ_2^*. The following generalization of Corollary 2.2 is one such application.

COROLLARY 4.2. *Let* $\Lambda = k\Gamma/I$ *and suppose that* Γ *contains the cycle (not necessarily directed)* $v_1 \xrightarrow{a_1} v_2 \xrightarrow{a_2} \cdots v_e \xrightarrow{a_e} v_1$. *If there exists some* a_i, $i = 1,...,e$, *such that*

i) a_i does not divide any support path in R or
ii) a_i divides only monomial relations in R or
iii) for each relation $r = \sum_{j=1}^{n} \alpha_j p_j \in R$, there exists a nonnegative integer m_r such that $a_i | p_j$ exactly m_r times for $j = 1,...,m_r$

then $H^1(\Lambda) \neq 0$.

Proof. If a_i does not divide any support path in R, then $\phi_2^*(\bar{a}_i) = 0$. If a_i divides only monomial relations in R, then ϕ_2^* sends $\bar{a}_i$ into the a_i position of any path p_j, $j = 1,...,m$, that a_i divides. So we just obtain scalar multiples of the paths that are zero relations. Finally, if for each relation $r = \sum_{j=1}^{n} \alpha_j p_j \in R$ there exists a nonnegative integer m_r such that $a_i | p_j$ exactly m_r times for $j = 1, ..., m_r$, then in the r-component of $\phi_2^*(\bar{a}_i) \in P_2^*$ we have $\sum_{j=1}^{n} m_r \alpha_j p_j = m_r r = 0$. In each of these cases $\bar{a}_i \in \ker \phi_2^*$ and $\alpha \notin \mathrm{im} \phi_1^*$. □

REMARK: If one examines the proof of proposition 4.1 and corollary 4.2, one can see that the items i) and ii) of proposition 4.2 are valid in any characteristic.

REFERENCES

[Ad] Assem, I, de la Peña, J.A, The Fundamental Group of a Triangular Algebra, Communications in Algebra **24** (1996) 187-208.

[ARS] Auslander, M, Reiten I, Smalo, S; Representation Theory of Artin Algebras, Cambridge Studies in Advanced Mathematics 36, Cambridge University Press, 1995.

[BG] Bardzell, M.J., Green, E.L.; An Invariant Characterization of Monomial Algebras; Communications in Algebra, 27(5), 2331-2344, 1999.

[BM] Bardzell, M.J., E.N. Marcos, Induced Boundary Maps for the Cohomology of Monomial and Auslander Algebras; Canadian Math Society Conference Proceedings, Volume 24, 47-54, 1998.

[BL] Buchweitz, R.O., Liu, S; Artin Algebras with Loops but no Outer Derivations, Preprint 232, University of Sherbrooke (1999).

[FFG] Farkas, D., Feustel, C., Green, E.L.; Synergy in the Theories of Gröbner Bases and Path Algebras, Canadian J. Math 45, 727-739 (1993).

[FGM] Farkas, D, Green, E.L., Marcos, E.; Diagonalizable Derivations of Finite Dimensional Algebras II, Trabalhos do Departmento de Matemática, preprint RT-MAT 99-04 1999.

[GS] Guil-Asensio, F, Saorin, M, The Automorphism Group and the Picard Group of a Monomial Algebra, Communications in Algebra **27** (1999) 857-887.

[Ha] Happel D., Hochschild Cohomology of Finite Dimensional Algebras, Springer Lecture Notes, 1404, 108-126, 1984.

[PS] de la Peña, J.A, Saorin, M, The First Cohomology Group of an Algebra, preprint 1999.

Tame tilted algebras with almost regular connecting components

GRZEGORZ BOBIŃSKI[1] Faculty of Mathematics and Informatics, Nicholas Copernicus University, ul. Chopina 12/18, 87-100, Toruń, Poland.
email: gregbob@mat.uni.torun.pl

ABSTRACT In the paper we classify the tame tilted algebras with almost regular connecting components in the terms of quivers and relations. We also use this classifications to describe the selfinjective algebras of Euclidean type which admit almost regular nonperiodic components in the Auslander–Reiten quiver.

Throughout the paper K denotes a fixed algebraically closed field. By an algebra we will mean a finite dimensional basic associative K-algebra. Finally, a module is always a finite dimensional right module.

A component of the Auslander–Reiten quiver Γ_A of an algebra A is called almost regular if it has only one nonstable τ_A-orbit which consists of exactly one vertex. It follows from the description of the Auslander–Reiten quiver of a tame tilted algebra presented in [3] that its connecting component cannot be stable. Our aim in this paper is to classify all representation-infinite tame tilted algebras with almost regular connecting components. Note that the field K is the unique representation-finite connected tilted algebra with almost regular connecting component.

The paper is organized as follows. In Section 1 we formulate our main result, in Section 2 we present the proof of it and in Section 3 we use the main result to classify selfinjective algebras of Euclidean type whose Auslander–Reiten quivers admit almost regular nonperiodic components.

1 THE MAIN RESULT

Before formulating the main result we introduce some families of algebras. For nonnegative integers u, v, w, t we denote by $\Delta(u,v,w,t)$ the following family of

---

[1]Supported by the Polish Scientific Grant KBN No. 2PO3A 012 14.

quivers

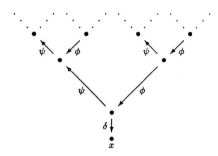

(∗)

where each vertex denoted by a square can be replaced by a connected finite subquiver of the following infinite quiver

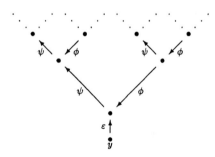

containing the vertex x, and the original vertex is then identified with the vertex x. Similarly, each vertex denoted by a circle can be replaced by a finite connected subquiver of the following quiver

containing the vertex y, and the original vertex is then identified with the vertex y. The integer u is the sum of p and the number of arrows in the quivers denoted by circles with a vertical bar in the picture (∗) and the integer v is the sum of q and the number of arrows in the quivers denoted by circles with a horizontal bar in the same picture. Analogously, using r (respectively, s) and the quivers denoted by squares with a vertical (respectively, horizontal) bar we define w (respectively, t). Similarly for positive integers u, v, w and t we denote by $\Delta'(u,v,w,t)$ the following family of quivers where the meaning of circles and squares is the same as above. We also define the numbers u, v, w, t in the same way as before. Finally, using the same conventions we define the family $\Delta''(u,v,w,t)$, $u,v \geq 2$, $w,t \geq 1$, as follows

[quiver diagram]

Let u, v, w and t be positive integers. We define the family of algebras $A(u,v,w,t)$ as the family of bound quiver algebras $K\Delta'/I'$, where the quiver Δ' belongs to the family $\Delta'(u,v,w,t)$ and the ideal I' is generated by $\alpha_p\rho_1$, $\beta_q\sigma_1$, $\alpha_1\cdots\alpha_p\sigma_1\cdots\sigma_s - \beta_1\cdots\beta_q\rho_1\cdots\rho_r$ and all elements of the form $\alpha_i\varepsilon$, $\beta_j\varepsilon$, $\delta\rho_k$, $\delta\sigma_l$, $\phi\psi$. Similarly, for positive integers u, v, w, t, $B(u,v,w,t)$ will denote the family of bound quiver algebras $K\Delta'/J'$, where Δ' belongs to the family $\Delta'(u,v,w,t)$ and the generators of J' are $\beta_q\sigma_1$, $\alpha_1\cdots\alpha_p\rho_1 - \beta_1\cdots\beta_q\rho_1$, $\alpha_p\rho_1\cdots\rho_r - \alpha_p\sigma_1\cdots\sigma_s$ and all elements of the form $\alpha_i\varepsilon$, $\beta_j\varepsilon$, $\delta\rho_k$, $\delta\sigma_l$, $\phi\psi$. For $u,v \geq 2$ and $w,t \geq 1$ the family of algebras of the form $K\Delta''/I''$, where Δ'' belongs to $\Delta''(u,v,w,t)$ and I'' is generated by $\alpha_1\cdots\alpha_p + \beta_1\cdots\beta_q + \gamma_1\gamma_2$, $\alpha_p\rho_1$, $\beta_q\sigma_1$, $\gamma_2\rho_1\cdots\rho_r - \gamma_2\sigma_1\cdots\sigma_s$ and all elements of the form $\alpha_i\varepsilon$, $\beta_j\varepsilon$, $\delta\rho_k$, $\delta\sigma_l$, $\phi\psi$, will be denoted by $C(u,v,w,t)$. Finally, for nonnegative integers u, w, v, t we have the family $D(u,v,w,t)$ consisting of the algebras of the form $K\Delta/I$, with Δ belonging to $\Delta(u,v,w,t)$ and the generators of I being $\alpha_1\cdots\alpha_p\eta_1\zeta_1\rho_1\cdots\rho_r - \beta_1\cdots\beta_q\eta_4\zeta_4\sigma_1\cdots\sigma_s$, $\eta_1\zeta_1 - \eta_2\zeta_2$, $\eta_3\zeta_3 - \eta_4\zeta_4$ and all elements of the form $\alpha_i\varepsilon$, $\beta_j\varepsilon$, $\delta\rho_k$, $\delta\sigma_l$, $\phi\psi$.

The main result of the paper is the following.

THEOREM 1. *Let A be a representation-infinite connected tame tilted algebra with almost regular connecting component. Then A or A^{op} belongs to one of the families $A(u,v,w,t)$, $u,v,w,t \geq 1$, $B(u,v,w,t)$, $u,w,v,t \geq 1$, $(u,w,v+t-1) = (2,2,n-2)$, $n \geq 4$, $(2,3,3)$, $(2,3,4)$, $(2,4,3)$, $(2,3,5)$, $(2,5,3)$, $C(u,v,w,t)$, $u,v \geq 2$, $w,t \geq 1$, $(u+w-1,v+t-1) = (2,n-2)$, $n \geq 4$, $(3,3)$, $(3,4)$, $(3,5)$, $D(u,v,w,t)$, $u,v,w,t \geq 0$, or to one of the families 1–14 defined below (each algebra from the families 1–14 is described as a bound quiver algebra with relations listed to the right of a quiver).*

Family 1.

[quiver diagram] $\eta\delta\alpha - \theta\varepsilon\beta$

             $\theta\varepsilon\beta - \iota\zeta\gamma$

Family 2.

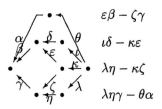

$\varepsilon\beta - \zeta\gamma$
$\iota\delta - \kappa\varepsilon$
$\lambda\eta - \kappa\zeta$
$\lambda\eta\gamma - \theta\alpha$

$\varepsilon\beta - \eta\delta\gamma$
$\iota\eta\delta - \kappa\zeta$
$\theta\alpha - \kappa\zeta\gamma$

Family 3.

$\iota\zeta\delta\alpha - \theta\gamma$
$\kappa\eta\varepsilon\beta - \theta\gamma$

Family 4.

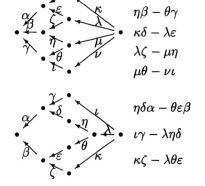

$\varepsilon\alpha - \zeta\beta$
$\eta\beta - \theta\gamma$
$\kappa\delta - \lambda\varepsilon$
$\lambda\zeta - \mu\eta$
$\mu\theta - \nu\iota$

$\theta\gamma - \iota\delta\alpha$
$\varepsilon\alpha - \zeta\beta$
$\mu\iota\delta - \kappa\varepsilon$
$\kappa\zeta - \lambda\eta$

$\eta\delta\alpha - \theta\varepsilon\beta$
$\iota\gamma - \lambda\eta\delta$
$\kappa\zeta - \lambda\theta\varepsilon$

Family 5.

$\delta\beta - \varepsilon\gamma$
$\iota\varepsilon - \lambda\theta\zeta$
$\kappa\eta\alpha - \lambda\theta\zeta\gamma$

Family 6.

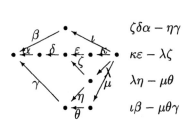

$\zeta\delta\alpha - \eta\gamma$
$\kappa\varepsilon - \lambda\zeta$
$\lambda\eta - \mu\theta$
$\iota\beta - \mu\theta\gamma$

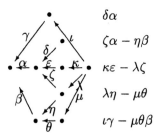

$\delta\alpha$
$\zeta\alpha - \eta\beta$
$\kappa\varepsilon - \lambda\zeta$
$\lambda\eta - \mu\theta$
$\iota\gamma - \mu\theta\beta$

Tame Tilted Algebras with Almost Regular Connecting Components

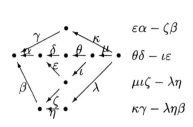
$\varepsilon\alpha - \zeta\beta$
$\theta\delta - \iota\varepsilon$
$\mu\iota\zeta - \lambda\eta$
$\kappa\gamma - \lambda\eta\beta$

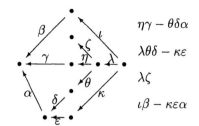
$\eta\gamma - \theta\delta\alpha$
$\lambda\theta\delta - \kappa\varepsilon$
$\lambda\zeta$
$\iota\beta - \kappa\varepsilon\alpha$

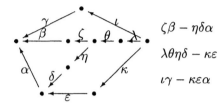
$\zeta\beta - \eta\delta\alpha$
$\lambda\theta\eta\delta - \kappa\varepsilon$
$\iota\gamma - \kappa\varepsilon\alpha$

Family 7.

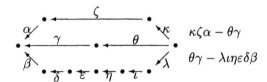
$\kappa\zeta\alpha - \theta\gamma$
$\theta\gamma - \lambda\iota\eta\varepsilon\delta\beta$

Family 8.

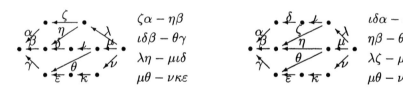

Left diagram:
$\zeta\alpha - \eta\beta$
$\iota\delta\beta - \theta\gamma$
$\lambda\eta - \mu\iota\delta$
$\mu\theta - \nu\kappa\varepsilon$

Right diagram:
$\iota\delta\alpha - \zeta\beta$
$\eta\beta - \theta\gamma$
$\lambda\zeta - \mu\eta$
$\mu\theta - \nu\kappa\varepsilon$

Bottom diagram:
$\varepsilon\alpha - \zeta\beta$
$\kappa\eta\delta\beta - \theta\gamma$
$\lambda\zeta - \mu\kappa\eta\delta$
$\mu\theta - \nu\iota$

Family 9.

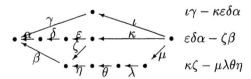
$\iota\gamma - \kappa\varepsilon\delta\alpha$
$\varepsilon\delta\alpha - \zeta\beta$
$\kappa\zeta - \mu\lambda\theta\eta$

Family 10.

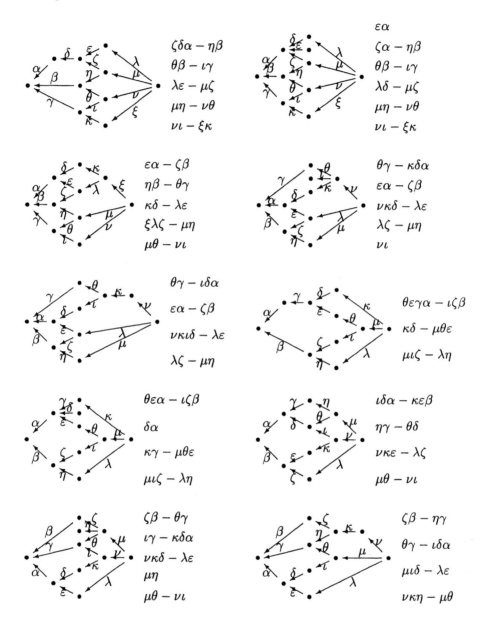

Family 11.

Tame Tilted Algebras with Almost Regular Connecting Components

Family 12.

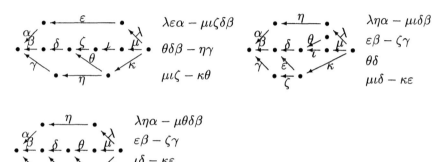

$\delta\alpha - \varepsilon\beta$

$\lambda\theta\varepsilon\beta - \mu\iota\zeta\gamma$

$\kappa\eta - \lambda\theta\delta$

Family 13.

Family 14.

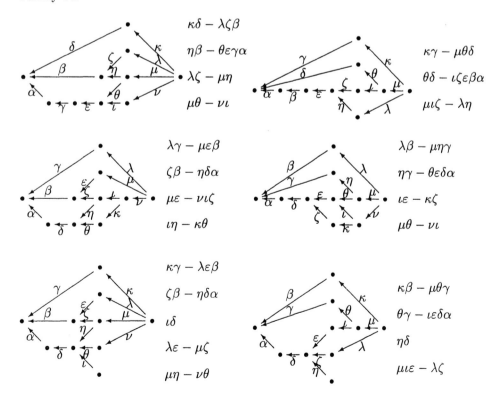

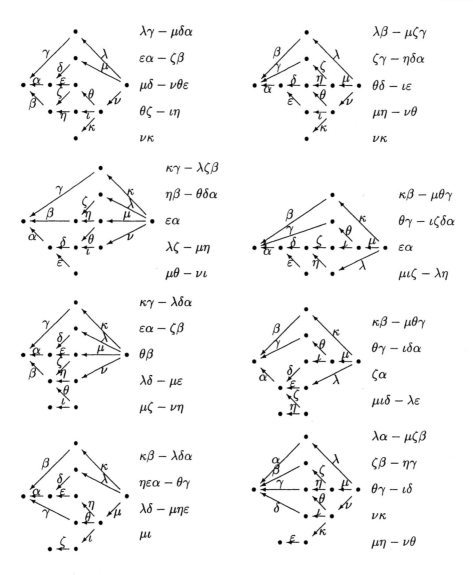

REMARK. It follows from the proof of the above theorem presented in the next section that the algebras from the families 14 and $A(u, v, w, t)$, $u, v, w, t \geq 1$, $B(u, v, w, t)$, $u, v, w, t \geq 1$, $(u, w, v + t - 1) = (2, 2, n - 2)$, $n \geq 4$, $(2, 3, 3)$, $(2, 3, 4)$, $(2, 4, 3)$, $(2, 3, 5)$, $(2, 5, 3)$, $C(u, v, w, t)$, $u, v \geq 2$, $w, t \geq 1$, $(u + w - 1, v + t - 1) = (2, n - 2)$, $n \geq 4$, $(3, 3)$, $(3, 4)$, $(3, 5)$, and their opposite algebras are all tilted algebras of extended Euclidean types with almost regular connecting components.

2 PROOF OF THE MAIN RESULT

Let A be a connected representation-infinite tame tilted algebra with almost regular connecting component. Denote by X the unique projective-injective A-module. It is well-known that $\operatorname{rad} X$ and $X/\operatorname{soc} X$ are indecomposable A-modules and we have the Auslander–Reiten sequence of the form

$$0 \longrightarrow \operatorname{rad} X \longrightarrow X \oplus \operatorname{rad} X/\operatorname{soc} X \longrightarrow X/\operatorname{soc} X \longrightarrow 0.$$

Let a and b be vertices of the ordinary quiver Q_A of A such that $X = P_A(a) = I_A(b)$. If we denote by B_1 the full subcategory of A formed by all objects except a, and by B_2 the full subcategory of A formed by all objects except b, then $A = B_1[\operatorname{rad} X] = [X/\operatorname{soc} X]B_2$. Moreover, if B denotes the full subcategory of A formed by all objects except a and b, then $B_1 = [\operatorname{rad} X/\operatorname{soc} X]B$ and $B_2 = B[\operatorname{rad} X/\operatorname{soc} X]$. It follows also that B_1 and B_2 are representation-infinite tilted algebras of Euclidean type, and B is a product of tilted algebras of Dynkin type. We will call B_1 the left end algebra of A. Similarly, B_2 will be called the right end algebra of A.

Thus our objective is to study the following situation. Let B be a product of tilted algebras of Dynkin type and R a B-module. We are asking when $B[R]$ and $[R]B$ are representation-infinite tilted algebras of Euclidean type. We will also denote by A the algebra

$$\begin{bmatrix} K & R & K \\ 0 & B & D(R) \\ 0 & 0 & K \end{bmatrix}$$

where multiplication is the usual multiplication of matrices up to rule $r \cdot \varphi = \varphi(r)$ for any $r \in R$ and $\varphi \in D(R)$. In our investigations we shall use vector space category methods. Details on vector space and subspace categories can be found in [5] and [6]. The facts necessary to follow the below considerations can be also found in [2].

We have to consider different cases which may occur. First assume that $B[R]$ is a tilted algebra of type $\widetilde{\mathbb{A}}_m$, $m \geq 1$. Then B has to be a tilted algebra of type $\mathbb{A}_m$ and according to [2, Proposition 3.5] we get that A belongs to some family $A(u, w, v, t)$, $u, v, w, t \geq 1$, $u + v + w + t = m + 3$.

Assume now that $B[R]$ is a tilted algebra of type $\widetilde{\mathbb{D}}_n$, $n \geq 4$. Again the case when B is a tilted algebra of type $\mathbb{D}_n$, $n \geq 4$, has been studied in [2] and it follows from [2, Proposition 3.5] that A or A^{op} has to be one of the algebras from the families $B(2, v, 2, t)$, $v, t \geq 1$, $v + t - 1 = n - 2$, $C(2, v, 1, t)$, $v \geq 2$, $t \geq 1$, $v + t - 1 = n - 2$. Hence we have to consider the case when B is a product of at least two tilted algebras of Dynkin type. It follows from [2, Lemma 3.3] that we can start from the situation when the unique projective-injective A-module is sincere. Then A is a tilted algebra of type Σ, where Σ is obtained from the following quiver

$k \geq 0$, by orienting edges. Hence, according to [4, Theorem 1], A is of the form $KQ(p, q, r, s)/I(p, q, r, s)$, $p, q, r, s \geq 0$, $p + q + r + s = n - 4$, where $Q(p, q, r, s)$ is

the following quiver

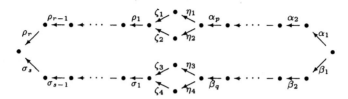

and the ideal $I(p,q,r,s)$ is generated by the relations

$$\alpha_1\cdots\alpha_p\eta_1\zeta_1\rho_1\cdots\rho_r - \beta_1\cdots\beta_q\eta_4\zeta_4\sigma_1\cdots\sigma_s,\ \eta_1\zeta_1 - \eta_2\zeta_2,\ \eta_3\zeta_3 - \eta_4\zeta_4.$$

If we omit the assumption that the unique projective-injective A-module is sincere, then the knowledge of modules lying on the mouths of tubes in Γ_{A_0}, where A_0 is one of the algebras above, leads to the conclusion that A is one of the algebras from the families $D(u,v,w,t)$, $u,v,w,t \geq 0$, $u+v+w+t = n-4$.

Let now $B[R]$ be a tilted algebra of type $\widetilde{\mathbb{E}}_6$. There are possible three situations: B can be a product of three tilted algebras of type $\mathbb{A}_2$, B can be a product of a tilted algebra of type $\mathbb{A}_5$ and a tilted algebra of type $\mathbb{A}_1$, and finally B can be a tilted algebra of type $\mathbb{E}_6$. In the first case the vector space category $\text{Hom}(R, \text{mod } B)$ has to be the following category

$$\begin{array}{c}
\bullet = \text{Hom}_B(R,Z_1) \\
\bullet = \text{Hom}_B(R,X_1) \\
\bullet = \text{Hom}_B(R,Z_2) \\
\bullet = \text{Hom}_B(R,X_2) \\
\bullet = \text{Hom}_B(R,Z_3) \\
\bullet = \text{Hom}_B(R,X_3)
\end{array},$$

where $R = X_1 \oplus X_2 \oplus X_3$. Similarly the dual category $\text{Hom}(\text{mod } B, R)$ is the following category

$$\begin{array}{c}
\bullet = \text{Hom}_B(Y_1,R) \\
\bullet = \text{Hom}_B(X_1,R) \\
\bullet = \text{Hom}_B(Y_2,R) \\
\bullet = \text{Hom}_B(X_2,R) \\
\bullet = \text{Hom}_B(Y_3,R) \\
\bullet = \text{Hom}_B(X_3,R)
\end{array}.$$

Hence X_1, X_2, X_3, Z_1, Z_2, Z_3 are injective B-modules and it follows that A has to be the unique algebra from family 1.

In the second case, when B is a product of two tilted algebras of types $\mathbb{A}_4$ and $\mathbb{A}_1$, the vector space category $\text{Hom}(R, \text{mod } B)$ is a full subcategory of the following category

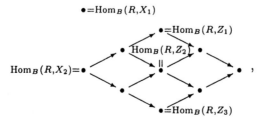

where $R = X_1 \oplus X_2$. Since the algebra $B[R]$ is representation-infinite it follows that $\operatorname{Hom}(R, \operatorname{mod} B)$ has to contain the objects $\operatorname{Hom}_B(R, X_1)$, $\operatorname{Hom}_B(R, Z_1)$, $\operatorname{Hom}_B(R, Z_2)$, $\operatorname{Hom}_B(R, Z_3)$. Dually, the vector space category $\operatorname{Hom}(\operatorname{mod} B, R)$ is a full subcategory of the following category

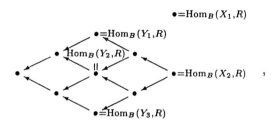

and has to contain the objects $\operatorname{Hom}_B(X_1, R)$, $\operatorname{Hom}_B(Y_1, R)$, $\operatorname{Hom}_B(Y_2, R)$, $\operatorname{Hom}_B(Y_3, R)$. Since the functions $f_1, f_2, f_3 : (\Gamma_B)_0 \to \mathbb{Z}$ given by $f_i(X) := \dim_K \operatorname{Hom}_B(Y_i, X)$ for $i = 1, 2, 3$ are additive on Γ_B and take nonnegative values it follows that the modules Z_1, Z_2 and Z_3 are injective. Of course, the module X_1 is also injective. Hence, the possible configurations (up to symmetry) of indecomposable injective B-modules are the following

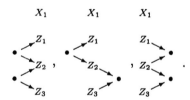

and $\dim_K \operatorname{Hom}_B(R, I) = 1$ for each indecomposable injective B-module. Thus, it easily follows that A or A^{op} is one of the algebras from the family 2.

The last case of B being a tilted algebra of type $\mathbb{E}_6$ has been studied in [2] and hence according to [2, Proposition 3.5] we get one of the algebras from the families $B(2, v, 3, t)$, $v, t \geq 1$, $v + t - 1 = 3$, $C(u, v, w, t)$, $u, v \geq 2$, $w, t \geq 1$, $(u + w - 1, v + t - 1) = (3, 3)$, or their opposite algebras.

Analogous considerations as above conducted in cases when $B[R]$ is tilted of type $\widetilde{\mathbb{E}}_7$ or $\widetilde{\mathbb{E}}_8$ give us the families 3–14 and $B(u, v, w, t)$, $u, v, w, t \geq 1$, $(u, w, v + t - 1) = (2, 3, 4), (2, 4, 3), (2, 3, 5), (2, 5, 3)$, $C(u, v, w, t)$, $u, v \geq 2$, $w, t \geq 1$, $(u + w - 1, v + t - 1) = (3, 4), (3, 5)$. Here we only list for each family the type of $B[R]$ and B, and if A is a tilted algebra of extended Euclidean type then also the type of A. If the algebra B is not connected then we list the types of blocks.

Family	Type of B	Type of $B[R]$	Type of A
Family 3	$\mathbb{A}_3, \mathbb{A}_3, \mathbb{A}_1$	$\widetilde{\mathbb{E}}_7$	
Family 4	$\mathbb{A}_7$	$\widetilde{\mathbb{E}}_7$	
Family 5	$\mathbb{A}_5, \mathbb{A}_2$	$\widetilde{\mathbb{E}}_7$	
Family 6	$\mathbb{D}_6, \mathbb{A}_1$	$\widetilde{\mathbb{E}}_7$	
$B(u,v,w,t)$, $u,v,w,t \geq 1$, $(u,w,v+t-1) = (2,3,4), (2,4,3)$ $C(u,v,w,t)$, $u,v \geq 2$, $w,t \geq 1$, $(u+w-1, v+t-1) = (3,4)$	$\mathbb{E}_7$	$\widetilde{\mathbb{E}}_7$	$\widetilde{\widetilde{\mathbb{E}}}_7$
Family 7	$\mathbb{A}_5, \mathbb{A}_2, \mathbb{A}_1$	$\widetilde{\mathbb{E}}_8$	
Family 8	$\mathbb{A}_8$	$\widetilde{\mathbb{E}}_8$	
Family 9	$\mathbb{A}_7, \mathbb{A}_1$	$\widetilde{\mathbb{E}}_8$	
Family 10	$\mathbb{D}_8$	$\widetilde{\mathbb{E}}_8$	
Family 11	$\mathbb{A}_4, \mathbb{A}_4$	$\widetilde{\mathbb{E}}_8$	
Family 12	$\mathbb{D}_5, \mathbb{A}_3$	$\widetilde{\mathbb{E}}_8$	
Family 13	$\mathbb{E}_6, \mathbb{A}_2$	$\widetilde{\mathbb{E}}_8$	
Family 14	$\mathbb{E}_7, \mathbb{A}_1$	$\widetilde{\mathbb{E}}_8$	$\widetilde{\widetilde{\mathbb{D}}}_8$
$B(u,v,w,t)$, $u,v,w,t \geq 1$, $(u,w,v+t-1) = (2,3,5), (2,5,3)$ $C(u,v,w,t)$, $u,v \geq 2$, $w,t \geq 1$, $(u+w-1, v+t-1) = (3,5)$	$\mathbb{E}_8$	$\widetilde{\mathbb{E}}_8$	$\widetilde{\widetilde{\mathbb{E}}}_8$

3 APPLICATION TO SELFINJECTIVE ALGEBRAS

An algebra A is called selfinjective if each projective A-module is injective. An important class of selfinjective algebras is formed by the selfinjective algebras of Euclidean type, that is algebras of the form $\hat{B}/G$, where $\hat{B}$ is the repetitive category of a tilted algebra B of Euclidean type and G is an admissible (infinite cyclic) group of K-linear automorphisms of $\hat{B}$ (for definitions of notions presented in this section we refer to [2], [7] and [9]). We may even assume that B is a domestic tubular extension of a tame concealed algebra. Precisely, for each tilted algebra B of Euclidean type there exists a domestic tubular extension B' of a tame concealed algebra such that $\hat{B} \simeq \hat{B}'$. The analogous fact holds for domestic tubular coextensions of tame concealed algebras. It has been proved by Skowroński in [7] that a connected selfinjective algebra which admits a simply connected Galois covering is of domestic representation type if and only if it is a selfinjective algebra of Euclidean type.

The connection between the Auslander–Reiten quivers $\Gamma_{\hat{B}/G}$ of $\hat{B}/G$ and $\Gamma_{\hat{B}}$ of $\hat{B}$ described in [7] and the reflection procedure of constructing the repetitive category for domestic tubular extensions of tame concealed algebras investigated in [1] allow to classify selfinjective algebras of Euclidean type whose Auslander–Reiten quivers admit almost regular nonperiodic components. Namely, we have the following theorem (see [2, Section 4] for arguments).

THEOREM 2. *Let A be a selfinjective algebra of Euclidean type. The Auslander–Reiten quiver Γ_A of A admits an almost regular nonperiodic component if and* only if $A \simeq \hat{B}/G$, where B is the left end (respectively, right end) algebra of a representation-infinite tame tilted algebra with almost regular connecting component and G is an admissible group of K-linear automorphisms of $\hat{B}$.

Following [8] a subquiver $\mathcal{C}$ of Γ_A is called generalized standard if for any two modules X and Y in $\mathcal{C}$ the infinite radical $\operatorname{rad}^\infty(X, Y)$ is zero. We have the following consequences of the above theorem and [9, Theorem 5.5, Corollary 5.6] (compare also [2, Theorems 2 and 3]). In the below corollaries $\nu_{\hat{B}}$ denotes the Nakayama automorphism.

COROLLARY 3. *Let A be a connected selfinjective algebra. The following conditions are equivalent.*

(i) *A is of Euclidean type, Γ_A has at least two nonperiodic components, and at least one of them is almost regular.*

(ii) *Γ_A admits an almost regular nonperiodic component and a generalized standard left stable full translation subquiver of Euclidean type which is closed under predecessors in Γ_A.*

(iii) *Γ_A admits an almost regular nonperiodic component and a generalized standard right stable full translation subquiver of Euclidean type which is closed under successors in Γ_A.*

(iv) *$A \simeq \hat{B}/(\varphi \nu_{\hat{B}})$, where B is the left end (respectively, right end) algebra of a representation-infinite tame tilted algebra with almost regular connecting component and φ is a positive automorphism of $\widetilde{B}$.*

COROLLARY 4. *Let A be a connected selfinjective algebra. The following conditions are equivalent.*

(i) *A is of Euclidean type, Γ_A has at least three nonperiodic components, and at least one of them is almost regular.*

(ii) *A is tame, Γ_A has at least one generalized standard almost regular nonperiodic component.*

(iii) *Γ_A contains a nonperiodic component $\mathcal{C}$ such that $A/\operatorname{ann}\mathcal{C}$ is a representation-infinite tame tilted algebra with almost regular connecting component $\mathcal{C}$.*

(iv) *$A = \hat{B}/(\varphi \nu_{\hat{B}})$, where B is the left end (respectively, right end) algebra of a representation-infinite tame tilted algebra with almost regular connecting component and φ is a strictly positive automorphism of $\hat{B}$.*

The arguments needed to prove the above results are similar to the ones presented in the proof of the main results of [2]. In [2] one can also find a characterization of selfinjective algebras of Euclidean type whose all nonperiodic components are almost regular.

REFERENCES

[1] I. Assem, J. Nehring and A. Skowroński, *Domestic trivial extensions of simply connected algebras*, Tsukuba J. Math. **13** (1989), 31–72.

[2] G. Bobiński and A. Skowroński, *Selfinjective algebras of Euclidean type with almost regular nonperiodic Auslander–Reiten components*, preprint, Toruń, 1999.

[3] O. Kerner, *Tilting wild algebras*, J. London Math. Soc. **39** (1989), 29–47.

[4] J. A. de la Peña, *The families of two-parametric tame algebras with sincere directing modules*, Canad. Math. Soc. Conf. Proc. **14** (1993), 361–392.

[5] C. M. Ringel, *Tame algebras and integral quadratic forms*, Lecture Notes in Math. **1099**, Springer, 1984.

[6] D. Simson, *Linear representations of partially ordered sets and vector space categories*, Algebra, Logic and Applications **4**, Gordon and Breach Science Publishers, 1992.

[7] A. Skowroński, *Selfinjective algebras of polynomial growth*, Math. Ann. **285** (1989), 177–199.

[8] A. Skowroński, *Generalized standard Auslander–Reiten components*, J. Math. Soc. Japan **46** (1994), 517–543.

[9] A. Skowroński and K. Yamagata, *Galois coverings of selfinjective algebras by repetitive algebras*, Trans. Amer. Math. Soc. **351** (1999), 715–734.

Reflexive modules are not closed under submodules

GABRIELLA D'ESTE Dipartimento di Matematica, Università di Milano, via Saldini 50, 20133 Milano, Italy, email: DESTE@mat.unimi.it

ABSTRACT We show that the two classes of reflexive modules with respect to a cotilting bimodule fail to be closed under submodules. More precisely, we show that any generalized Kronecker algebra A of infinite dimension has the following property: $_AA_A$ is a cotilting bimodule, and any faithful module M such that M is reflexive with respect to $_AA_A$ has a non reflexive socle.

1 INTRODUCTION

The first remark of Colpi in his paper [C] on cotilting bimodules and their dualities says the following: "The main difference between our and Colby's setting is that we are not assuming the further hypothesis that the class of reflexive modules is closed under submodules". The example presented in this note shows that the situation studied by Colpi in [C] is much more general than that considered by Colby in [Cb1] and [Cb2] for several reasons, concerning the shape and the size of both the rings and the Δ-reflexive modules involved. (Before we recall all the useful definitions, we point out that Δ-reflexive module means W-reflexive module, in the sense of [AF], with respect to a cotilting bimodule W.) In the following we construct a cotilting bimodule $_RW_S$ such that even the most obvious Δ-reflexive left R-modules (resp. right S-modules) [AF, Propositions 20.13, 20.14 and Corollary 20.16], namely the summands of both $_RR$ and $_RW$ (resp. S_S and W_S), may have a submodule which is not Δ-reflexive. More precisely, given any infinite cardinal d, we construct an algebra A of dimension d over an algebraically closed field K, with the following properties:

- $_AA_A$ is a cotilting bimodule (Lemma 2.2).

- Both the classes of Δ-reflexive modules fail to be closed under submodules (Theorem 2.5 (ii)).

In our example it actually occurs that the Δ-reflexive modules are as few as possible, i.e. coincide with the finitely generated projectives modules (Lemma 2.3). Moreover, the Δ-reflexive modules admitting only Δ-reflexive submodules are as small as possible, i.e. coincide with the Δ-reflexive modules of finite dimension over K (Lemma 2.4). Before we describe the last part of the paper, we recall some definitions, and we fix the notation used in the sequel. First of all, we say that a left (resp. right) module W over a ring R is a *cotilting module* [CDT1] if W satisfies the following conditions:

1. inj $\dim_R(W) \leq 1$;

2. $Ext^1_R(W^\alpha, W) = 0$ for any cardinal α;

3. $\mathrm{Ker}\, Hom_R(-, W) \cap \mathrm{Ker}\, Ext^1_R(-, W) = 0$.

Next, we say that a faithfully balanced bimodule ${}_RW_S$ is a *cotilting bimodule* [C] if both ${}_RW$ and W_S are cotilting modules. As usually, for any ring A, we denote by A-Mod (resp. Mod-A) the category of all left (resp. right) A-modules. Moreover, given a cotilting bimodule ${}_RW_S$, we simply denote by Δ both the contravariant functors

$$Hom_R(-, W): R - \mathrm{Mod} \to \mathrm{Mod} - S, \mathrm{Hom}_S(-, W): \mathrm{Mod} - S \to R - \mathrm{Mod}.$$

In the following, for any left R-module (resp. right S-module) M, the *evaluation morphism* $\delta_M: M \to \Delta^2(M)$ is defined by the formula $(\delta_M(x))(\xi) = \xi(x)$ for any $x \in M$ and $\xi \in \Delta(M)$. If δ_M is an isomorphism, i.e. if M is W-reflexive in the sense of [AF], we say that M is Δ-*reflexive*. Finally, we simply denote by Γ both the contravariant functors

$$Ext^1_R(-, W): R - \mathrm{Mod} \to \mathrm{Mod} - S, \mathrm{Ext}^1_S(-, W): \mathrm{Mod} - S \to R - \mathrm{Mod}.$$

With this notation, we point out other surprising properties of our example. First of all, even indecomposable left (resp. right) modules with a very easy structure belong to $\mathrm{Ker}\, \Delta^2 \cap \mathrm{Ker}\, \Gamma^2$. Indeed, for any positive integer n, we exhibit (Theorem 2.5 (iv)) an indecomposable module M, of dimension n over K, such that

- $\Delta(M) = 0$;

- $\Gamma(M)$ and $\Delta\Gamma(M)$ are free modules of uncountable rank.

As we shall see, these modules M are of the form X/Y, where X is an indecomposable projective module, hence a Δ-reflexive module, and Y is a semisimple projective module which is not Δ-reflexive. Hence, by just dealing with algebras and cotilting bimodules of infinite but countable dimension, even a simple module M which is countably presented does not admit an exact sequence of the form

(+) $$0 \to \Gamma^2(M) \to M \to \Delta^2(M) \to 0.$$

We recall that, by the Cotilting Theorem proved by Colpi [C, Theorem 6], every module which is the quotient of two Δ-reflexive modules admits an exact sequence as in (+). We also recall that the results obtained by Tonolo in [T] explain the relationships among the functors Γ^2, Δ^2 and the identity functor, that is the three functors in (+). More precisely, by [T, Theorem 1.2], a derived functor has a "key role to relate" these three functors. Secondly, using a Δ-reflexive module whose socle is not Δ-reflexive, that is a Δ-reflexive module which is not finitely cogenerated, we construct (Theorem 2.5 (iii)) infinitely many pairwise non-isomorphic indecomposable modules X such that

- X is isomorphic to $\Gamma^2(X)$;

- $\Delta(X) = 0$ and X is the quotient of two indecomposable Δ-reflexive modules.

Consequently, even a cotilting bimodule admitting only finitely many indecomposable Δ-reflexive modules may admit infinitely many indecomposable Γ-reflexive modules in the sense of [C]. For a new homological definition of Γ-reflexive modules, we refer to [T], where Tonolo addressed and solved the problem of a good notion of Γ-reflexive modules with respect to the so-called weakly cotilting bimodules. Finally, the cotilting bimodule of our example suggests that the asymmetry between the dualities induced by Δ and Γ [C, Theorem 6] and [T, Corollary 2.9] does not depend only on how many indecomposable modules are involved. Indeed, the behaviour of Γ (resp. Γ^2) is as bad (resp. as good) as possible on all finitely generated modules belonging to Ker Δ which are not finitely presented (see (a) and (c) in Corollary 2.8). As a partial symmetry between Δ and Γ, we show that the countably generated modules M which are Δ-reflexive (resp. such that $\Delta(M) = 0$ and M is isomorphic to $\Gamma^2(M)$) are just the finitely presented modules (Lemma 2.3; Corollaries 2.8 and 2.10). However, also by looking at finitely presented modules and by dealing with a regular cotilting bimodule $_AA_A$ such that there exists an isomorphism $f : A \to A^{op}$, the functor Γ seems to act as a kind of concealed reflection. More precisely, the functor Γ used in our example acts in an easy and geometric way on infinitely many quotients of an indecomposable Δ-reflexive module (Remark 2.11 (a)). However, in the same example the action of Γ is much more complicated even on the quotient of an indecomposable Δ-reflexive module with respect to a two-dimensional Δ-reflexive submodule (Remark 2.11 (b)).

2 PROOFS AND REMARKS

Throughout the paper, we always assume that K is an algebraically closed field and that d is an infinite cardinal. Moreover, we say that A is the *generalized Kronecker algebra of dimension d over K* (compare with [HU, page 182]) if A is the K-algebra given by the quiver depicted in Figure 1, where the arrows, say α_j, from 1 to 2 are indexed by a set J of cardinality d. Hence, following the terminology of [R], A is the *one-point extension* of K by a vector space V of dimension d over K (i.e. A is isomorphic to $\begin{bmatrix} K & 0 \\ V & K \end{bmatrix}$, the set of all matrices $\begin{pmatrix} a & 0 \\ v & b \end{pmatrix}$, with $a, b \in K$ and $v \in V$, subject to the usual addition and multiplication of matrices). Finally, given a generalized Kronecker algebra A, we denote by e_1 (resp. e_2) the primitive idempotent of A corresponding to the vertex 1 (resp. 2), and we denote by P, Q, $\tilde{P}$, $\tilde{Q}$ the following indecomposable modules:

$$P = Ae_1, \quad Q = Ae_2, \quad \tilde{P} = e_1 A, \quad \tilde{Q} = e_2 A$$

Keeping all this notation, we recall some properties of direct products of projective modules used in the sequel.

LEMMA 2.1. *Let A be the generalized Kronecker algebra of infinite dimension d, let P (resp. $\tilde{Q}$) be the indecomposable faithful projective left (resp. right) A-module. If m is an infinite cardinal, then the following facts hold:*

(i) *P^m (resp. $\tilde{Q}^m$) admits a decomposition of the form $X \oplus Y$, where X is isomorphic to the direct sum of $|K^m|$ copies of P (resp. $\tilde{Q}$) and Y is a semisimple projective module of dimension $|A^m|$ over K.*

(ii) *P^m (resp. $\tilde{Q}^m$) is free if and only if $|K^m| \geq d$.*

Proof. The proof of [D, Lemmas 2.1 and 2.2] shows that the left A-module P^m satisfies (i) and (ii). On the other hand, there is an isomorphism $f : A \to A^{op}$ satisfying $e_1 \mapsto e_2$, $e_2 \mapsto e_1$ and $\alpha_j \mapsto \alpha_j$ for any arrow α_j from 1 to 2. Since $\tilde{Q}$ is the right A-module obtained by means of f from the right A^{op}-module P [J, page 26], a dual argument shows that also the right A-module $\tilde{Q}^m$ satisfies (i) and (ii). □

The next lemma shows that the modules P and $\tilde{Q}$ are subspaces of codimension one of a cotilting bimodule.

LEMMA 2.2. *Let A be a generalized Kronecker algebra of infinite dimension. Then A is coherent and perfect on both sides and ${}_AA_A$ is a cotilting bimodule.*

Proof. An argument similar to that used in the first part of [CDT1, Example 5.3 (c)] (see also the correction [CDT2] of the last part of (c)) shows that A is right coherent and left perfect and that P is a cotilting module. Since P is a summand of ${}_AA$, we obviously have

$$\operatorname{Ker} Hom_A(-, {}_AA) \cap \operatorname{Ker} Ext^1_A(-, {}_AA) = 0.$$

Since A is left hereditary, it follows that ${}_AA$ is a cotilting module. On the other hand, Lemma 2.1 and a dual proof show that A is left coherent and right perfect, and that both $\tilde{Q}$ and A_A are cotilting modules. Since ${}_AA_A$ is faithfully balanced, this implies that ${}_AA_A$ is a cotilting bimodule. □

It is easy to see that the regular bimodule constructed in Lemma 2.2 admits as few as possible Δ-reflexive modules.

LEMMA 2.3. *Let A be a generalized Kronecker algebra of infinite dimension, and let M be an A-module. Then the following conditions are equivalent:*

(i) *M is projective and finitely generated.*

(ii) *M is Δ-reflexive with respect to $_AA_A$.*

Proof. (i) $\Rightarrow$ (ii). This is well-known [AF, Proposition 20.13 and Corollary 20.16]. (ii) $\Rightarrow$ (i). Since A is hereditary on both sides, we deduce from Lemma 2.2 and [C, Lemma 2 (b)] that A is semiperfect on both sides and that any Δ-reflexive A-module is projective. By the characterization of projective modules over semiperfect rings [AF, Theorem 27.11], this implies that any Δ-reflexive A-module is a direct sum of indecomposable projective modules. Moreover, we clearly have $\Delta(P) \simeq \tilde{P}$ and $\Delta(Q) \simeq \tilde{Q}$. It follows that

(1) Δ interchanges indecomposable faithful projective modules and simple projective modules.

To end the proof, let L be an indecomposable projective module, let m be an infinite cardinal, and let X denote the direct sum of m copies of L. Then we clearly have

(2) $\Delta(X) \simeq \Delta(L)^m$.

Assume first L is simple. Then, putting (1) and (2) together, we deduce from Lemma 2.1 that $\Delta(X)$ has a non-zero free summand. Consequently, also $\Delta^2(X)$ has a non-zero free summand, and so X is not isomorphic to $\Delta^2(X)$. This means that

(3) *A semisimple projective A-module of infinite dimension is not Δ-reflexive.*

Suppose now L is faithful. Then, by (1), (2) and an obvious remark (see i.e. [D, Remark 2.3 (iii)]), $\Delta(X)$ is a semisimple projective module of uncountable dimension over K. This observation and (3) imply that $\Delta(X)$ is not Δ-reflexive. Therefore, by [AF, Proposition 20.14 (3)], X is not Δ-reflexive. Thus any Δ-reflexive module is finitely generated. This result completes the proof of the lemma. $\square$

As the following lemma shows, the property of admitting only Δ-reflexive submodules may be very restrictive.

LEMMA 2.4. *Let A be a generalized Kronecker algebra of infinite dimension, and let M be a Δ-reflexive module with respect to $_AA_A$. Then the following conditions are equivalent:*

(a) *Every submodule L of M is Δ-reflexive.*

(b) *$dim_K M$ is finite.*

(c) *M is artinian.*

(d) *M is finitely generated semisimple.*

Proof. By Lemma 2.3, it suffices to note that every faithful projective module has an infinite dimensional socle. $\square$

We are now ready to prove

THEOREM 2.5. *Let A be the generalized Kronecker algebra of infinite dimension d over K. Then $_AA_A$ is a cotilting bimodule with the following properties:*

(i) There are only finitely many indecomposable Δ-reflexive modules, up to isomorphism.

(ii) Both the classes of Δ-reflexive modules are not closed under submodules.

(iii) There are infinitely many pairwise non-isomorphic indecomposable A-modules X such that $\Delta(X) = 0$ and $\Gamma^2(X)$ is isomorphic to X.

(iv) For every cardinal c such that $1 \leq c \leq d$, there is an indecomposable cyclic A-module Y such that $\dim_K Y = c$, $\Delta(Y) = 0$ and $\Gamma(Y)$ is a free module of uncountable rank.

Proof. We first note that P and Q (resp. $\tilde{P}$ and $\tilde{Q}$) are the only indecomposable projective left (resp. right) A-modules, up to isomorphism. Consequently, (i) and (ii) follow from Lemmas 2.2, 2.3 and 2.4.

To prove (iii), fix an arrow α_j from 1 to 2, and let X denote the left (resp. right) A-module $P/A\alpha_j$ (resp. $\tilde{Q}/\alpha_j A$). Then we clearly have $\Delta(X) = 0$ and $A\alpha_j \simeq Q$ (resp. $\alpha_j A \simeq \tilde{P}$). Hence, either a direct calculation, or an application of [C, Theorem 6] shows that $\Gamma^2(X)$ is isomorphic to X. Since the annihilator of X is the subspace generated by α_j, it follows that the modules $P/A\alpha_j$ (resp. $\tilde{Q}/\alpha_j A$) are pairwise non-isomorphic. Thus also (iii) holds. Finally, take a cardinal c such that $1 \leq c \leq d$. Since the arrows α_j with $j \in J$ are a base of $soc\, P$ and $|J| = d$, we can fix a subspace L of $soc\, P$ such that L is generated by d arrows and $\dim_K P/L = c$. Next, let $i : L \to P$ denote the canonical inclusion, and let Y denote the module P/L. Then there is an exact sequence in Mod-A of the form

$$0 \longrightarrow \Delta(P) \xrightarrow{\Delta(i)} \Delta(L) \longrightarrow \Gamma(Y) \longrightarrow 0. \tag{1}$$

For brevity, let F denote the module $\Delta(L)$, and let V_1 and V_2 denote the subspaces Fe_1 and Fe_2 respectively. Since L is isomorphic to a direct sum of d copies of Q, we have $F \simeq \tilde{Q}^d$. By Lemma 2.1, this implies that

$$F = \Delta(L) \quad \text{is a free module of rank} \quad |K^d|. \tag{2}$$

Let T denote the submodule of F generated by V_2, that is let $T = V_2 A$. Then T is a summand of F. We also note that

$$F = T \oplus U \quad \text{for any subspace} \quad U \quad \text{of} \quad V_1 \tag{3}$$
$$\text{such that} \quad V_1 = Te_1 \oplus U.$$

Fix any $t \in T$. Since $T = V_2 A = (Fe_2)A$, there exist finitely many elements $f_1, \cdots, f_n \in F$ and $a_1, \cdots, a_n \in e_2 A$ such that we may write $t = f_1 a_1 + \cdots + f_n a_n$. Now let W denote the K-vector space generated by the subset $\{a_1, \cdots, a_n\}$. Then W is a left ideal of A of finite dimension over K. Hence our hypotheses on t and the structure [AF, Proposition 4.4] of the right A-module $F = Hom_A(L, {}_AA_A)$ imply that $t(L) \subseteq W$. It follows that

$$\dim_K t(L) \quad \text{is finite for any } t \in T. \tag{4}$$

On the other hand, let $g : L \to A$ denote the canonical inclusion. Since $g(x) = x = xe_1$ for any $x \in L$, we obtain

$$g = ge_1 \in V_1 = Fe_1 \quad \text{and} \quad \dim_K g(L) = d. \tag{5}$$

Putting (4) and (5) together, we conclude that the subspace of F generated by g, which coincides with $Im\,\Delta(i)$, is a subspace of V_1 such that $T \cap Im\,\Delta(i) = 0$. Therefore, we may choose a (semisimple) module U containing $Im\,\Delta(i)$ such that F has a decomposition of the form $F = T \oplus U$ as in (3). Thus we deduce from (1) and (2) that $\Gamma(Y)$ is a free right module of rank $|K^d|$, as claimed in (iv). Our assumptions on L guarantee that L is also a submodule of $\tilde{Q}$ with the property that $\dim_K \tilde{Q}/L = c$ and $\Delta(\tilde{Q}/L) = 0$. Hence a dual proof shows that $\Gamma(\tilde{Q}/L)$ is a free left module of rank $|K^d|$. This remark completes the proof of (iv). $\square$

Before we point out an application of the previous results, concerning Δ-reflexive modules and generalized linearly compact modules, we recall some definitions and results. Following the terminology of [CF] suggested by [GGW], given a cotilting module ${}_RW$, we say that a left R-module M is W-*torsionless linearly compact* if $M \in \text{Ker}\,\Gamma$ and, for any inverse system of morphisms $\{p_\lambda : M \to M_\lambda\}$ with $M_\lambda \in \text{Ker}\,\Gamma$ and $\text{Coker}\,p_\lambda \in \text{Ker}\,\Delta$ for all λ's, we have $\text{Coker}(\varprojlim p_\lambda) \in \text{Ker}\,\Delta$. Let us recall two facts used in the sequel concerning Δ-reflexive modules and torsionless linearly compact modules with respect to a cotilting bimodule W (see also [Mü] and [X, Theorem 4.1]).

- Every W-torsionless linearly compact module is Δ-reflexive [C, Proposition 10].

- A Δ-reflexive module M is W-torsionless linearly compact if and only if every submodule of $\Delta(M)$ is Δ-reflexive [CF, Theorem 1.8].

Surprisingly enough, a very easy property, namely having only infinite dimensional indecomposable summands, may characterize the Δ-reflexive modules which are torsionless linearly compact.

COROLLARY 2.6. *Let A be a generalized Kronecker algebra of infinite dimension, and let M be a Δ-reflexive module with respect to ${}_AA_A$. Then the following conditions are equivalent:*

(a) M is A-torsionless linearly compact.

(b) Every indecomposable summand of M is faithful.

(c) M does not have an indecomposable summand of finite dimension over K.

Proof. By the proof of Lemma 2.3, any Δ-reflexive left (resp. right) A-module M is isomorphic to $P^r \oplus Q^s$ (resp. $\tilde{Q}^r \oplus \tilde{P}^s$) for some natural numbers r and s. Consequently, we have $\Delta(M) \simeq \tilde{P}^r \oplus \tilde{Q}^s$ (resp. $\Delta(M) \simeq Q^r \oplus P^s$). Moreover, by Lemma 2.4, every submodule of $\Delta(M)$ is Δ-reflexive if and only if $soc\,\Delta(M)$ is Δ-reflexive, that is if and only if $s = 0$. This remark and the characterization of torsionless linearly compact modules given in [CF, Theorem 1.8] complete the proof of the corollary. □

REMARK 2.7. Given a cotilting bimodule ${}_RW_S$, we know from [M, Proposition 1.6] that a nice property, i.e. the property that $\Delta\Gamma(M) = 0$ for any left R-module M, implies that the class of Δ-reflexive S-modules is closed under submodules. On the other hand, Γ induces a duality between the modules X such that $\Delta(X) = 0$ and X is the quotient of two Δ-reflexive modules [C, Proposition 5 (d); Theorem 6 (c)]. By Lemma 2.3, this implies that

(∗) $Ext_A^1(\cdot, A)$ induces a duality between the finitely presented modules belonging to Ker $Hom_A(\cdot, A)$ for any generalized Kronecker algebra A of infinite dimension.

Moreover, by condition (iv) of Theorem 2.5, a module in the image of $\Delta\Gamma$ may be extremely big. The next corollary shows that the behaviour of Γ and $\Delta\Gamma$ on certain finitely generated modules is quite different from that of Γ^2.

COROLLARY 2.8. *Let ${}_AA_A$ be the regular cotilting bimodule over a generalized Kronecker algebra of infinite dimension, and let M be a finitely generated A-module such that $\Delta(M) = 0$. Then the following facts hold:*

(a) $\Gamma(M)$ is finitely generated if and only if M is finitely presented.

(b) If M is not finitely presented, then $\Gamma(M)$ and $\Delta\Gamma(M)$ have a free summand of uncountable rank.

(c) $\Gamma^2(M)$ is finitely presented.

Proof. Let M be a non-zero left A-module as in the hypotheses. Then there is an exact sequence of the form

(1) $$0 \longrightarrow L \xrightarrow{i} P^n \longrightarrow M \longrightarrow 0,$$

where n is a positive integer and L is a submodule of $soc\,P^n$. Assume M is finitely presented. Then the results of [C] mentioned in Remark 2.7 guarantee that $\Gamma(M)$ is finitely presented. Now suppose M is not finitely presented. Then L is isomorphic to the direct sum of infinitely many copies of Q. Hence, by (1) and Lemma 2.1, there is an exact sequence of the form

(2) $$0 \longrightarrow \Delta(P^n) \xrightarrow{\Delta(i)} \Delta(L) \longrightarrow \Gamma(M) \longrightarrow 0,$$

where $\Delta(L)$ is a projective module admitting a free summand of uncountable rank. Since $\Delta(P^n) \simeq \tilde{P}^n$, this implies that $\Gamma(M)$ has a decomposition of the form $X \oplus X'$ with the following properties:

(3) X is finitely presented and $\Delta(X) = 0$.

(4) X' is a projective module admitting a free summand of uncountable rank.

Reflexive Modules 61

Hence we deduce from (4) that $\Gamma(M)$ is not finitely generated, and that $\Delta\Gamma(M)$ has a free summand of uncountable rank. Thus (a) and (b) hold for any left module satisfying the hypotheses of the corollary. Moreover, by (3) and (4), we obviously have $\Gamma^2(M) \simeq \Gamma(X)$. This remark and (a) guarantee that $\Gamma^2(M)$ is finitely presented, as claimed in (c). This completes the proof for any finitely generated left A-module belonging to $\operatorname{Ker} \Delta$. A dual argument shows that (a), (b) and (c) hold also for any finitely generated right A-module belonging to $\operatorname{Ker} \Delta$. The corollary is proved. □

REMARK 2.9. As in Corollary 2.8, let A be a generalized Kronecker algebra, and let Δ and Γ be the contravariant functors induced by the cotilting bimodule ${}_AA_A$. Then the structure of projective A-modules (see the proof of Lemma 2.3) guarantees that Δ carries finitely generated modules to finitely generated modules. This observation and Lemma 2.3 imply that

(∗) ${}_AA$ and A_A are finitely cotilting modules and Colby-modules in the sense of [An].

Hence, by (∗) and [An, Remark 4.5], the existence of a finitely generated A-module M such that $\Gamma(M)$ is not finitely generated (Theorem 2.5 (iv), Corollary 2.8 (b)) follows also from the fact that A is neither left nor right noetherian.

The next partial result gives some information on the images under Γ and Γ^2 of the non-finitely generated modules belonging to $\operatorname{Ker} \Delta$.

COROLLARY 2.10. *Let A be the generalized Kronecker algebra of infinite dimension d over K, and let M be an A-module such that $\Delta(M) = 0$ and M is not finitely generated. If m is the smallest cardinality of a set of generators of M, then the following facts hold:*

(a) Any set of generators of $\Gamma(M)$ has at least $|K^m|$ elements.

(b) Either $\Gamma^2(M)$ is finitely generated, or $\Gamma^2(M)$ is not countably generated.

Proof. Let M be a left A-module as in the hypotheses. Then there is an exact sequence of the form

(1) $$0 \longrightarrow L \xrightarrow{i} L' \longrightarrow M \longrightarrow 0,$$

where L' is isomorphic to the direct sum of m copies of P, while L is a submodule of $\operatorname{soc} L'$. Let $l = \dim_K L$. Since M does not have a non-zero projective summand and the dimension of $\operatorname{soc} L'$ is equal to dm, it follows that $l \geq m$. Hence L is isomorphic to the direct sum of infinitely many copies of Q. Thus, by (1) and Lemma 2.1, there is an exact sequence in Mod-A of the form

(2) $$0 \longrightarrow \Delta(L') \xrightarrow{\Delta(i)} \Delta(L) \longrightarrow \Gamma(M) \longrightarrow 0,$$

where $\Delta(L)$ has a free summand of rank equal to $|K^l|$. Since $Im\, \Delta(i)$ is a submodule of $\Delta(L)e_1$, it follows that any set of generators of $\Gamma(M)$ has a least $|K^l|$ elements. Therefore (a) holds. On the other hand, $\Gamma(M)$ has a decomposition of the form $X \oplus Y$, where $\Delta(X) = 0$ and Y is projective. Thus we obviously have

(3) $$\Gamma^2(M) \simeq \Gamma(X).$$

Assume first X is finitely presented. This hypothesis and (∗) in Remark 2.7 guarantee that

(4) $\Gamma(X)$ is finitely presented.

Suppose now X is finitely generated, but not finitely presented. Then condition (b) of Corollary 2.8 guarantees that

(5) $\Gamma(X)$ has a free summand of uncountable rank.

Suppose finally X is not finitely generated. In this case, by simply replacing M by X in the first part of this proof, we see that

(6) $\Gamma(X)$ is not countably generated.

By combining (3), (4), (5) and (6), we obtain (b). The proof is finished. □

We end with a remark on the behaviour of Γ on infinitely many very special finitely presented modules defined over a generalized Kronecker algebra.

REMARK 2.11. Let A be a generalized Kronecker algebra of infinite dimension. As in the proof of Lemma 2.1, let $f : A \to A^{op}$ be the isomorphism such that $f(e_1) = e_2$, $f(e_2) = e_1$ and $f(\alpha_j) = \alpha_j$ for any arrow α_j from 1 to 2. Since A-Mod (resp. Mod-A) is isomorphic to Mod-A^{op} (resp. A^{op}-Mod) in an obvious way [J, page 26], we may use f to obtain an isomorphism $M \mapsto M'$ between A-Mod and Mod-A (resp. Mod-A and A-Mod). For any arrow α_j, and let X denote the module $P/A\alpha_j$ (resp. $\tilde{Q}/\alpha_j A$). Next, let $i : Q \to P$ (resp. $i : \tilde{P} \to \tilde{Q}$) denote the right (resp. left) multiplication by α_j. Then there is an exact sequence of the form

$$0 \longrightarrow \Delta(P) \xrightarrow{\Delta(i)} \Delta(Q) \longrightarrow \Gamma(X) \longrightarrow 0$$

$$(\text{resp. } 0 \longrightarrow \Delta(\tilde{Q}) \xrightarrow{\Delta(i)} \Delta(\tilde{P}) \longrightarrow \Gamma(X) \longrightarrow 0).$$

Thus $\Gamma(X)$ is isomorphic to Coker $\Delta(i)$, and so $\Gamma(X)$ has a base of the form $\{v, v_i | i \neq j\}$ such that $ve_2 = v$, $v\alpha_j = 0$, $v\alpha_i = v_i$ (resp. $e_1 v = v$, $\alpha_j v = 0$, $\alpha_i v = v_i$) for any $i \neq j$. This means that $\Gamma(X)$ is isomorphic to X'. Hence, by condition (iii) of Theorem 2.5, we conclude that

(a) $\Gamma(M) \simeq M'$ for infinitely many indecomposable modules M such that $M \simeq \Gamma^2(M)$ and $\Delta(M) = 0$.

Finally, fix two different arrows α_j and α_l, and let Y denote the module $P/(A\alpha_j \oplus A\alpha_l)$ (resp. $\tilde{Q}/(\alpha_j A \oplus \alpha_l A)$). Then we obviously have

(1) $\dim_K Y' e_2 = 1$ (resp. $\dim_K e_1 Y' = 1$).

Now, let $i : Q \oplus Q \to P$ (resp. $i : \tilde{P} \oplus \tilde{P} \to \tilde{Q}$) be a morphism such that $Im\, i$ is the submodule of P (resp. $\tilde{Q}$) generated by α_j and α_l. Then we have an exact sequence of the form

$$0 \longrightarrow \Delta(P) \xrightarrow{\Delta(i)} \Delta(Q \oplus Q) \longrightarrow \Gamma(Y) \longrightarrow 0$$

$$(\text{resp. } 0 \longrightarrow \Delta(\tilde{Q}) \xrightarrow{\Delta(i)} \Delta(\tilde{P} \oplus \tilde{P}) \longrightarrow \Gamma(Y) \longrightarrow 0).$$

Consequently, $\Gamma(Y)$ is isomorphic to Coker $\Delta(i)$, and so we clearly have

(2) $\dim_K \Gamma(Y)e_2 = 2$ (resp. $\dim_K e_1\Gamma(Y) = 2$).

Therefore, by (1) and (2), $\Gamma(Y)$ is not isomorphic to Y'. This remark and another application of [C, Theorem 6] guarantee that

(b) $\Gamma(M) \not\simeq M'$ for infinitely many indecomposable modules M such that $M \simeq \Gamma^2(M)$ and $\Delta(M) = 0$.

ACKNOWLEDGEMENTS

I would like to thank the referee for his suggestions "to improve the readability" of my paper. He also pointed out that we may proceed as in the proof of Lemma 2.2 to obtain the following more general result:
"Every ring A which is hereditary and perfect on both sides is a cotilting bimodule".
In fact, by a well-known result of Chase [Theorem 3.3, Trans. Amer. Math. Soc. 97 (1960), 457-473], both $_AA$ and A_A are product-complete modules in the sense of Krause-Saorin [Theorem 3.8, Proceedings of the Seattle Conference]. I take the opportunity to mention that Professor K.R. Fuller made a similar remark at the Ohio Algebra Conference (Athens, March 1999), during his conversations on dualities with R. Colpi, F. Mantese, E. Gregorio, A. Tonolo and myself.

REFERENCES

[AF] ANDERSON F.W. - FULLER K.R., Rings and categories of modules, 2nd ed. GTM 13, Springer-Verlag (1992).

[An] ANGELERI HÜGEL L., Finitely cotilting modules, Comm. Algebra 28 (4) (2000), 2147-2172.

[Cb1] COLBY R.R., A generalization of Morita duality and the tilting theorem, Comm. Algebra 17 (7) (1989), 1709-1722.

[Cb2] COLBY R.R., A cotilting theorem for rings, Methods in Module Theory 140, M. Dekker (1993), 33-37.

[C] COLPI R., Cotilting bimodules and their dualities, to appear in Murcia Conference Proceedings 1998, M. Dekker.

[CDT1] COLPI R. - D'ESTE G. - TONOLO A., Quasi-tilting modules and counter equivalences, J. Algebra 191 (1997), 461-494.

[CDT2] COLPI R. - D'ESTE G. - TONOLO A., Corrigendum, J. Algebra 206 (1998), 370-370.

[CF] COLPI R. - FULLER K.R., Cotilting modules and bimodules, Pacific J. Math. 192 (2) (2000), 275-291.

[D] D'ESTE G., Free modules obtained by means of infinite direct products, to appear in Ohio Conference Proceedings.

[GGW] GÓMEZ PARDO J.L. - GUIL ASENSIO P.A. - WISBAUER R., Morita dualities induced by the M-dual functors, Comm. Algebra 22 (1994) 5903-5934.

[HU] HAPPEL D. - UNGER L., *A family of infinite dimensional non self-extending bricks for wild hereditary algebras*, CMS Conference Proceedings 114 (1991), 181-189.

[J] JACOBSON N., *Basic Algebra II*, W.H. Freeman and C., San Francisco (1980).

[M] MANTESE F., *Hereditary cotilting modules*, J. Algebra, to appear.

[Mü] MÜLLER B.J., *Linear compactness and Morita duality*, J. Algebra 16 (1970), 60-66.

[R] RINGEL C.M., *Tame algebras and integral quadratic forms*, Springer LMN 1099 (1984).

[T] TONOLO A., *Generalizing Morita duality: a homological approach*, J. Algebra, to appear.

[X] XUE W., *Rings with Morita duality*, Springer LMN 11523 (1992).

Fibre sum functors and the bimodule Ext

PETER DRÄXLER Fakultät für Mathematik, Universität Bielefeld, POBox 100131, D-33501 Bielefeld, Germany

ABSTRACT Representations of the bimodule $\operatorname{Ext}_A^1(-,-)$ and fibre sum functors both provide techniques for the investigation of module categories for finite-dimensional algebras. We clarify the relation between these two constructions.

1 INTRODUCTION

It is a classical technique in the representation theory of finite-dimensional algebras to consider the A-modules as extensions of modules over smaller subcategories $\mathcal{F}$ and $\mathcal{T}$ thus identifying the module category with the representation category of a bimodule $\operatorname{Ext}_A^1(-,-)$ acting on $\mathcal{F} \times \mathcal{T}$. If $\mathcal{T} = \operatorname{add} S$ for a simple module S, then the category of representations of this bimodule can be identified with the subspace category of a vector space category (see [Ri1]).

In [Dr1] the fibre sum functor with respect to a module P is introduced. We will also use properties of the fibre sum construction which were established in [Dr2]. The fibre sum functor relates the category of A-modules with the category of representations of the bimodule $\operatorname{Hom}_A(-,-)$ acting on $\mathcal{T} \times \mathcal{K}$ for appropriate subcategories $\mathcal{T}$, $\mathcal{K}$. In case the endomorphism algebra of P is a field or more specially if P is a simple module, this leads to a vector space category as well.

The aim of this note is to analyse the relation between these two reduction processes to vector space categories. After recalling the first reduction concept in the next section, in the final section we will clarify the relation completely.

For simplicity we consider only finite-dimensional algebras over an algebraically closed field k which we assume to be basic. We use the term algebra for this concept. For notation and background we refer to [GR] and [Ri2].

2 REDUCTION TO THE BIMODULE $\text{Ext}^1_A(-,-)$

2.1 Let us start out by recalling the concept of a bimodule. Following [GR] an aggregate is a k-additive category with finite-dimensional morphism spaces such that each object is the direct sum of subobjects with local endomorphism algebras. A typical example for an aggregate is the category $A-\text{mod}$ of finite-dimensional (left) modules over an algebra A.

A bimodule over two aggregates $\mathcal{F}$ and $\mathcal{T}$ is a k-linear bifunctor $H: \mathcal{F} \times \mathcal{T} \to k-\text{mod}$ which is covariant in the second and contravariant in the first argument. The category $\text{rep}(H)$ of representations of H has as objects the triples (X, h, Y) where $X \in \mathcal{F}, Y \in \mathcal{T}$ and $h \in H(X, Y)$. A morphism from (X, h, Y) to (X', h', Y') is a pair (s, t) of morphisms $s: X \to X'$ in $\mathcal{F}$ and $t: Y \to Y'$ in $\mathcal{T}$ such that $H(X, t)(h) = H(s, Y')(h')$. The category $\text{rep}(H)$ is again an aggregate.

2.2 The classical example for studying the category $A-\text{mod}$ of an algebra A by representations of a bimodule is the following: Let $\mathcal{F}$ and $\mathcal{T}$ be two subaggregates of $A-\text{mod}$ and $G: A-\text{mod} \to A-\text{mod}$ a subfunctor of the identity functor such that $G(X) \in \mathcal{T}$ and $X/G(X) \in \mathcal{F}$ for all X in $A-\text{mod}$. We consider the bimodule $H = \text{Ext}^1_A(-,-)$ acting on $\mathcal{F} \times \mathcal{T}$ and obtain a full functor $R: A-\text{mod} \to \text{rep}(H)$ by mapping the module X to its canonical exact sequence:

$$0 \to G(X) \xrightarrow{\iota_X} X \xrightarrow{\pi_X} X/G(X) \to 0$$

For illustration we mention that the zero object in $\text{rep}(H)$ is the triple (X, h, Y) with $X = Y = 0$. In general, the functor R is neither dense nor faithful. Its kernel consists of the morphisms $f: X \to Y$ which factorise as $f = \iota_Y g \pi_X$. Since this kernel is contained in the Jacobson radical of the aggregate $A-\text{mod}$, we see that $A-\text{mod}$ is representation equivalent to its image category inside $\text{rep}(H)$. Thus we have 'reduced' the study of $A-\text{mod}$ to the study of its image. It turns out that in many cases representations of a bimodule are easier to handle than the module category itself.

Let us provide some examples for choices of $\mathcal{F}, \mathcal{T}$ and G. We first start with a full subaggregate $\mathcal{T}$ of $A-\text{mod}$. For an A-module X we define $G(X) = G_{\mathcal{T}}(X)$ as the trace of $\mathcal{T}$ in X i.e. the sum over all images $f(Z)$ for all Z in $\mathcal{T}$ and $f \in \text{Hom}_A(Z, X)$. For $\mathcal{F}$ we choose a full subaggregate of $A-\text{mod}$ containing all modules $X/G(X)$. Another class of examples arises by considering an ideal I of A. We take $\mathcal{T}$ as the subcategory of A/J-modules and $\mathcal{F}$ as the subcategory of A/Imodules of $A-\text{mod}$ where J is the left annihilator of I in A. The functor $G = G_I$ is defined as $G_I(X) = IX$. Dually, we can take $\mathcal{T} = A/I-\text{mod}, \mathcal{F} = A/J-\text{mod}$, and $G = G^I$ as the annihilator of I in X where J is the right annihilator of I in A (this means $J = G^I(I)$).

Note, that the lack of density of R is repaired if one assumes that the pair $(\mathcal{T}, \mathcal{F})$ is a torsion theory or in other words G is a radical subfunctor of the identity functor i.e. $G(X/G(X)) = 0$ for all A-modules X. If in addition to $(\mathcal{T}, \mathcal{F})$ being a torsion theory one also assumes that $\text{Hom}_A(\mathcal{F}, \mathcal{T}) = 0$ or that $(\mathcal{F}, \mathcal{T})$ is also a torsion theory, then R is faithful. Being a torsion class forces $\mathcal{T}$ to be extension closed. Therefore, if $\mathcal{T} = \text{add}\, S$, then $\text{Ext}^1_A(S, S) = 0$.

Fibre Sum Functors 67

2.3 We will look in more detail a the situation that $H : \mathcal{F} \times \mathcal{T} \to k-\text{mod}$ is a bimodule such that $\mathcal{T}$ is an aggregate of the shape $\mathcal{T} = \text{add}\, S$ for some object S of $\mathcal{T}$. Additionlly we assume $\mathcal{T}(S,S) \cong k$. In this case $\text{rep}(H)$ can be rewritten as a subspace category. Let us recall the relevant notation.

A vector space category is a k-additive functor $M: \mathcal{F} \to k-\text{mod}$ where $\mathcal{F}$ is an aggregate. Its subspace category $\breve{\mathcal{U}}(M)$ is an aggregate which has as objects the triples $U = (U_\omega, \gamma_U, U_0)$ where $U_\omega \in k-\text{mod}$, $U_0 \in \mathcal{F}$ and $\gamma_U \in \text{Hom}_k(U_\omega, M(U_0))$. Morphisms $U \to U'$ in this category are pairs $f = (f_\omega, f_0)$ such that $f_\omega : U_\omega \to U'_\omega$ is k-linear, $f_0: U_0 \to U'_0$ is a morphism in $\mathcal{F}$ and $M(f_0)\gamma_U = \gamma_{U'} f_\omega$.

If now H is a bimodule such that $\mathcal{T} = \text{add}\, S$ and $\mathcal{T}(S,S) \cong k$ as considered above, then H is completely determined by the contravariant functor $M := H(-,S): \mathcal{F} \to k-\text{mod}$ which we consider as covariant functor $\mathcal{F}^{op} \to k-\text{mod}$. Moreover, any object of $\text{rep}(H)$ lying in $H(X, S^n)$ may by the Yoneda lemma be identified with a morphism in $\text{Hom}_{\mathcal{T}(S,S)}(\mathcal{T}(S^n, S), H(X, S))$. This identification yields an antiequivalence $\text{rep}(H) \to \breve{\mathcal{U}}(M)$.

Dually, one can transform $\text{rep}(H)$ into a subspace category if $\mathcal{F} = \text{add}\, S$. This happens e.g. for G^I introduced above for an ideal I satisfying $A/J \cong k$. This is used in [GNRSV] where the considered functor R usually will not be dense, but the image is calculated precisely.

3 FIBRE SUM FUNCTORS

3.1 Let $(\mathcal{T}, \mathcal{F})$ be a torsion theory in $A-\text{mod}$. We define $\mathcal{K}_\mathcal{T}$ to be the full sub-aggregate of $A-\text{mod}$ whose objects are the modules V satisfying $\text{Ext}^1_A(V, \mathcal{T}) = 0$. The bimodule L acts on $\mathcal{T} \times \mathcal{K}$ as $\text{Hom}_A(-,-)$. We put $\text{rep}_{mon}(L)$ to be the full subcategory of $\text{rep}(L)$ given by all monomorphisms $h : X \to V$. Then the definition of $\mathcal{K}_\mathcal{T}$ implies that the functor $\widetilde{F} : \text{rep}_{mon}(L) \to A-\text{mod}$ which sends h to its cokernel is full.

As a torsion class $\mathcal{T}$ is closed under factor modules. Let us assume that $\mathcal{T}$ has a cover P i.e. $\mathcal{T} = \text{fac}\, P$ for some module P in $A-\text{mod}$. Then we can even assume that P is a minimal cover and therefore $\text{Ext}^1_A(P, \text{fac}\, P) = 0$. It follows from [AR, 1.4] (see also [Dr2, 2.2]) that $\widetilde{F}$ is dense.

The category $\text{rep}_{mon}(L)$ seems to be hard to understand. To improve the situation we assume additionally that also $\text{Ext}^1_A(P, \text{sub}\, P) = 0$. Let C be the bimodule which acts as $\text{Hom}_A(-,-)$ on $\text{add}\, P \times \mathcal{K}_\mathcal{T}$. Then the functor $\Psi : \text{rep}(C) \to \text{rep}_{mon}(L)$ which sends $f : U \to V$ to the inclusion of $\text{Im}\, f$ into V is full and dense. Altogether we have derived the following result from [Dr2] where we put $F_P = \widetilde{F}\Psi$ which is said to be the fibre sum functor with respect to P.

PROPOSITION. *Suppose the S-module P satisfies $\text{Ext}^1_A(P, \text{sub}\, P) = 0$ and $\text{Ext}^1_A(P, \text{fac}\, P) = 0$. Then the fibre sum functor F_P is full and dense.*

It is calculated in [Dr2, 1.4] that the kernel of F_P consists of the morphisms factoring through an object $f : U \to V$ in $\text{rep}(C)$ which is an epimorphism. As shown in [Dr3, 2.5] there are interesting cases such that there exist up to isomorphism only finitely many indecomposable objects of this shape in $\text{rep}(C)$.

3.2 Now we consider the special case that $\operatorname{End}_A(P) \cong k$. Then we can identify rep(C) with $\check{\mathcal{U}}(N)$ where N is the functor $\operatorname{Hom}_A(P,-)$ acting on $\mathcal{K}_\mathcal{T}$. Namely, a homomorphism f in $\operatorname{Hom}_A(P \otimes_k k^n, V)$ is mapped to its adjoint homomorphism $\check{f}$ in $\operatorname{Hom}_k(k^n, \operatorname{Hom}_A(P,V))$.

If in addition $P = S$ is a simple module, then two functors relating A–mod with the subspace category of a vector space category were introduced namely $F_S : \check{\mathcal{U}}(N) \to A$–mod and the antiequivalence $R : A$–mod $\to \check{\mathcal{U}}(M)$ from the previous section. It is our final aim to calculate the composition RF_S. Note, that R is already an equivalence whereas F_P in general cannot be an equivalence because $F_P(k,0,0) = 0 = F(k,\operatorname{id}_k,S)$. Erasing these two objects makes also F_S into an equivalence. More precisely, we consider the full subcategory $\mathcal{K}'_\mathcal{T}$ of $\mathcal{K}_\mathcal{T}$ whose objects do not admit a summand isomorphic to S and denote by N' the restriction of N to $\mathcal{K}'_\mathcal{T}$. Considering only $\check{\mathcal{U}}(N')$ excludes direct summands of the form $(k,\operatorname{id}_k,S)$. Unfortunately, it cancels also direct summands of the form $(0,0,S)$ but, as we will see below, this is the price we have to pay to get satisfactory results. Furthermore, we replace $\check{\mathcal{U}}(N')$ by its full subcategory $\mathcal{U}(N')$ having only those objects U such that γ_U is a monomorphism. In this way we get rid of direct summands of the shape $(k,0,0)$. Now the restriction $F_S : \mathcal{U}(N') \to A$–mod$'$ is an equivalence where A–mod$'$ is the full subaggregate of modules X admitting now direct summand S. On the other hand, R maps A–mod$'$ onto $\mathcal{U}(M)^{op}$. Thus RF_S becomes an antiequivalence $\mathcal{U}(N') \to \mathcal{U}(M)$.

The best that can happen for such an anti equivalence of subspace categories is that it is induced by a suitable antiequivalence of the corresponding vector space categories. We first provide an equivalence $\mathcal{K}'_\mathcal{T} \to \mathcal{F}$.

LEMMA. *Let G' be the functor sending X in A–mod to $X/G(X)$. Then G' induces an equivalence $\mathcal{K}'_\mathcal{T} \to \mathcal{F}$.*

Proof. For given V in $\mathcal{K}'$ the adjoint of the inclusion of $G(V)$ into V is an object $U = (k^n, \gamma_U, V)$ satisfying $F_S(U) = G'(V)$. Therefore the density and fullness of F_S implies the required density and fullness of G'. That G' acts faithfully on $\mathcal{K}'_\mathcal{T}$ follows easily because there do not exist non-zero maps from $\mathcal{K}'_\mathcal{T}$ to S. But note that $G'(S) = 0$. Here it pays out that we passed from $\mathcal{K}_\mathcal{T}$ to $\mathcal{K}'_\mathcal{T}$. □

Furthermore we need an natural isomorphism between the involved functors. As usual $\operatorname{D} = \operatorname{Hom}_k(-,k)$ denotes the dual space functor.

LEMMA. *There is a natural isomorphism $\varphi : \operatorname{D}\operatorname{Hom}_A(S,-) \to \operatorname{Ext}^1_A(G'(-),S)$ of functors $\mathcal{K}'_\mathcal{T} \to k$–mod.*

Proof. Let us consider V in $\mathcal{K}'_\mathcal{T}$. The canonical exact sequence

$$0 \to G(V) \xrightarrow{\iota_V} V \xrightarrow{\pi_V} G'(V) \to 0$$

induces an exact sequence

$\operatorname{Hom}_A(G'(V),S) \xrightarrow{\operatorname{Hom}_A(\pi_V,S)} \operatorname{Hom}_A(V,S) \xrightarrow{\operatorname{Hom}_A(\iota_V,S)} \operatorname{Hom}_A(G(V),S) \longrightarrow$

$\longrightarrow \operatorname{Ext}^1_A(G'(V),S) \xrightarrow{\operatorname{Ext}^1_A(\pi_V,S)} \operatorname{Ext}^1_A(V,S) = 0$

where $\operatorname{Hom}_A(\pi_V,S)$ is actually an isomorphism because $\operatorname{Hom}_A(V,S)\iota_V = 0$. Hence we obtain an isomorphism $\operatorname{Hom}_A(G(V),S) \cong \operatorname{Ext}^1_A(G'(V),S)$. As $G(V)$ is in $\mathcal{T} =$

add S, we obtain $\mathrm{D}\operatorname{Hom}_A(G(V), S) \cong \operatorname{Hom}_A(S, G(V))$. Finally, $\operatorname{Hom}_A(S, G(V)) \cong \operatorname{Hom}_A(S, V)$ by the definition of G. The composition of all these isomorphisms yields the desired natural isomorphism φ. $\square$

Using the two lemmas above and calculating RF_S we obtain:

THEOREM. *If $P = S$ is a simple projective A-module, then the functor $\mathcal{U}(N') \to \mathcal{U}(M)$ sending an object $U = (U_\omega, \gamma_U, U_0)$ to $(\mathrm{D}\, U'_\omega, \varphi_{U_0}^{-1}\, \mathrm{D}\, \gamma'_U, G'(U_0))$ is an antiequivalence which is isomorphic to RF_S.*

REFERENCES

[AR] M. Auslander, I. Reiten, *Applications of contravariantly finite subcategories*, Adv. Math. 86 (1991), 111-152.

[C-B] W. Crawley-Boevey, *On tame algebras and bocses*, Proc. London Math. Soc. (3) 56 (1988), 451-483.

[Dr1] P. Dräxler, *$\mathcal{U}$-Fasersummen in darstellungsendlichen Algebren*, J. Algebra 113 (1988), 430-437.

[Dr2] P. Dräxler, *On the density of fiber sum functors*, Math. Z. 216 (1994), 645-656.

[Dr3] P. Dräxler, *Generalized one-point extensions*, Math. Ann. 304 (1996), 645-667.

[Dd1] Y.A. Drozd, *Matrix problems and categories of matrices*, Zap. Nauchn. Sem. LOMI 28 (1972), 144-153

[Dd2] Ju. A. Drozd: *Tame and wild matrix problems*, Lecture Notes in Math. 832 (1980), 242-258.

[GNRSV] P. Gabriel, L.A. Nazarova, A.V. Roiter, V.V. Sergejchuk, D. Vossieck, *Tame and wild subspace problems*, Ukr. Math. J. 45 (1993), 313-352.

[GR] P. Gabriel, A.V. Roiter, *Representations of finite-dimensional algebras*, Encyclopedia of the Mathematical Sciences, Vol. 73, Algebra VIII, A.I. Kostrikin and I.V. Shafarevich (Eds.), Berlin, Heidelberg, New York, 1992.

[NR] L.A. Nazarova, A.V. Roiter, *Kategorielle Matrizenprobleme und die Brauer-Thrall-Vermutung*, Mitt. Math. Sem. Giessen 115 (1975).

[Ri1] C.M. Ringel, *Report on the Brauer-Thrall conjectures*, Lecture Notes in Math. 831 (1980), 104-136.

[Ri2] C.M. Ringel, *Tame algebras and integral quadratic forms*, Lecture Notes in Math. 1099 (1984).

Smooth Automorphism Group Schemes

DANIEL R. FARKAS Department of Mathematics, Virginia Polytechnic Institute and State University, Blacksburg, VA 24061

CHRISTOF GEISS Instituto de Matemáticas, UNAM, Ciuadad Universitaria, C.P. 04510, Mexico D. F., Mexico

EDUARDO N. MARCOS Departamento de Matemática, I.M.E., Universidade de São Paulo, CP 66281, São Paulo, SP05389.970, Brasil

ABSTRACT Smoothness for the automorphism group scheme of a finite-dimensional algebra in positive characteristic can be interpreted as a property of the Hopf algebra representing the scheme. With this approach, it is proved that the scheme is smooth if and only if all derivations of the original finite-dimensional algebra are integrable. This criterion is applied to commutative monomial algebras and used, as well, to establish a general Morita invariance theorem.

The most naive way to understand a finite-dimensional associative algebra is to find a basis and analyze its multiplication table. In the modern incarnation, one considers the scheme of all associative n-dimensional algebras over the field k as a subscheme of affine space $\operatorname{Hom}_k(k^n \otimes k^n, k^n)$. Then $GL_n(k)$ acts on the k-rational points of the scheme so that orbits can be interpreted as isomorphism classes of n-dimensional algebras.

The stabilizer of a point can be identified with the automorphism group (scheme) of the corresponding algebra. The geometry at a point seems to behave particularly well when the automorphism group scheme is smooth. For example, Gabriel ([Ga], 2.4) proves that if the algebra A corresponds to the point μ then $\mathrm{H}^2(A)$ is isomorphic to the tangent space at μ in the entire scheme modulo the tangent space at μ in its GL_n-orbit. This result requires smoothness of the stabilizer, as first explicitly pointed out in [Maz].

The automorphism group scheme is automatically smooth when char $k = 0$ by the classical characterization of cocommutative connected Hopf algebras. The situation in positive characteristic has been more mysterious. The main contribution of this paper is to provide a simple, user-friendly reformulation of smoothness. We prove that the automorphism group scheme of A is smooth if and only if every k-derivation of A is integrable. Here we mean that D is *integrable* if it is a member of a sequence of k-endomorphisms of A,

$$D^{(0)} = I,\ D^{(1)} = D,\ D^{(2)}, D^{(3)}, \ldots$$

such that

$$D^{(m)}(ab) = \sum_{i+j=m} D^{(i)}(a) D^{(j)}(b)$$

for all $a, b \in A$. The notion of integrability (which also appears in the literature under the name "higher derivations") is far from new although we believe that this application is novel. The proof of our criterion is essentially Hopf algebraic and found in the first section.

The second section reviews known properties of integrable derivations. It also includes a generalization of the well known fact that a derivation of a finite-dimensional algebra over a field of characteristic zero sends the Jacobson radical into itself.

Next, a particular class of examples is studied. Using our criterion, we present a clumsy but algorithmically tractable description of those commutative monomial algebras whose automorphism group scheme is smooth. We obtain both expected results (e.g., smoothness follows when relations "avoid" the characteristic) and bizarre examples.

In the fourth and last section, we prove that the property of having a smooth automorphism group scheme is a Morita invariant. Indeed, it is shown more generally that integrable derivations contribute to a Morita invariant piece of the first Hochschild cohomology group.

1 SWEEDLER'S THEOREM

We begin by deriving a transparent, intrinsic condition on a finite-dimensional algebra which is equivalent to its having a smooth automorphism group scheme. Two different proofs are presented. The first is a leisurely algebraic exposition which depends on classical Hopf algebra constructions. The second proof is short and geometric. This time the real work is hidden in several standard lemmas.

Until further notice, we let H denote a commutative affine Hopf algebra over the field k. If H represents an affine group scheme then the scheme is smooth precisely when H is reduced, i.e., when H has no nonzero nilpotent elements. It is well known that H is always reduced when char $k = 0$ ([Wa]). In case the characteristic of k is positive and k is perfect, Sweedler ([Sw]) has found a characterization of reduced Hopf algebras which we wish to apply. This result depends on the analysis of a certain k-coalgebra, the *hyperalgebra*, associated with H.

Let ϵ_H be the augmentation map for H and let $\mathcal{M}$ be its kernel. $\mathbf{Hyp}(H)$ is the subcoalgebra of the dual H° consisting of all linear functionals which vanish on some power of $\mathcal{M}$. It is possible to prove that $\mathbf{Hyp}(H)$ is the irreducible component of H° containing ϵ_H ([Abe], p.198).

Suppose C is a coalgebra with counit ϵ_C. Given $d \in C$, an *infinite sequence of divided powers lying over* d is a sequence $d_0, d_1, d_2, \ldots$ of elements in C such that

$$\Delta d_n = \sum_{i=0}^{n} d_i \otimes d_{n-i} \text{ for all } n, \ \epsilon_C(d_n) = 0 \text{ for } n > 0, \text{ and } \epsilon_C(d_0) = 1$$

with $d_1 = d$. Equivalently, we may regard an infinite sequence of divided powers in C as a coalgebra morphism from the coalgebra of divided powers

$$B = kx^{(0)} + kx^{(1)} + \cdots$$

to C. Now suppose $d_0, d_1, \ldots$ is an infinite sequence of divided powers in $\mathbf{Hyp}(H)$. Notice that $d_0(1) = 1$ and $d_n(1) = 0$ for $n > 0$. Since d_0 must be group-like, we have $d_0 = \epsilon_H$. Moreover, consider any infinite sequence of divided powers $d_0, d_1, \ldots$ in H° such that $d_0 = \epsilon_H$. If $a, b \in \mathcal{M}$ then

$$d_1(ab) = d_0(a)d_1(b) + d_1(a)d_0(b) = 0 .$$

Continuing by induction, we see that $d_n(\mathcal{M}^{n+1}) = 0$. Hence the sequence lies in $\mathbf{Hyp}(H)$. Thus we may identify the collection of all infinite divided powers in $\mathbf{Hyp}(H)$ with the subgroup (under convolution) of $\mathbf{coalg}(B, H^\circ)$ consisting of those α with $\alpha(x^{(0)}) = \epsilon_H$.

Recall that an element a in a bialgebra is *primitive* when $\Delta a = 1 \otimes a + a \otimes 1$. Since 1 in H° is identified with ϵ_H, we see that any term d_1 belonging to an infinite sequence of divided powers in $\mathbf{Hyp}(H)$ must be primitive. (In this context, a linear functional $d \in H^\circ$ with

$$d(ab) = \epsilon_H(a)d(b) + \epsilon_H(b)d(a) \text{ for all } a, b \in H$$

is also called an ϵ-derivation.)

THEOREM 1.1 ([Sw]). *Assume H is an affine commutative Hopf algebra over a perfect field k of positive characteristic. Then H is reduced if and only if there is an infinite sequence of divided powers in $\mathbf{Hyp}(H)$ lying over each primitive element.*

In order to apply this theorem when H represents the automorphism group scheme of a finite-dimensional k-algebra A, we need to interpret infinite sequences of divided powers intrinsically for A. This will be done in a series of steps which are more or less standard. We begin by reminding the reader that if B is the k-coalgebra of divided powers then B^* can be identified with $k[[t]]$ by sending $f \in B$ to $\sum f(x^{(n)})t^n$.

LEMMA 1.1. $\mathbf{coalg}(B, H^\circ) \simeq \mathbf{alg}(H, B^*)$ *as groups.*

Proof. We have the obvious group homomorphism

$$\Phi : \mathbf{coalg}(B, H^\circ) \to \mathbf{alg}(H, B^*)$$

given by $\Phi(\alpha)(h)(b) = \alpha(b)(h)$ for $h \in H$ and $b \in B$. We first argue that Φ is surjective. Indeed, let $\theta \in \mathbf{alg}(H, B^*)$. Let $\pi_m : H \to k[[t]]/(t^m)$ be the composition of θ with the obvious projection. The kernel of this algebra map is a two-sided ideal I_m of cofinite dimension in H. Let c_m be the linear functional in H^* which sends $h \in H$ to the coefficient of t^m in $\theta(h)$. Then $I_{m+1} \subseteq \text{Ker } c_m$, whence $c_m \in H^\circ$. If $\alpha \in \textbf{coalg}(B, H^\circ)$ is defined by $\alpha(x^{(m)}) = c_m$ then $\Phi(\alpha) = \theta$.

Next, we compute the kernel of Φ,

$$\{\alpha : B \to H^\circ \mid \Phi(\alpha) = \eta_{B^*} \epsilon_H\}$$

where η denotes the unit. Since $\Phi(\alpha)(h) = \sum \alpha(x^{(m)})(h) t^m$ we see that if $\alpha \in \text{Ker} \Phi$ then

$$\alpha(x^{(0)}) = \epsilon_H \quad \text{and} \quad \alpha(x^{(m)}) = 0 \text{ for } m \geq 1 .$$

We have described the identity element for $\textbf{coalg}(B, H^\circ)$. $\square$

For the remainder of this section, we shall assume that H represents the automorphism group scheme $\textbf{Aut}_A$ of A. That is, if R is any commutative k-algebra then

$$\textbf{Aut}_A(R) = \textbf{alg}(H, R) .$$

Of course, $\textbf{Aut}_A(k) = \text{Aut}_k(A)$. The action of this automorphism group on A can be described via an H-comodule algebra structure on A: there is a coaction

$$\lambda : A \to H \otimes A$$

making A a left H-comodule so that

$$\lambda(ab) = \sum a_{(0)} b_{(0)} \otimes a_{(1)} b_{(1)} \quad \text{and} \quad \lambda(1) = 1 \otimes 1$$

for $a, b \in A$. (See [Mo], section 4.1.) The explicit isomorphism from $\textbf{alg}(H, k)$ to $\text{Aut}_k(A)$ sends θ to the automorphism $\widetilde{\theta}$ where

$$\widetilde{\theta}(a) = \sum \theta(a_{(0)}) a_{(1)} .$$

If R is any commutative k-algebra then the group $\textbf{Aut}_A(R)$ is isomorphic to $\text{Aut}_R(R \otimes_k A)$ under the extension of the comodule algebra action to $R \otimes H$ on $R \otimes A$.

We are particularly interested in the case that $R = B^*$. Since A is finite-dimensional, we have

$$B^* \otimes A \simeq k[[t]] \otimes A \simeq A[[t]] .$$

Again, since A is finite-dimensional, a $k[[t]]$-automorphism of $A[[t]]$ is determined by its effect on elements of A.

A k-algebra map $\delta : A \to A[[t]]$ is a *higher derivation* of A provided that for all $a \in A$, the constant term of the power series $\delta(a)$, is simply a. Clearly, higher derivations of A are in one-to-one correspondence with $k[[t]]$-automorphisms of $A[[t]]$ which "preserve constant terms". Alternatively, we may regard a higher derivation as a sequence of linear endomorphisms of A, say $D^{(0)} = I, D^{(1)}, D^{(2)}, \ldots$, such that

$$D^{(n)}(ab) = \sum_{i+j=n} D^{(i)}(a) D^{(j)}(b)$$

for all $a, b \in A$ and $n \geq 0$. (The point is to expand $\delta(c) = \sum_{n=0}^\infty D^{(n)}(c) t^n$ for $c \in A$.)

LEMMA 1.2. *There is a one-to-one correspondence between infinite sequences of divided powers in* $\mathbf{Hyp}(H)$ *and higher derivations of* A. *The map sends* $\epsilon_H = d_0, d_1, \ldots$ *to* $I = D^{(0)}, D^{(1)}, \ldots$ *where*

$$D^{(m)}(a) = \sum d_m(a_{(0)})a_{(1)}$$

for all $a \in A$.

Proof. By virtue of the previous lemma and our discussion so far, there is a group isomorphism

$$\mathbf{coalg}(B, H^\circ) \simeq \mathrm{Aut}_{k[[t]]}(A[[t]]) \ .$$

The isomorphism sends $\alpha \in \mathbf{coalg}(B, H^\circ)$ to the automorphism

$$\sum c_n t^n \mapsto \sum_n \sum_{i+j=n} [\alpha(x^{(i)})(c_j)_{(0)}(c_j)_{(1)}]t^n \ .$$

In particular, this automorphism sends $c \in A$ to $\sum_n [\sum \alpha(x^{(n)})(c_{(0)})c_{(1)}]t^n$.

Note that if $\alpha(x^{(0)}) = \epsilon_H$ then the associated automorphism preserves constants. Conversely, we argue that if $f = \alpha(x^{(0)}) \in H^\circ$ and $\sum f(c_{(0)})c_{(1)} = c$ for all $c \in A$ (i.e., the automorphism preserves constants) then $f = \epsilon_H$. But f is group-like, so $f \in \mathbf{alg}(H, k)$. The claim follows from our isomorphism $\mathbf{alg}(H, k) \simeq \mathrm{Aut}_k(A)$.

The lemma is now a consequence of restricting the group isomorphism to coalgebra maps from B to H° which send $x^{(0)}$ to ϵ_H. □

It is easy to see that the $D^{(1)}$-term of a higher derivation is always an ordinary derivation. We say that a derivation $D \in \mathrm{Der}_k(A)$ is *integrable* provided there exists a higher derivation $D^{(0)} = I, D^{(1)}, D^{(2)}, \ldots$ such that $D^{(1)} = D$.

THEOREM 1.2. *Let* A *be a finite-dimensional* k-*algebra and assume that* H *represents the affine group scheme* Aut_A. *Every* ϵ-*derivation of* H *has an infinite sequence of divided powers in* $\mathbf{Hyp}(H)$ *lying over it if and only if every derivation of* A *is integrable.*

Proof. It is easy to check directly that if d is an ϵ-derivation of H then the linear endomorphism D of A given by

$$D(a) = \sum d(a_{(0)})a_{(1)}$$

is a derivation. It is well known that this map from ϵ-derivations to $\mathrm{Der}_k(A)$ is an isomorphism ([Wa]). (This can also be seen by replacing $k[[t]]$ with $k[t]/t^2$ in the arguments we have just presented.) Apply the lemma. □

If the characteristic of k is zero then it is a well known consequence of the Leibniz rule (see [Hu], p.8) that any derivation D can be integrated to the higher derivation

$$I, D, \frac{1}{2}D^2, \ldots, \frac{1}{n!}D^n, \ldots \ .$$

Thus integrability of derivations is only an issue when $\mathrm{char}\, k > 0$, which brings us back to Sweedler's Theorem. We summarize our discussion for this section.

COROLLARY 1.1. *Assume that A is a finite-dimensional algebra over the perfect field k. The affine group scheme Aut_A is smooth if and only if every k-derivation of A is integrable.*

Gerstenhaber observes (see [GS]) that if the second Hochschild cohomology group $\mathrm{H}^2(A, A)$ vanishes then all derivations of A are integrable. Though this is only a sufficient condition, it does suggest that a detailed study of H^2 might pinpoint the precise obstruction.

As promised, we outline a second proof. Again, assume that H is an affine cocommutative Hopf algebra over k with augmentation ϵ. Set $\widetilde{H} = H/\mathrm{rad}(H)$, so $\widetilde{H}$ is reduced. The tangent space at unity for the algebraic group scheme associated to H can be described by using "dual numbers", which we identify with $k[[t]]/(t^2)$:

$$\mathcal{T}_H = \{\alpha \in \mathbf{alg}(H, k[[t]]/(t^2)) \mid \alpha(h) \equiv \epsilon(h) \pmod{t} \text{ for all } h \in H\}.$$

It is easy to see that $\alpha \in \mathcal{T}_H$ if and only if $\alpha - \epsilon$ is an ϵ-derivation of H. Thus when H represents the automorphism group scheme of A, our earlier discussion identifies $\mathcal{T}_H$ with $\mathrm{Der}_k(A)$.

The quotient map $\rho : H \to \widetilde{H}$ induces an injective k-linear map

$$d\rho^* : \mathcal{T}_{\widetilde{H}} \to \mathcal{T}_H$$

by sending β to $\beta \circ \rho$. The key technical lemma we need (see [Wa]) is that H is reduced if and only if $d\rho^*$ is an isomorphism. Our characterization can now be expressed in the following form.

THEOREM 1.3. *Assume that k is a perfect field. Then H is reduced if and only if for each $\alpha \in \mathcal{T}_H$ there exists a k-algebra map $\check{\alpha} : H \to k[[t]]$ with*

$$\check{\alpha}(h) \equiv \alpha(h) \pmod{t^2}$$

for all $h \in H$.

Proof. First suppose that each $\alpha \in \mathcal{T}_H$ can be lifted to $\check{\alpha}$. Then $\check{\alpha}(\mathrm{rad}(H)) = 0$ because $k[[t]]$ is an integral domain. Hence $\alpha(\mathrm{rad}(H)) = 0$. This says that α factors through a map in $\mathcal{T}_{\widetilde{H}}$.

Conversely, assume H is reduced. The local ring $H_{\mathrm{Ker}(\epsilon)}$ is regular, so the Cohen Structure Theorem implies its completion C is isomorphic to a power series algebra $k[[t_1, \ldots, t_m]]$ with m the Krull dimension of H. (It is at this point that we use the assumption that k is perfect.) The natural ring homomorphism $\eta : H \to C$ induces an isomorphism $d\eta^* : \mathcal{T}_C \to \mathcal{T}_H$. Suppose $\alpha \in \mathcal{T}_H$. Choose $\gamma \in \mathcal{T}_C$ such that $\alpha = \gamma \circ \eta$. Certainly γ lifts to some $\check{\gamma}$. Then α lifts to $\check{\gamma} \circ \eta$. □

It is clear from the second argument that the restriction to perfect fields is only required for one direction. In general, if all derivations are integrable then the automorphism group scheme is smooth. Here is an application: any finite-dimensional hereditary algebra has a smooth automorphism group scheme because its second Hochschild cohomology group vanishes ([Ha]). We will give an alternative proof at the end of the paper.

2 INTEGRABLE DERIVATIONS

A derivation of a finite-dimensional algebra whose scalar field has characteristic zero always sends the radical into itself. We shall see that integrable derivations extend this behavior.

LEMMA 2.1. *If S is a semiprime ring then so is $S[[t]]$.*

Proof. We must show that if $\alpha \in S[[t]]$ is nonzero then $\alpha S[[t]]\alpha \neq 0$. But $\alpha S \alpha \neq 0$, as can be seen by looking at the lowest term of α. □

As a consequence, if R is any ring and I is a nilpotent ideal of $R[[t]]$ then $I \subseteq (\text{prime rad}\,(R))[[t]]$.

THEOREM 2.1. *Let A be a finite-dimensional algebra. If the algebra map $\phi : A \to A[[t]]$ is a higher derivation then $\phi(\operatorname{rad} A) \subseteq (\operatorname{rad} A)[[t]]$.*

Proof. It suffices to prove that $\phi(\operatorname{rad} A)$ lies in a nilpotent ideal of $A[[t]]$. Let $\widetilde{A}$ denote the algebra generated by $\phi(A)$ and t. (We do not ask that $\widetilde{A}$ be closed.) The condition that $\phi(a) = a+$"higher terms" implies that $\widetilde{A}$ is dense in $A[[t]]$ with respect to the (t)-adic topology.

Let n be the index of nilpotence for $\operatorname{rad} A$ and choose $w_j \in \operatorname{rad} A$. Choose $r_j, s_j \in \widetilde{A}$ for $j = 1, \ldots, n$. Because ϕ is an algebra map and t is central,

$$(r_1\phi(w_1)s_1)(r_2\phi(w_2)s_2)\cdots(r_n\phi(w_n)s_n) = 0 .$$

Each member of $A[[t]]$ is the limit of a sequence of elements in $\widetilde{A}$. Thus, by continuity, the identity above extends to all $r_j, s_j \in A[[t]]$. We conclude that the ideal in $A[[t]]$ generated by $\phi(\operatorname{rad} A)$ is nilpotent. □

COROLLARY 2.1. *Let A be a finite-dimensional k-algebra. If $D \in \operatorname{Der}_k(A)$ is integrable then $D(\operatorname{rad} A) \subseteq \operatorname{rad} A$.*

Proof. Let $I = D^{(0)}, D = D^{(1)}, D^{(2)}, \ldots$ be a higher derivation. According to the theorem, $D^{(m)}(\operatorname{rad} A) \subseteq \operatorname{rad} A$ for all m. □

We shall see, when we examine monomial algebras, that it is possible for every derivation of a finite-dimensional algebra to leave the radical invariant even though its automorphism group scheme is not smooth. Nonetheless, the corollary does provide a useful test.

THEOREM 2.2. *Assume that k is a perfect field of characteristic p and G is a non-trivial finite p-group. Then the group algebra $k[G]$ never has a smooth automorphism group scheme.*

Proof. Since G/G' is not trivial, there is a nonzero additive character $\lambda \in \operatorname{Hom}(G, k^+)$. Define $D : k[G] \to k[G]$ by linearly extending the function $D(g) = \lambda(g)g$ with $g \in G$. It is easy to see that D is a derivation.

Choose $h \in G$ with $\lambda(h) \neq 0$. Then $h-1$ lies in the augmentation ideal of $k[G]$, which coincides with the radical. But

$$D(h-1) = \lambda(h)h$$

so $D(h-1)$ is not in the radical. $\square$

It is tempting to conjecture that $k[G]$ does not have a smooth automorphism group scheme whenever p divides the order of G. However, we will see in a few moments that inner derivations are always integrable. Thus a "prerequisite" to the conjecture is the knowledge that such group algebras possess outer derivations. The good news is that this weaker assertion is true ([FJL]). The bad new is that the only known proof requires the classification of finite simple groups.

We record some well known properties of integrable derivations for future use. (See, e.g., [Mat].) Let $\mathcal{Z}(\ )$ denote the center of a ring.

PROPOSITION 2.1. *The integrable derivations of the k-algebra A constitute a $\mathcal{Z}(A)$-submodule of all derivations.*

Proof. If D and E are integrable derivations then there exist ϕ and ψ, constant preserving $k[[t]]$-automorphisms of $A[[t]]$, such that

$$\phi(a) = a + D(a)t + \cdots \quad \text{and} \quad \psi(a) = a + E(a)t + \cdots$$

for all $a \in A$. Then the composition $\psi\phi$ is an automorphism which preserves constants. Explicitly, if $D^{(0)}, D^{(1)} = D, D^{(2)}, \ldots$ and $E^{(0)}, E^{(1)} = E, E^{(2)}, \ldots$ are the corresponding higher derivations then we have constructed a new higher derivation whose m^{th} term is $\sum_{i+j=m} E^{(i)} D^{(j)}$. In particular, the $m = 1$ term is $D + E$. Thus the collection of integrable derivations is closed under addition.

For any central $\lambda \in A$, the sequence

$$\lambda^0 D^{(0)}, \lambda^1 D^{(1)}, \lambda^2 D^{(2)}, \ldots$$

is also a higher derivation. $\square$

PROPOSITION 2.2. *Every inner derivation of the algebra A is integrable.*

Proof. Let $a \in A$. Conjugation by the unit $1 - at$ is an algebra automorphism of $A[[t]]$ and for any $r \in A$,

$$(1-at)^{-1} r (1-at) = r + (ar - ra)t + \text{higher terms}.$$

Thus ad a is integrable. $\square$

It is well known that a diagonalizable derivation of a k-algebra A is equivalent to a grading of A by the additive group k^+. Indeed, the eigenspaces of the derivation are the homogeneous components for the grading. Such gradings can be difficult to deal with when the characteristic of k is positive; it would be nice to lift $\mathbf{Z}/(p)$-gradings to $\mathbf{Z}$-gradings. This goal is encoded in the following definition. We say that a higher derivation $D^{(0)}, D^{(1)}, \ldots$ is diagonalizable when the $D^{(m)}$ are simultaneously diagonalizable k-endomorphisms of A. If $\phi : A \to A[[t]]$ is the algebra map version of the higher derivation then diagonalizability means that there is a basis $v_1, \ldots, v_n$ of A so that $\phi(v_i) = f_i v_i$ for some $f_i \in k[[t]]$. Moreover, the fact that ϕ preserves constants tells us that $f_i \in \mathcal{U}_1(k[[t]])$, the multiplicative group of units in $k[[t]]$ with constant term 1. With very little additional work, we have

PROPOSITION 2.3. *There is a one-to-one correspondence between diagonalizable higher derivations of the finite-dimensional k-algebra A and $\mathcal{U}_1(k[[t]])$-gradings of A.*

Observe that the group $\mathcal{U}_1(k[[t]])$ is always torsion free, no matter what the characteristic of k is. Thus if a diagonalizable derivation of A lifts to a diagonalizable higher derivation then a k^+-grading lifts to a grading by a torsion free abelian group. The converse is more valuable. Since $\mathcal{U}_1(k[[t]])$ is uncountable, it is abelian of infinite rank. As a consequence, every finitely generated torsion free abelian group embeds in $\mathcal{U}_1$. We conclude that if D is a diagonalizable derivation of A whose grading lifts to a second grading via a (finitely generated) torsion free abelian group then the D is integrable (and is the $D^{(1)}$-term of a diagonalizable higher derivation).

3 COMMUTATIVE MONOMIAL ALGEBRAS

We regard monomial algebras as a rich source of elementary examples. Our study of this family of rings begins with a more or less computable criterion for integrability in this case.

Recall (cf. [FGGM]) that if I is an ideal of the polynomial algebra $k[X_1, \ldots, X_n]$ then every derivation of

$$R = k[X_1, \ldots, X_n]/I$$

lifts to a derivation of $k[X_1, \ldots, X_n]$ which stabilizes I. If I is a monomial ideal then every derivation of R is a linear combination of images of such derivations with the special form $m\frac{\partial}{\partial X_j}$ for some monomial m.

In this section, we will always assume that I has finite codimension in the polynomial algebra.

THEOREM 3.1. *Let I be a monomial ideal of $k[X, Y_1, \ldots Y_n]$ and set*

$$R = k[X, Y_1, \ldots, Y_n]/I .$$

Assume that m is a monomial which does not involve X such that $m\frac{\partial}{\partial X}$ stabilizes I. Then the derivation D it induces on R is integrable if and only if for each monomial $X^e \nu \in I$, where ν does not involve X,

$$\binom{e}{j} X^{e-j} m^j \nu \in I \text{ for } j = 0, 1, \ldots, e.$$

Proof. Choose an automorphism ϕ of $R[[t]]$ such that

$$\phi \mid_R = I + Dt + \cdots .$$

Underline to denote the image of a polynomial in R. Suppose $d \leq e$. Since $\phi(\underline{X}) = \underline{X} + \underline{m}t + \cdots$, the coefficient of t^d in $\phi(\underline{X})^e$ has the form

$$\binom{e}{d} \underline{X}^{e-d} \underline{m}^d + \underline{X}^{e-d+1} s_d$$

for some $s_d \in R$. (The pigeon-hole principle is at work here: no more than d of the factors $\phi(\underline{X})$ in $\phi(\underline{X})^e$ can contribute a term rt^i for $i \geq 1$.) Similarly,

$$\phi(\underline{\nu}) = \sum_{h \geq 0} a_h t^h \text{ with } a_0 = \underline{\nu} .$$

Hence for $j \leq e$, the coefficient of t^j in $\phi(\underline{X})^e \phi(\underline{\nu})$ has the form

$$\sum_{d=0}^{j} (\binom{e}{d} \underline{X}^{e-d} \underline{m}^d + \underline{X}^{e-d+1} s_d) a_{j-d} = \binom{e}{j} \underline{X}^{e-j} \underline{m}^j \underline{\nu} + \underline{X}^{e-j+1} s$$

for some $s \in R$. On the other hand,

$$\phi(\underline{X})^e \phi(\underline{\nu}) = \phi(\underline{X}^e \underline{\nu}) = 0 \,.$$

Since R is strongly graded by monomials, we see from the powers of $\underline{X}$ in our expression for 0 that

$$\binom{e}{j} \underline{X}^{e-j} \underline{m}^j \underline{\nu} = 0 \,.$$

This proves one direction of the theorem.

As to the converse, assume that for a set of generating relations $\underline{X}^e \underline{\nu} \in I$ we have $\binom{e}{j} \underline{X}^{e-j} \underline{m}^j \underline{\nu} = 0$ in R for $j = 0, \ldots, e$. Consider the assignments

$$\psi(\underline{X}) = \underline{X} + \underline{m}t \quad \text{and} \quad \psi(\underline{Y}_i) = \underline{Y}_i$$

for $i = 1, \ldots, n$. Since $\psi(\underline{X})^e \psi(\underline{\nu}) = 0$, ψ extends to an algebra map from R to $R[t]]$. It is easy to check that the coefficient of t in the expansion of ψ agrees with D on the generators $\underline{X}, \underline{Y}_1, \ldots, \underline{Y}_n$ of R. Hence D is integrable. □

The previous theorem only handles images of $m \frac{\partial}{\partial X}$ when X does not appear in m. Fortunately, the remaining "monomial" derivations are always integrable.

THEOREM 3.2. *Let I be a monomial ideal of $k[X, Y_1, \ldots Y_n]$ and set*

$$R = k[X, Y_1, \ldots, Y_n]/I \,.$$

Assume that m is a monomial which involves X such that $m \frac{\partial}{\partial X}$ stabilizes I. Then the derivation D it induces on R is integrable.

Proof. According to Proposition 2.1, it suffices to show that the image D of $X \frac{\partial}{\partial X}$ is integrable. Define $\phi(\underline{X}) = \underline{X} + \underline{X}t$ and $\phi(\underline{Y}_j) = \underline{Y}_j$ for $j = 1, \ldots n$. If $X^e \nu$ is a monomial in I such that X does not appear in ν then

$$(X + Xt)^e \nu = X^e \nu (1+t)^e \in I \,.$$

Thus ϕ extends to an algebra map from R to $R[[t]]$. Its t term agrees with D. □

We use the previous two theorems to illustrate the metatheorem that an algebra whose relations do not interfere with the characteristic has a smooth automorphism group.

THEOREM 3.3. *Assume that k is a field of characteristic $p > 0$. Let I be a monomial ideal of $k[X_1, \ldots X_n]$ and set*

$$R = k[X_1, \ldots, X_n]/I \,.$$

If no minimal monomial in I has positive degree in any X_j which is divisible by p then the automorphism group scheme of R is smooth.

Smooth Automorphism Group Schemes

Proof. By virtue of the previous theorem and Proposition 2.1, we need only prove that if m is a monomial which does not involve X_s and $m\frac{\partial}{\partial X_s}$ stabilizes I then its image derivation of R is integrable. We apply Theorem 3.1. It suffices to show that if $X_s^e \nu$ is a monomial in I such that X_s does not appear in ν then

$$X_s^{e-j} m^j \nu \in I \quad \text{for} \quad j = 0, \ldots, e \,.$$

By induction, we are reduced to the case $j = 1$. Since $m\frac{\partial}{\partial X_s}$ stabilizes I,

$$eX_s^{e-1}\nu m \in I \,.$$

We are done unless $p|e$. Suppose this is the case. Now $X_s^e \nu$ is divisible by some minimal monomial relation μ. But the X_s-degree of μ is either zero or a positive integer not divisible by p. In either event, we must have $X_s^{e-1}\nu \in I$. □

The hope is to look at the minimal monomials generating an ideal and immediately tell whether the corresponding monomial algebra has a smooth automorphism group scheme. Since we do not yet know how to do this, we offer a more modest result.

THEOREM 3.4. *Assume that k is a field of characteristic $p > 0$. Let I be a monomial ideal of $k[X_1, \ldots X_n]$ and set*

$$R = k[X_1, \ldots, X_n]/I \,.$$

Every derivation of R stabilizes the radical if and only if for each j there exists a minimal monomial $\mu_j \in I$ such that the X_j-degree of μ_j is not divisible by p.

Proof. First assume that every minimal monomial in I has the form $X_1^{ps}\alpha$ where α is a monomial not involving X_1. Then

$$\frac{\partial}{\partial X}(X_1^{ps}\alpha) = \frac{\partial}{\partial X}(X_1^{ps})\alpha + X_1^{ps}\frac{\partial}{\partial X}(\alpha) = 0 \,.$$

Thus $\frac{\partial}{\partial X}$ stabilizes I. We see that $\frac{\partial}{\partial X}$ induces a derivation of R which sends the image of X_1, which is in the radical, to 1.

Conversely, assume that I has minimal generators as described in the theorem. We must show that if D is derivation of $k[X_1, \ldots, X_n]$ and $D(I) \subseteq I$ then $D(X_j) \in (X_1, \ldots, X_n)$ for $j = 1, \ldots, n$. Choose a minimal monomial $X_j^f \beta \in I$ such that β does not involve X_j and p does not divide f.

$$fX_j^{f-1}D(X_j)\beta + X_j^f D(\beta) = D(X_j^f \beta) \in I$$

Write $D(X_j) = c + H$ where $c \in k$ and $H \in (X_1, \ldots, X_n)$. Then

$$cfX_j^{f-1}\beta + fX_j^{f-1}\beta H + X_1^f D(\beta) \in I \,.$$

If we write the second term as a nonredundant linear combination of monomials then each monomial which appears has length greater than the length of $X_j^{f-1}\beta$. Each monomial in the support of the third term has X_j-degree at least f. Thus the first monomial term cannot be in the support of $fX_j^{f-1}\beta H + X_1^f D(\beta)$. The fact that I is a monomial ideal implies now that

$$cfX_j^{f-1}\beta \in I.$$

But $X_j^{f-1}\beta \notin I$ by minimality and f is not zero in k. We conclude that $c = 0$, i.e., $D(X_j) \in (X_1, \ldots, X_n)$. □

As promised, we can now construct many examples of finite-dimensional algebras all of whose derivations stabilize the radical, but which do not have smooth automorphism group schemes.

For example, suppose that k is a perfect field of characteristic p, that $n \geq 2$, and

$$p < e(1) \leq e(2) \leq \cdots \leq e(n).$$

Let I be the ideal of $k[X_1, \ldots X_n]$ generated by

$$X_1^{e(1)}, \ldots, X_n^{e(n)}, \text{ and } X_1^p X_2 \cdots X_n.$$

We first observe that if p is relatively prime to $e(1)$ then $R = k[X_1, \ldots, X_n]/I$ does not have a smooth automorphism group scheme. Set $w = X_2 \cdots X_n$. It is easy to check that $w\frac{\partial}{\partial X_1}$ induces a derivation of R. We argue that this derivation is not integrable. Otherwise, we may apply theorem 3.1 to conclude that

$$X_1 X_2^{e(1)-1} \cdots X_n^{e(1)-1} = \binom{e(1)}{e(1)-1} w^{e(1)-1} \in I.$$

However, $e(1) \leq e(j)$ for all j, so

$$e(1) - 1 < e(j) \text{ for all } j \geq 2.$$

We have created an element of I which is not divisible by any of the defining generators for I.

If we assume, in addition, that all $e(j)$ are relatively prime to p then Theorem 3.4 tells us that every derivation of R sends the radical into itself. The curious aspect of this example is that when $e(1) \geq 2p$ we can add the relation $(X_2 \cdots X_n)^2$ and the new algebra has a smooth automorphism group scheme.

4 MORITA INVARIANCE

We prove that having a smooth automorphism group scheme is a Morita invariant for finite-dimensional algebras over perfect fields. In fact, we establish a stronger result with no restriction on the scalar field. It is well known that Hochschild cohomology is a Morita invariant. According to Proposition 2.2, for any finite-dimensional algebra A it makes sense to define the subspace

$$\int \mathrm{HH}^1(A) = \text{integrable derivations of } A \text{ / inner derivations of } A$$

in $\mathrm{HH}^1(A)$. We show that $\int \mathrm{HH}^1$ is a Morita invariant in the sense that if A and B are Morita equivalent finite-dimensional algebras then there is an isomorphism from $\mathrm{HH}^1(A)$ to $\mathrm{HH}^1(B)$ which carries $\int \mathrm{HH}^1(A)$ isomorphically to $\int \mathrm{HH}^1(B)$. Although this subspace is the correct one for our strategy of proof, there are good reasons to "cointerpret" it. As a consequence of our theorem, the obstruction to integrability

$$\mathrm{Der}_k(A) \text{ / integrable derivations of } A$$

is a Morita invariant.

To prove the result as originally stated, we need a particular explicit map from $\mathrm{HH}^1(A)$ to $\mathrm{HH}^1(B)$. Throughout this section, we adopt the following set-up. A is a finite-dimensional k-algebra and P is a left progenerator for A. We may assume that the algebra Morita equivalent to A has the form

$$B = (\mathrm{End}_A(P))^{\mathrm{op}} \ .$$

(Henceforth, we ignore the presence of the opposite ring in the description of B. The concerned reader may move all of our functions from the left of their arguments to the right.) Let

$$\{(f_i, p_i) \in \mathrm{Hom}_A(P, A) \times P \mid 1 \leq i \leq m\}$$

be a projective basis for P. If $D \in \mathrm{Der}_k(A)$, denote by $D^* \in \mathrm{End}_k(P)$ the function given by

$$D^*(x) = \sum_{i=1}^m Df_i(x)p_i \ .$$

A linear map $\nu \in \mathrm{End}_k(P)$ is said to be *D-twisted* provided that

$$\nu(ap) = a\nu(p) + D(a)p$$

for all $a \in A$ and $p \in P$. It is easy to check that D^* is a D-twisted map.

LEMMA 4.1.

(1) If $\nu \in \mathrm{End}_k(P)$ is a D-twisted map then

$$\mathrm{ad}\,\nu \in \mathrm{Der}_k(B) \ .$$

(2) If ν_1 and ν_2 are D-twisted then $\mathrm{ad}\,(\nu_1 - \nu_2)$ is an inner derivation of B.

(3) If D is inner then $\mathrm{ad}\,D^*$ is inner.

Proof. To prove (1), it suffices to demonstrate that if $\beta \in B = \mathrm{End}_A(P)$ then $\nu \circ \beta - \beta \circ \nu \in B$. (It is a standard observation that $\mathrm{ad}\,\nu$ is a derivation of $\mathrm{End}_k(P)$.) Suppose that $a \in A$ and $p \in P$.

$$\begin{aligned}
\mathrm{ad}\,\nu(\beta)(ap) &= \nu(\beta(ap)) - \beta(\nu(ap)) \\
&= \nu(a\beta(p)) - \beta(a\nu(p) + D(a)p) \\
&= a\nu(\beta(p)) + D(a)\beta(p) - a\beta(\nu(p)) - D(a)\beta(p) \\
&= a\,\mathrm{ad}\,\nu(\beta)(p).
\end{aligned}$$

Hence $\mathrm{ad}\,\nu(\beta)$ is an A-module map.

As to (2), notice that $\nu_1 - \nu_2$ is an A-module map; that is, it is already some element $\gamma \in B$. Therefore
$$\operatorname{ad}(\nu_1 - \nu_2) = \operatorname{ad}\gamma .$$

Finally, suppose that $D = \operatorname{ad} w$ for some $w \in A$. Let λ_w denote left multiplication by w, thought of as a function in $\operatorname{End}_k(P)$. For $a \in A$ and $p \in P$,

$$\begin{aligned}(D^* - \lambda_w)(ap) &= D^*(ap) - w(ap) \\ &= aD^*(p) + \operatorname{ad}(w)(a)p - wap \\ &= aD^*(p) + wap - awp - wap \\ &= a(D^* - \lambda_w)(p) .\end{aligned}$$

We conclude that $D^* - \lambda_w \in B$. Hence
$$\operatorname{ad} D^* - \operatorname{ad} \lambda_w = \operatorname{ad}(D^* - \lambda_w)$$
is inner. But if $\beta \in B$ then β is A-linear. In other words, $\lambda_w \circ \beta - \beta \circ \lambda_w = 0$. It follows that $\operatorname{ad} \lambda_w = 0$. Thus $\operatorname{ad} D^*$ is inner. $\square$

Parts one and three of the lemma tells us that the k-linear map which sends D to $\operatorname{ad} D^*$ induces a linear map
$$\Phi : \operatorname{HH}^1(A) \to \operatorname{HH}^1(B) .$$

THEOREM 4.1. Φ *is an isomorphism.*

Proof. Since both cohomology groups are finite-dimensional and Morita equivalence is symmetric, it suffices to prove that the kernel of Φ is zero. This amounts to showing that if $\operatorname{ad} D^*$ is inner then D is inner. So assume that there is some $\gamma \in B$ so that
$$D^* \circ \beta - \beta \circ D^* = \gamma \circ \beta - \beta \circ \gamma$$
for all $\beta \in B$.

We focus on those β of the following type:
$$\beta(x) = h(x)y$$
for a fixed $h \in \operatorname{Hom}_A(P, A)$ and a fixed $y \in P$. In this case,
$$\begin{aligned}(D^* \circ \beta - \beta \circ D^*)(x) &= D^*(h(x)y) - h(D^*(x))y \\ &= h(x)D^*(y) + D(h(x))y - h(D^*(x))y \\ &= h(x)\sum Df_i(y)p_i + D(h(x))y - \sum Df_i(x)h(p_i)y .\end{aligned}$$

By assumption, the last expression is also equal to
$$h(x)\gamma(y) - h(\gamma(x))y .$$

We apply this equality in two situations. Recall that P is a progenerator. It satisfies a trace condition of the following form: there exist $g_t \in \operatorname{Hom}_A(P, A)$ and $\pi_t \in P$ such that
$$\sum_{t=1}^r g_t(\pi_t) = 1 .$$

First set $y = \pi_t$, apply g_t to both sides of the equality, and then sum over t.

$$h(x)\sum_{i,t} Df_i(\pi_t)g_t(p_i) + D(h(x)) - \sum_i Df_i(x)h(p_i)$$
$$= h(x)\sum_t g_t(\gamma(\pi_t)) - h(\gamma(x)) .$$

Rearranging terms and using A-linearity, we obtain

(†) $\quad D(h(x)) = h(x)[\sum_t g_t(\gamma(\pi_t)) - \sum_{i,t} Df_i(\pi_t)g_t(p_i)]$
$$- h(\gamma(x) - \sum_i Df_i(x)p_i) .$$

Next set $h = g_t$, set $x = \pi_t$, and then sum over all t.

$$\sum_i Df_i(y)p_i + 0 - \sum_{i,t} Df_i(\pi_t g_t(p_i)y = \gamma(y) - \sum_t g_t(\gamma(\pi_t))y .$$

Replace y with x and rearrange.

$$\gamma(x) - \sum_i Df_i(x)p_i = \sum_t g_t(\gamma(\pi_t))x - \sum_{i,t} Df_i(\pi_t)g_t(p_i)x .$$

Finally, apply h to each side.

(††) $\quad h(\gamma(x) - \sum_i Df_i(x)p_i) =$
$$[\sum_t g_t(\gamma(\pi_t))x - \sum_{i,t} Df_i(\pi_t)g_t(p_i)]h(x) .$$

Notice that the expression in square brackets is an element of A independent of h and x. Call this expression c and substitute (††) in the second long term on the right side of equality (†).

$$D(h(x)) = h(x)c - ch(x) .$$

Since h and x in (†) are arbitrary, we can invoke the trace condition one last time to see that
$$D(a) = \text{ad}(-c)(a)$$
for all choices $a \in A$. $\square$

The remainder of the argument is inspired by [GAS]. We shall prove that if $D \in \text{Der}_k(A)$ is integrable then $\text{ad}\, D^*$ is integrable. Assume that

$$\theta = I + Dt + D^{(2)}t^2 + \cdots$$

is a $k[[t]]$-automorphism of $A[[t]]$. The $k[[t]]$-vector space $P[[t]]$ can be given two $A[[t]]$-module structures. The first is extended from the action of A on P in the obvious way; we denote this module $P[[t]]$. A second action comes from the automorphism. If $a \in A[[t]]$ and $p \in P[[t]]$ define

$$a * p = \theta(a)p .$$

The second module will be written ${}^\theta P[[t]]$. Both modules are finitely generated projective $A[[t]]$-modules.

LEMMA 4.2. $P[[t]] \simeq {}^\theta P[[t]]$ as $A[[t]]$-modules.

Proof. Since A is finite-dimensional, $A[[t]]$ is semiperfect. Hence finitely generated $A[[t]]$-modules have unique projective covers.

The key is that $\theta \equiv I(\mathrm{mod}\ t)$. This means that

$$P[[t]]/tP[[t]] \simeq {}^\theta P[[t]]/t^\theta P[[t]]$$

as A-modules. It is immediate that they are isomorphic as $A[[t]]$-modules. But $tA[[t]] \subseteq \mathrm{rad}\, A[[t]]$ so

$$P[[t]]/(\mathrm{rad}\, A[[t]])P[[t]] \simeq {}^\theta P[[t]]/(\mathrm{rad}\, A[[t]])^\theta P[[t]]$$

as $A[[t]]$-modules. Their respective projective covers are $P[[t]]$ and ${}^\theta P[[t]]$. $\square$

Let $\psi : P[[t]] \to {}^\theta P[[t]]$ be an $A[[t]]$-isomorphism guaranteed by the lemma. We will regard ψ as a member of $\mathrm{End}_{k[[t]]} P[[t]]$ which satisfies

$$\psi(ap) = \theta(a)\psi(p)$$

for all $a \in A[[t]]$ and $p \in P[[t]]$. Since ψ is determined by its behavior on those $p \in P$, we can write

$$\psi = \psi_0 + \psi_1 t + \psi_2 t^2 + \cdots$$

where $\psi_j \in \mathrm{End}_k(P)$ for all j. Once again we use the fact that $\theta \equiv I(\mathrm{mod}\ t)$ to see that $\psi_0 \in \mathrm{End}_A(P)$. By repeating this observation for the inverse of ψ we conclude that ψ_0 is an A-module isomorphism of P. Extend it in the obvious way to an $A[[t]]$-module isomorphism of $P[[t]]$, and call the new map ψ_0 as well. It is easy to check that

$$\psi_0^{-1}\psi : P[[t]] \to {}^\theta P[[t]]$$

is also an $A[[t]]$-isomorphism. Thus we may replace ψ with $\psi_0^{-1}\psi$ and thereby assume that $\psi_0 = I$.

LEMMA 4.3. ψ_1 is a D-twisted k-endomorphism of P.

Proof. We know that

$$\psi(ap) = \theta(a)\psi(p)$$

whenever $a \in A$ and $p \in P$. As a "power series", this means

$$ap + \psi_1(ap)t + \psi_2(ap)t^2 + \cdots$$
$$= (a + D(a)t + D^{(2)}t^2 + \cdots)(p + \psi_1(p)t + \psi_2(p)t^2 + \cdots) .$$

Read off the coefficient of t. $\square$

LEMMA 4.4 ([GAS]). *If we identify $B[[t]]$ with $\text{End}_{A[[t]]}(P[[t]])$ then*

$$\psi \circ s \circ \psi^{-1} \in B[[t]]$$

for all $s \in B[[t]]$.

Proof. For $a \in A[[t]]$ and $p \in P[[t]]$, we have

$$\begin{aligned}
(\psi \circ s \circ \psi^{-1})(ap) &= (\psi \circ s)(\psi^{-1}(ap)) \\
&= (\psi \circ s)\left(\psi^{-1}(\theta(\theta^{-1}(a))p)\right) \\
&= (\psi \circ s)(\psi^{-1}(\theta^{-1}(a) * p)) \\
&= (\psi \circ s)(\theta^{-1}(a)\psi^{-1}(p)) \\
&= \psi\left(\theta^{-1}(a)(s \circ \psi^{-1})(p)\right) \\
&= \theta(\theta^{-1}(a))(\psi \circ s \circ \psi^{-1})(p) \ .
\end{aligned}$$

$\square$

THEOREM 4.2.
$$\Phi\left(\int \text{HH}^1(A)\right) = \int \text{HH}^1(B)$$

Proof. By symmetry and finite-dimensionality, it suffices to show that

$$\Phi\left(\int \text{HH}^1(A)\right) \subseteq \int \text{HH}^1(B) \ .$$

According to the previous lemma, the map $s \mapsto \psi \circ s \circ \psi^{-1}$ defines an automorphism

$$\widehat{\theta} \in \text{Aut}_{k[[t]]}(B[[t]]) \ .$$

Since this automorphism is determined by its behavior on B, it has a power series description,
$$\widehat{\theta} = \widehat{\theta_0} + \widehat{\theta_1}t + \widehat{\theta_2}t^2 + \cdots \ .$$
Computing $\widehat{\theta}(\beta) = \psi \circ \beta \circ \psi^{-1}$ when $\beta \in B$, we obtain

$$\widehat{\theta}(\beta) = \beta + (\psi_1 \circ \beta - \beta \circ \psi_1)t + \cdots \ .$$

Thus $\widehat{\theta_0} = I$ and $\widehat{\theta_1} = \text{ad}\,\psi_1$. It follows from Lemma 4.1 that $\text{ad}\,\psi_1 \in \text{Der}_k(B)$ and that this derivation is integrable (to $\widehat{\theta}$). The same lemma says that

$$\text{ad}\,\psi_1 \equiv \text{ad}\,D^*$$

as a member of $\text{HH}^1(B)$. But inner derivations are always integrable. and integrable derivations constitute a vector space. Therefore $\text{ad}\,D^*$ is integrable. $\square$

As promised earlier, we provide a second proof that hereditary algebras over an algebraically closed field have a smooth automorphism group scheme.

THEOREM 4.3. *All derivations of a finite-dimensional hereditary algebra over an algebraically closed field are integrable.*

Proof. A theorem due to Gabriel ([ARS]) states that one of these hereditary algebra is Morita equivalent to a path algebra $k\Gamma$ whose underlying graph Γ has no oriented cycles. The main result of this section tells us that we may take $A = k\Gamma$.

It is well known that every derivation of $k\Gamma$ is the sum of an inner derivation and a derivation which vanishes on all vertices. (A stronger version of this observation can be found in [FGGM], Theorem 4.1). A derivation of the second kind is determined by its behavior on arrows. Notice that if D vanishes on vertices and a is an arrow from vertex e to vertex f then

$$D(a) = D(eaf) = eD(a)f \ .$$

It follows that every derivation of the second kind is a linear combination of derivations $m\frac{\partial}{\partial a}$ where m is a path which shares the same origin and the same terminus as a. (Here $m\frac{\partial}{\partial a}$ sends a to m and all other arrows to zero.)

We are reduced to proving that if $D = m\frac{\partial}{\partial a}$ then D is integrable. It is at this stage that we take advantage of the hypothesis that Γ has no oriented cycles by noting that

$$D(A)D(A) = 0 \ .$$

Indeed, if $y \in D(A)$ then any path in the support of y has a subpath which begins at the origin of a and ends at the terminus of a. No two paths like this can be concatenated.

It follows immediately that $I + Dt$ is an automorphism of $A[[t]]$. □

REFERENCES

[Abe] E. Abe, *Hopf Algebras*, Cambridge Univ. Press, Cambridge, 1977.

[FGGM] D. R. Farkas, Ch. Geiss, E. L. Green, E. N. Marcos, Diagonalizable derivations of finite-dimensional algebras I, *Israel Journal*, to appear.

[FJL] P. Fleischmann, I. Janiszczak, and W. Lempken, Finite groups have non-Schur centralizers, *Manuscr. Math.* **80** (1993), 213-224.

[Ga] P. Gabriel, Finite representation type is open, in *Representations of Algebras*, Springer Lect. Notes **488**, Berlin, 1975, 132-155.

[GS] M. Gerstenhaber and S. D. Schack, Algebraic cohomology and deformation theory, in *Deformation Theory of Algebras and Structures and Applications*, NATO ASI Ser. C **247**, Kluwer, Dordrecht, 1988, 11-264.

[GAS] F. Guil-Asensio and M. Saorin, On automorphism groups induced by bimodules, *Archiv. der Math Basel*, to appear.

[Ha] D. Happel, Hochschild cohomology of finite dimensional algebras, in *Séminaire d'Algèbre*, Springer Lect. Notes **1404**, Berlin, 1984, 108-126.

[Hu] J. E. Humphreys, *Introduction to Lie Algebras and Representation Theory*, Springer-Verlag Grad. Texts in Math. **9**, New York, 1980.

[Mat] H. Matsumura, Integrable derivations, *Nagoya Math. J.* **87** (1982), 227-245.

[Maz] G. Mazzola, The algebraic and geometric classification of associative algebras of dimension five, *Manuscr. Math.* **27** (1997), 81-101.

[Mo] S. Montgomery, *Hopf Algebras and Their Actions on Rings*, CBMS **82** Amer. Math. Soc., Providence, 1993.

[Sw] M. E. Sweedler, Hopf algebras with one group-like element, *Trans. AMS* **127** (1967), 515-26.

[Ta] M. Takeuchi, Tangent coalgebras and hyperalgebras I, *Japanese Jour. of Math.* **42** (1974), 1-143.

[Wa] W. C. Waterhouse, *Introduction to Affine Group Schemes*, Springer-Verlag Grad. Texts in Math. **66**, New York, 1979.

A combinatorial characterization of hereditary categories containing simple objects

DIETER HAPPEL Fakultät für Mathematik, Technische Universität Chemnitz, D-09107 Chemnitz, GERMANY

IDUN REITEN Department of Mathematical Sciences, Norwegian University of Science and Technology, 7491 Trondheim, Norway

ABSTRACT Let $\mathcal{H}$ be a connected hereditary abelian category with finite dimensional homomorphism and extension spaces. If $\mathcal{H}$ contains a tilting object then $\mathcal{H}$ has almost split sequences and if $\mathcal{H}$ is not a module category this yields an equivalence τ. We consider here τ−invariant additive functions on $\mathcal{H}$ and show as a main result that $\mathcal{H}$ contains simple objects iff there exists a nonnegative additive τ−invariant function.

Let k be an algebraically closed field and $\mathcal{H}$ a connected hereditary abelian k-category with finite dimensional homomorphism and extension spaces. Assume that $\mathcal{H}$ has a tilting object, that is, an object T such that $\mathrm{Ext}^1_{\mathcal{H}}(T,T) = 0$ and such that $\mathrm{Hom}_{\mathcal{H}}(T, X) = 0 = \mathrm{Ext}^1_{\mathcal{H}}(T, X) = 0$ implies $X = 0$ (see [HRS],[H],[HR]).

It was proved in [HR] that if $\mathcal{H}$ has a simple object and $\mathcal{H}$ is not equivalent to mod Λ for a finite dimensional hereditary k-algebra Λ, then $\mathcal{H}$ must be derived equivalent to a category coh $\mathbb{X}$ of coherent sheaves on some weighted projective line $\mathbb{X}$ in the sense of Geigle-Lenzing [GL1]. However, not all hereditary categories derived equivalent to some coh $\mathbb{X}$ have a simple object. In this paper we give a combinatorial characterization of the hereditary abelian k-categories $\mathcal{H}$ with tilting object, which are not equivalent to mod Λ for a finite dimensional hereditary k-algebra Λ, and which have some simple object.

When $\mathcal{H}$ is a hereditary abelian k-category with tilting object, then $\mathcal{H}$ has almost split sequences [HRS]. If $\mathcal{H}$ is not equivalent to mod Λ for some finite dimensional hereditary k-algebra Λ, then $\mathcal{H}$ has no nonzero projective object (see [H]). There is then an equivalence $\tau \colon \mathcal{H} \to \mathcal{H}$ such that for each indecomposable object C in $\mathcal{H}$

there is an almost split sequence $0 \to \tau C \to B \to C \to 0$ [HRS].

A function $l: \mathcal{H} \to \mathbb{N}_0$ (the nonnegative integers) is **additive** if for each exact sequence $0 \to A \to B \to C \to 0$ we have $l(B) = l(A) + l(C)$. When we have such an additive function, there is induced a function $\bar{l}: K_0(\mathcal{H}) \to \mathbb{Z}$, where $\mathbb{Z}$ denotes the integers and $K_0(\mathcal{H})$ denotes the Grothendieck group of $\mathcal{H}$ (modulo exact sequences). Under our assumptions $K_0(\mathcal{H})$ is free abelian of finite rank [HRS].

Assume that our $\mathcal{H}$ is not equivalent to some mod Λ for a finite dimensional hereditary k-algebra Λ. A function $l: \mathcal{H} \to \mathbb{N}_0$ is τ-**invariant** if $l(C) = l(\tau C)$ for each C in $\mathcal{H}$. Our main result is that $\mathcal{H}$ has some simple object if and only if there is a nonnegative additive τ-invariant function on $\mathcal{H}$.

Note that if $\mathcal{H}$ is mod Λ for a finite dimensional hereditary k-algebra Λ, then there can be no nonzero τ-invariant additive function l on $\mathcal{H}$. For if P is projective, then $l(P) = 0$ since $\tau P = 0$ and for every C there is an exact sequence $0 \to P \to P' \to C \to 0$ with P and P' projective.

We would like to thank the referee for helpful comments.

1 COHERENT SHEAVES AND MODULE CATEGORIES

In this section we recall some background material on the category coh $\mathbb{X}$ of coherent sheaves on the weighted projective line $\mathbb{X}$, and on hereditary categories derived equivalent to coh $\mathbb{X}$, from the point of view of existence of simple objects and of nonnegative additive τ-invariant functions. Through this we give motivating examples for the main result of this paper. We also discuss hereditary categories derived equivalent (but not equivalent) to mod Λ for some finite dimensional hereditary k-algebra Λ.

Let $\underline{p} = (p_1, \cdots, p_t)(t \geq 3)$ be a weight sequence, that is, a sequence of positive integers and $\underline{\lambda} = (\lambda_3, \cdots, \lambda_t)$ a sequence of pairwise distinct nonzero elements of the field k. Let $\mathbb{X} = \mathbb{X}(\underline{p}, \underline{\lambda})$ be the associated weighted projective line, and coh $\mathbb{X}$ the corresponding category of coherent sheaves. This category has simple objects, and here simple objects give naturally rise to additive functions, as we now show.

Let S be a simple object in a homogenous tube of coh $\mathbb{X}$, so that $\tau S \simeq S$. Then we have the rank function $r = <-, S> = \dim_k \operatorname{Hom}(-, S) - \dim_k \operatorname{Ext}^1(-, S)$ on coh $\mathbb{X}$ (see [GL1]). This function is clearly additive and τ-invariant, and is also nonnegative. It takes value 0 on the objects of finite length, and is positive on each indecomposable object of infinite length.

With a weight sequence $\underline{p} = (p_1, \cdots, p_t)$ is associated the number $h = (t-2)p - \sum_{i=1}^t p/p_i$, where p is the least common multiple of $p_1, \cdots, p_t$. When $h \neq 0$, that is, coh $\mathbb{X}$ is not of tubular type, the hereditary categories $\mathcal{H}$ derived equivalent to coh $\mathbb{X}$ are obtained by "tilting" with respect to a split torsion pair $(\mathcal{T}, \mathcal{F})$, where $\mathcal{T}$ consists of objects of finite length (see [LS, H]). The rank function r on coh $\mathbb{X}$ gives rise to a function $\bar{r}$ on $K_0(\text{coh } \mathbb{X}) \simeq K_0(D^b(\text{coh } \mathbb{X})) \simeq K_0(\mathcal{H})$. In this way we obtain an additive τ-invariant function on $\mathcal{H}$ when $\mathcal{H}$ is not equivalent to some mod Λ for a hereditary k-algebra Λ. This function is nonnegative since r takes value 0 on $\mathcal{T}$, and hence also on $\mathcal{T}[-1]$.

In the tubular case, that is, $h = 0$, we have families of tubes $\mathcal{C}_a$ indexed by $Q_+ \cup \{\infty\}$, where Q_+ denotes the positive rational numbers. If $a \neq b$ there is a map from an object in the family $\mathcal{C}_a$ to an object in the family $\mathcal{C}_b$ if and only if $a < b$. Let $(\mathcal{T}, \mathcal{F})$ be a split torsion pair where the torsion class $\mathcal{T}$ consists of the union of families $\mathcal{C}_a$ where $a > s$ for a fixed irrational number s. Tilting with respect to $(\mathcal{T}, \mathcal{F})$ we obtain a hereditary category $\mathcal{H}$ having no simple object. For it is known from [HR] that the simple objects must lie "furthest to the right" or "furthest to the left". As a consequence of our main result it will follow that there is no nonzero nonnegative additive τ-invariant function on $\mathcal{H}$.

When $\mathcal{H}$ is derived equivalent (but not equivalent) to mod Λ for some wild indecomposable finite dimensional hereditary k-algebra Λ, we know that $\mathcal{H}$ has no simple object since there are no tubes. For the structure of such $\mathcal{H}$, see [H]. We now show that in this case there is also no additive function of the desired type.

PROPOSITION 1.1. *Let $\mathcal{H}$ be a connected hereditary category derived equivalent (but not equivalent) to mod Λ for a finite dimensional wild hereditary k-algebra Λ. Then there is no nonzero nonnegative additive τ-invariant function on $\mathcal{H}$.*

Proof. We know that $\mathcal{H}$ has a component of type $Z\vec{\Delta}$, where $\vec{\Delta}$ is a connected quiver which is not Dynkin or extended Dynkin. Assume to the contrary that there is some nonzero nonnegative additive τ-invariant function l on $\mathcal{H}$. There are two different types of such $\mathcal{H}$. In one case, when the component $Z\vec{\Delta}$ is on the left hand side, it is easy to see that there is for each C in $\mathcal{H}$ on exact sequences $0 \to B_1 \to B_0 \to C \to 0$ where the indecomposable summands of B_1 and B_0 belong to $Z\vec{\Delta}$. In the second case there is for each C in $\mathcal{H}$ an exact sequence $0 \to C \to B_0 \to B_1 \to 0$ where the indecomposable summands of B_0 and B_1 belong to $Z\vec{\Delta}$. Hence l is nonzero also on $Z\vec{\Delta}$. Since l is τ-invariant, there is induced a nonzero nonnegative function on the underlying graph Δ of $\vec{\Delta}$. We claim that this function is positive. For if not, let x be a vertex with $l(x) = 0$, having a neighbour y with $l(y) > 0$. Then the additivity formula at x gives a contradiction. Since Δ is a wild graph, this is impossible. Hence there is no nonzero nonnegative additive τ-invariant function on $\mathcal{H}$. □

2 THE MAIN RESULT

The aim of this section is to prove our main result. We also show that a nonzero nonnegative τ-invariant function must be positive on a central class of objects, which will be used for proving in which sense such functions are unique.

Recall that an object E in $\mathcal{H}$ is exceptional if $\text{Ext}^1_{\mathcal{H}}(E, E) = 0$. The object E is torsionable if it is the factor of a direct sum of copies of some tilting object. We denote by $E^\perp$ the perpendicular category, whose objects are the C in $\mathcal{H}$ with $\text{Hom}(E, C) = 0 = \text{Ext}^1(E, C)$.

We have the following main result.

THEOREM 2.1. *Let $\mathcal{H}$ be a connected hereditary abelian k-category with tilting object, where k is an algebraically closed field, and assume that $\mathcal{H}$ is not equivalent to mod Λ for a finite dimensional hereditary k-algebra Λ.*

There is some nonzero nonnegative additive τ-invariant function l on $\mathcal{H}$ if and only $\mathcal{H}$ has some simple object.

Proof. Assume that $\mathcal{H}$ is not equivalent to mod Λ for a finite dimensional hereditary k-algebra Λ.

Assume first that $\mathcal{H}$ has some simple object. Then there is a split torsion pair $(\mathcal{T}, \mathcal{F})$ in $\mathcal{H}$ with all objects in $\mathcal{F}$ of finite length, such that when we tilt with respect to this pair we obtain some category coh $\mathbb{X}$ [HR, Section 3, Th 6.1]. Then $(\mathcal{F}(1), \mathcal{T})$ is a split torsion pair for coh $\mathbb{X}$. Since the rank r is 0 on $\mathcal{F}$, it follows that the induced τ-invariant additive function on $\mathcal{H}$ is also nonzero and nonnegative.

Assume now that $\mathcal{H}$ has some desired function l, but no simple object. By Proposition 1.1 it follows that $\mathcal{H}$ is not derived equivalent to mod Λ for a wild finite dimensional hereditary k-algebra Λ. Further $\mathcal{H}$ is not derived equivalent to mod Λ where Λ is tame hereditary, since $\mathcal{H}$ would then have a simple object because we are in the situation $h \neq 0$ discussed in section 1. If Λ was hereditary of finite type, then any $\mathcal{H}$ derived equivalent to mod Λ would have a nonzero projective object, which is impossible by the assumption that $\mathcal{H}$ is not equivalent to any mod Λ where Λ is a finite dimensional hereditary k-algebra (see [HRS]). Hence $\mathcal{H}$ is not equivalent to mod Λ for any finite dimensional hereditary k-algebra Λ.

Then there must be some object of infinite length in $\mathcal{H}$. For otherwise all objects are of finite length, and since there are no nonzero projectives, the AR-quiver for $\mathcal{H}$ is a union of tubes (see [L]), with no nonzero maps between objects in different tubes. Since $\mathcal{H}$ is connected, there is only one tube. The rank of the tube is the rank of $K_0(\mathcal{H})$. This contradicts the fact that $\mathcal{H}$ has a tilting object, since a tube of rank t does not contain an exceptional object which is the direct sum of t nonisomorphic exceptional objects.

Since $\mathcal{H}$ has some object of infinite length, there is by [H] a tilting object of infinite length, and hence some indecomposable torsionable exceptional object E of infinite length. We choose E such that $l(E) = a$ is minimal. As pointed out before, the perpendicular category $E^{\perp}$ is equivalent to mod H for some finite dimensional (basic) hereditary k-algebra H. Consider the almost split sequence $0 \to \tau E \to M \to E \to 0$. Since $\mathcal{H}$ is not equivalent to a hereditary module category, it follows that M is a sincere H-module [H].

Let now n be the rank of the Grothendieck group $K_0(\mathcal{H})$. The algebras $\text{End}_{\mathcal{H}}(T)^{op}$ where T is a tilting object are the quasitilted algebras [HRS]. A quasitilted algebra has no oriented cycles [HRS], and hence must be hereditary if there are at most two simple modules. This contradicts the assumption on $\mathcal{H}$, so that we must have $n \geq 3$. Since $T = H \oplus E$ is a tilting object, $K_0(\text{mod } H)$ has rank $n - 1$.

Let $S_1, \cdots, S_{n-1}$ be the nonisomorphic simple H-modules. We have $[M] = \sum_{i=1}^{n-1} t_i [S_i]$ in $K_0(\text{mod } H)$, where all t_i are nonzero since M is sincere. Each S_i is exceptional, and S_i is torsionable since $T = H \oplus E$ is a tilting object. Since by assumption $\mathcal{H}$ has no simple object, each S_i has infinite length. Then we have $l(S_i) \geq a$ by the definition of a. We now obtain the inequality $2a \geq a(\sum_{i=1}^{n-1} t_i) \geq (n-1)a$. It follows that $n = 3$.

Since $n = 3$, we must have $t_1 = 1 = t_2$, so that $[M] = [S_1] + [S_2]$ in $K_0(\mathcal{H})$. The quiver of H is then

$$1 \bullet \xrightarrow{\vdots} \bullet 2$$

where $m \geq 1$ denotes the number of arrows from the vertex v_1 at 1 to the vertex v_2 at 2. The one-point extension $H[M]$ is a quasitilted algebra derived equivalent to $\mathcal{H}$ [HR], and it is not tilted since $\mathcal{H}$ is not derived equivalent to some mod Λ for a finite dimensional hereditary k-algebra Λ. Then it follows from [HRS] that M is indecomposable.

For the algebra $H[M]$ we have the quiver

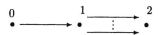

with m arrows from v_1 to v_2, and the space of relations of paths from v_0 to v_2 has dimension $m - 1$. We view this algebra as the one-point coextension $[N]H_1$, where H_1 is the path algebra of the quiver $0 \bullet \longrightarrow \bullet 1$. Since the indecomposable injective $H[M]$-module associated with the vertex v_2 has dimension vector (1m1), the H_1-module N has dimension vector (1m). Since there is no arrow from vertex v_0 to vertex v_2, the indecomposable summands of N must be one copy of $k \rightrightarrows k$, together with $m - 1$ copies of $0 \to k$. Since the modules are on a slice, it follows from [HRS] that $[N]H$ is tilted, and this gives a contradiction. □

Now we show that a nonzero nonnegative τ-invariant function must be positive on some class of exceptional objects.

PROPOSITION 2.2. *Assume that the hereditary abelian k-category $\mathcal{H}$ with tilting object is not equivalent to mod Λ for some finite dimensional hereditary k-algebra Λ. Let l be a nonzero nonnegative τ-invariant additive function on $\mathcal{H}$.*

If E is an indecomposable torsionable exceptional object of infinite length, then $l(E) > 0$.

Proof. Assume to the contrary that E is indecomposable torsionable exceptional and $l(E) = 0$. Since E has infinite length and is torsionable exceptional, we have $E^\perp = \text{mod } H$ for some finite dimensional hereditary k-algebra H [HR].

Consider the almost split sequence $0 \to \tau E \to M \to E \to 0$. Then we know that M is in $E^\perp = \text{mod } H$ (see [HR]). Since l is τ-invariant, we have $l(\tau E) = 0$. As already mentioned it follows that M is a sincere H-module [H]. The restriction $l|E^\perp \colon E^\perp \to \mathbb{N}_0$ is still additive. We have $l(M) = l(E) + l(\tau E) = 0$, and since $l \geq 0$, we have $l(S) = 0$ for each simple composition factor S of M. Since M is a sincere H-module, it follows that $l(S) = 0$ for each simple H-module, and hence $l|E^\perp = 0$. We know that $T = H \oplus E$ is a tilting object in $\mathcal{H}$ [HR], and hence gives rise to a basis for $K_0(\mathcal{H})$. Then we get $l = 0$, which is a contradiction. □

We end the section by pointing out that when we have a nonzero nonnegative τ-invariant additive function, then it is essentially unique.

PROPOSITION 2.3. *Let $\mathcal{H}$ be a hereditary abelian k-category with tilting object not equivalent to mod Λ for a finite dimensional hereditary k-algebra Λ, and having some nonzero nonnegative additive τ-invariant function l. Then any other function is a (rational) multiple of l.*

Proof. It follows from Theorem 2.1 that $\mathcal{H}$ has a simple object, and is hence derived equivalent to some coh $\mathbb{X}$.

We know that $\mathcal{H}$ is obtained from coh $\mathbb{X}$ by tilting with respect to a split torsion pair $(\mathcal{T}, \mathcal{F})$, where $\mathcal{T}$ consists of objects of finite length. Let l be a nonnegative τ-invariant additive function on $\mathcal{H}$.

We first show that l is zero on the objects of finite length. Let S be a simple object with $\operatorname{Hom}(S, X) = 0$ for each X indecomposable of infinite length, and assume $l(S) > 0$. There is some indecomposable Y of infinite length with $(Y, S) \neq 0$. Using the lifting property for almost split sequences, as in [HR], we get an epimorphism $f_n \colon Y \to A_n$, where A_n is uniserial of length n with S on the top. Since $l(T) \geq 0$ for each simple composition factor T of A_n, we get that $l(A_n) > l(Y)$. Hence we get $l(\ker f_n) < 0$, which is a contradiction. Similarly, if S is a simple object with $\operatorname{Hom}(X, S) = 0$ for each X indecomposable of infinite length, there is some indecomposable object Y of infinite length with $\operatorname{Hom}(S, Y) \neq 0$. Then we get a monomorphism $B_n \to Y$, where B_n is uniserial of length n with socle S, and for n large enough we get the contradiction that $l(Y/B_n) < 0$.

It follows from the above that if l is a τ-invariant additive function which is nonnegative on $\mathcal{H}$, which is derived equivalent to coh $\mathbb{X}$, then the induced function on coh $\mathbb{X}$ is also nonnegative on coh $\mathbb{X}$. Hence it is sufficient to consider coh $\mathbb{X}$.

We have the rank function r, which is positive on indecomposable objects of infinite length. Let C be indecomposable of infinite length and let E be an exceptional object of rank 1. Then for some i there is a nonzero map $g \colon E \to \tau^i C$ [LP], which must be a monomorphism since E has rank 1. Hence we have $l(C) = l(\tau^i C) \geq l(E)$, so that $l(C) > 0$ by Proposition 2.2.

Let A be indecomposable of infinite length, with $l(A) = a > 0$ minimal. If $r(A) > 1$, we have an exact sequence $0 \to A_1 \to A \to A_2 \to 0$ with $r(A_1) > 0$ and $r(A_2) > 0$. Since then $l(A_1) > 0$ and $l(A_2) > 0$, we get a contradiction, showing that $r(A) = 1$.

Assume $r(B) = 1$ and $l(B) > a$. With A chosen as above, there is for some i a nonzero map $g \colon A \to \tau^i B$, which must be a monomorphism since A has rank 1. Then $l(\tau^i B/A) > 0$ and $r(\tau^i B/A) = 0$, so that $\tau^i B/A$ has finite length. This gives a contradiction, and hence $l(B) = a$.

It now follows that $l = a \cdot r$ on coh $\mathbb{X}$, and hence on $\mathcal{H}$, and we are done. $\square$

3 POSITIVE ADDITIVE FUNCTIONS

When $\mathcal{H} = \operatorname{mod} \Lambda$ for a finite dimensional hereditary k-algebra Λ, we have a positive additive function given by ordinary length. However, no other hereditary k-category with tilting object has this property, as we now show. Note that we do not assume that our functions are τ-invariant.

PROPOSITION 3.1. *Let $\mathcal{H}$ be a hereditary abelian k-category with finite dimensional homomorphism and extension spaces, and having a tilting object.*

Then there is a positive additive function on $\mathcal{H}$ if and only if $\mathcal{H}$ is equivalent to $\operatorname{mod} \Lambda$ for some finite dimensional hereditary k-algebra Λ.

Proof. Assume that $\mathcal{H}$ is not equivalent to $\operatorname{mod} \Lambda$ for some finite dimensional hereditary k-algebra Λ, and assume that there is some positive additive function l on $\mathcal{H}$.

Assume first that $\mathcal{H}$ has some object of finite length. Then $\mathcal{H}$ is derived equivalent to some category coh $\mathbb{X}$. If S is a simple object with $\operatorname{Hom}(S, X) = 0$ for each indecomposable object X of infinite length, then choose Y indecomposable of infinite length with $\text{Hom}(Y,S) \neq 0$. As in section 2 we have an epimorphism $Y \to A_n$, where A_n is uniserial of length n and with top S. For n large enough we have $l(A_n) = n > l(Y)$, which gives a contradiction. If there is no such simple object S, then there is some simple object T with $\text{Hom}(X,T) = 0$ for each indecomposable object X of infinite length. In a similar way as above we get a contradiction. $\square$

We can now assume that $\mathcal{H}$ has no simple object. Let X be indecomposable with $l(X) = a$ minimal. Let Y be a proper indecomposable subobject of X. Then we have $l(Y) = a$, and hence $l(X/Y) = 0$, which is a contradiction.

REFERENCES

[GL1] W. Geigle and H. Lenzing, *A class of weighted projective curves arising in representation theory of finite dimensional algebras*, in: Singularities, representations of algebras and vector bundles, Springer Lecture Notes **1273** (1987), 265–297.

[GL2] W. Geigle and H. Lenzing, *Perpendicular categories with applications to representations and sheaves*, J. of Alg. **144** (1991), 339–389.

[H] D. Happel, *Quasitilted algebras*, Adv. in Proc.ICRA VIII (Trondheim), CMS Conf.proc., Vol. 23, Algebras and modules I (1998), 55–83.

[HR] D. Happel and I. Reiten, *Hereditary categories with tilting object*, Math. Zeitschr.**232** (1999) 559–588.

[HRS] D. Happel, I. Reiten and S. O. Smalø, *Tilting in abelian categories and quasitilted algebras*, Mem. Amer. Math. Soc. **575** (1996).

[L] H. Lenzing, *Hereditary noetherian categories with a tilting complex*, Adv. in Proc. AMS **125** (1997), 1893–1901.

[LP] H. Lenzing and J. A. de la Pena, *Wild canonical algebras*, Math.Z. **224** (1997) no.3, 403–425.

[LS] H. Lenzing and A. Skowronski, *Quasitilted algebras of canonical type*, Colloquium Mathematicum, Vol **71** (1996), 161–181.

Symmetric Quasi-schurian Algebras

OCTAVIO MENDOZA HERNÁNDEZ Departamento de Matemática, Universidad Nacional del Sur, 8000 Bahía Blanca, Argentina, E-mail: omendoza@criba.edu.ar [1]

ABSTRACT Let k denote an algebraically closed field. We say that a finite dimensional k-algebra Λ is quasi-schurian, if it satisfies the following two conditions:

QS1) $dim_k Hom_\Lambda(P,Q) \leq 1$ if P, Q are not isomorphic indecomposable projective Λ-modules.

QS2) $dim_k End_\Lambda(P) = 2$ for each indecomposable projective Λ-module P.

An important class of quasi-schurian algebras is the trivial extensions of finite representation type.

In this paper, we give necessary and sufficient conditions for a given quasi-schurian algebra Λ to be weakly-symmetric or symmetric. These conditions are given in a combinatorial approach using a graph $GS(\Lambda)$ associated to Λ, and a function $\phi_\Lambda : Ch(GS(\Lambda)) \to k$ where $Ch(GS(\Lambda))$ is the set of chains of the graph $GS(\Lambda)$. Finally we give some connections between symmetric quasi-schurian algebras and trivial extensions of algebras.

1 INTRODUCTION

Throughout this paper, we let k denote a fixed algebraically closed field. By algebra is always meant a finite dimensional associative k-algebra with an identity, which we assume moreover to be basic and connected, and by module is meant a finitely generated left A-module.

Let A be a schurian triangular algebra. It is well known that the trivial extension $T(A)$ of A satisfies $dim_k Hom_{T(A)}(P,Q) \leq 1$ and $dim_k End_{T(A)}(P) = 2$ where P, Q are non isomorphic indecomposable projective $T(A)$-modules. In this way, we are interested in the class of algebras Λ satisfying the above property. Thus, we say that an algebra Λ is quasi-schurian if it satisfies the following two conditions:

---

[1] Supported by a fellowship from CONICET, Argentina. The author gratefully acknowledges a grant from CONICET, Argentina.

QS1) $dim_k Hom_\Lambda(P,Q) \leq 1$ if P, Q are not isomorphic indecomposable projective Λ-modules.

QS2) $dim_k End_\Lambda(P) = 2$ for each indecomposable projective Λ-module P.

The aim of this paper is both to give necessary and sufficient conditions for a given quasi-schurian algebra to be weakly-symmetric or symmetric, and to say when a symmetric quasi-schurian algebra arises from a trivial extension of a schurian triangular algebra.

Let $\Lambda = kQ_\Lambda/I$ where Q_Λ is the ordinary quiver associated with Λ and I is an admissible ideal. If δ is a path in the quiver Q_Λ we will denote by $\underline{\delta}$ the sub quiver of Q_Λ having as vertices and arrows those which belong to δ, this $\underline{\delta}$ is called the support of δ. Let C be an oriented cycle. Each vertex j in the support $\underline{C}$ of C determines a cycle with origin j which we call $C(j)$.

Finally we denote by $\overline{\gamma}$ the congruence class $\gamma + I$ in $\Lambda = kQ_\Lambda/I$.

In section 3 we prove the following theorems

THEOREM. *Let $\Lambda = kQ_\Lambda/I$ be a quasi-schurian algebra. Then the following conditions are equivalent*

I) Λ is weakly-symmetric.

II) For every non zero path γ there exists a path δ such that $\delta\gamma$ is a non zero minimal oriented cycle.

III) For each non zero f in $Hom_\Lambda(P,Q)$ the induced morphism

$Hom_\Lambda(Q, f) : Hom_\Lambda(Q, P) \to End_\Lambda(Q)$ *is non zero, if P and Q are indecomposable non isomorphic projective Λ-modules.*

IV) Λ satisfies the following conditions

  a) If a minimal oriented cycle C is non zero, then $\overline{C(t)} \neq 0$ for each vertex t in the support $\underline{C}$ of C.

  b) Let $\{\underline{C_1}, \underline{C_2} \cdots, \underline{C_m}\}$ be the set of supports corresponding to the non zero oriented cycles. Then $Q_\Lambda = \cup_{i=1}^m \underline{C_i}$.

THEOREM. *Let $\Lambda = kQ_\Lambda/I$ be a quasi-schurian weakly-symmetric algebra. Let $\{\underline{C_1}, \underline{C_2} \cdots, \underline{C_m}\}$ be the set of supports of the non zero minimal oriented cycles. The following statements are equivalent:*

I) Λ is a symmetric algebra.

II) There are non zero elements $a_1, \cdots, a_m$ in the field k such that, for each i and j with $(\underline{C_i})_0 \cap (\underline{C_j})_0 \neq \emptyset$ the following condition holds

$$\overline{C_i(t)} = a_i a_j^{-1} \overline{C_j(t)} \quad \forall t \in (\underline{C_i})_0 \cap (\underline{C_j})_0.$$

In section 4 we give a combinatorial approach to the above last theorem using a graph $GS(\Lambda)$ associated to Λ, and a function $\phi_\Lambda : Ch(GS(\Lambda)) \to k$ where $Ch(GS(\Lambda))$ is the set of chains of the graph $GS(\Lambda)$. In this way, the existence of the non zero constants $a_1, \cdots, a_m$ which are required in the last theorem, is very closely related with the structure of the graph $GS(\Lambda)$ and with the function $\phi_\Lambda : Ch(GS(\Lambda)) \to k$. In fact, we prove that the quasi-schurian Weakly-Symmetric k-algebra Λ is symmetric if either the graph $GS(\Lambda)$ is a tree or ϕ_Λ satisfies $\phi_\Lambda(C) = 1$ for each minimal cycle C in $GS(\Lambda)$ with at least three vertices.

In section 5 we give a connexion between symmetric quasi-schurian algebras and trivial extensions of algebras, which we state next.

THEOREM. *Let Λ be basic connected finite dimensional k-algebra. The following statements are equivalent*

1) There exists a schurian basic triangular algebra Λ' such that $\Lambda \simeq T(\Lambda')$.

2) Λ is symmetric quasi-schurian, and there exists a set $\mathcal{C}(\Lambda)$ consisting of exactly one arrow in each non zero minimal oriented cycle, such that $Q_\mathcal{C}$ has non oriented cycles, where $Q_\mathcal{C}$ is the quiver obtained from Q_Λ by deleting the arrows in $\mathcal{C}(\Lambda)$.

If these conditions hold, then $\Lambda' \simeq \Lambda/I_\mathcal{C}$ where $I_\mathcal{C}$ is the ideal generated by $\mathcal{C}(\Lambda)$ in Λ.

In the case that Q is an oriented tree and $\Lambda = T(kQ)$ we can always choose a set $\mathcal{C}(\Lambda)$ as in 2) in the theorem. Moreover, we prove that for any such choice the factor algebra $\Lambda/I_\mathcal{C}$ is iterated tilted of type Q. This is a useful approach to obtain iterated tilted algebras of a given tree class.

2 PRELIMINARIES

It is well known that each basic finite dimensional algebra Λ over an algebraically closed field k is isomorphic to k-algebra kQ/I where Q is the finite quiver associated with Λ and I is an admissible ideal of the path algebra kQ.

Let Q be a quiver. We will denote by Q_0 the set of vertices and by Q_1 the set of arrows of Q. Given an arrow $\alpha \in Q_1$, we say it starts at the vertex $o(\alpha)$ and ends at $e(\alpha)$. A path in the quiver Q is either an oriented sequence of arrows $p = \alpha_n \cdots \alpha_1$ with $e(\alpha_t) = o(\alpha_{t+1})$ for $1 \leq t < n$, or the symbol e_i for $i \in Q_0$. We call the paths e_i trivial paths and we define $o(e_i) = e(e_i)$. For a nontrivial path $p = \alpha_n \cdots \alpha_1$ we define $o(p) = o(\alpha_1)$ and $e(p) = e(\alpha_n)$. If δ is a path in Q, we will denote by $\underline{\delta}$ the support of δ in Q. Thus, $\underline{\delta}$ is a sub quiver of Q having as vertices and arrows those which belong to δ. A nontrivial path p is said to be an oriented cycle if $o(p) = e(p)$.

Let $\mathcal{C} = \alpha_n \alpha_{n-1} \cdots \alpha_2 \alpha_1$ be an oriented cycle in Q. We will call $\mathcal{C}$ minimal oriented cycle if $n = 1$ or all the vertices $o(\alpha_1), o(\alpha_2), \cdots, o(\alpha_n)$ are different in case $n > 1$. Let j be a vertex in the support $\underline{\mathcal{C}}$ of $\mathcal{C}$, then the arrows of $\underline{\mathcal{C}}$ determine a cycle with origin j, which we call $\mathcal{C}(j)$. That is, $\mathcal{C}(j) = \alpha_{r-1} \cdots \alpha_2 \alpha_1 \alpha_n \cdots \alpha_{r+1} \alpha_r$ where $j = o(\alpha_r)$ is the origin of α_r.

Let Λ be a finite dimensional k-algebra, we denote by $mod(\Lambda)$ the category of finitely generated left-Λ modules, by Q_Λ the ordinary quiver associated with Λ, by $S(a)$ the simple Λ-module corresponding to the vertex a in $(Q_\Lambda)_0$, by $P(a)$ the projective cover, and by $I(a)$ the injective envelope of $S(a)$. Let γ be a path in Q_Λ. By $\overline{\gamma}$ we denote the congruence class $\gamma + I$ in $\Lambda = kQ_\Lambda/I$. We will say that the path γ is zero if $\overline{\gamma} = \overline{0}$.

DEFINITION: An algebra Λ is called quasi-schurian, if it satisfies the following two conditions:

QS1) $dim_k Hom_\Lambda(P,Q) \leq 1$ if P,Q are non isomorphic indecomposable projective Λ-modules.

QS2) $dim_k End_\Lambda(P) = 2$ for each indecomposable projective Λ-module.

An important class of quasi-schurian algebras consists of the trivial extensions of Cartan type D, with D a Dynkin quiver. These algebras are closely related with the iterated tilted algebras of Dynkin type D, see [1],[2]. More generally, consider a schurian algebra Λ such that Q_Λ has no oriented cycles. Then the trivial extension $T(\Lambda)$ of Λ will be quasi-schurian.

2.1 Symmetric algebras. Let Λ be a k-algebra. We denote by D_Λ the usual duality

$$Hom_k(-,k) : mod(\Lambda) \to mod(\Lambda^{op}).$$

The algebra Λ is called symmetric if there exists an isomorphism $\varphi : \Lambda \xrightarrow{\sim} D_\Lambda(\Lambda)$ as $\Lambda - \Lambda$ bimodules. It is well known that Λ is symmetric if and only if there is a non-degenerate Λ-balanced symmetric k-bilinear mapping $\theta : \Lambda \times \Lambda \to k$, see [4]. We will point out the following equivalent version of the above property.

PROPOSITION 1. *Let Γ be a finite dimensional k-algebra and $f \in D_\Gamma(\Gamma)$. Then there exists a $\Gamma - \Gamma$ bimodule isomorphism $\varphi : \Gamma \xrightarrow{\sim} D_\Gamma(\Gamma)$ such that $\varphi(1) = f$ if and only if f satisfies:*

α) For each $\gamma_1, \gamma_2 \in \Gamma$ we have that $\gamma_2 \gamma_1 = 0$ is equivalent to $\gamma_1 \Gamma \gamma_2 \subseteq Ker f$.

β) $\gamma_1 \gamma_2 - \gamma_2 \gamma_1 \in Ker f$ for every $\gamma_1, \gamma_2 \in \Gamma$.

Proof. straightforward calculations. □

REMARKS:

1) The condition α) may be changed by one of the following conditions

α') If $\gamma \Gamma \subseteq Ker f$, then $\gamma = 0$.

α'') If $\Gamma \gamma \subseteq Ker f$, then $\gamma = 0$.

2) Let $\{e_1, \cdots, e_n\}$ be a complete family of orthogonal idempotents in Γ. Then the condition α) implies that

i) $f(e_j \Gamma e_i) = 0$ for $i \neq j$.

ii) $f(e_i \Gamma e_i) \neq 0$ for each i.

2.2 The Supplement Property for quasi-schurian algebras.

DEFINITION: Let $\Lambda = kQ_\Lambda/I$ be a quasi-schurian algebra. We will say that Λ satisfies the Supplement Property if for every non zero path γ there exists a non zero minimal oriented cycle $\mathcal{C}$ such that

1) $o(\gamma) = o(\mathcal{C})$.

2) All the arrows in γ lie in the support $\underline{\mathcal{C}}$ of the cycle $\mathcal{C}$.

The path δ such that $\delta \gamma = \mathcal{C}$ is called the supplement of γ in the cycle $\mathcal{C}$.

E. Fernández and M.I. Platzeck proved that this property holds for the trivial extension $T(\Lambda)$ of a schurian algebra Λ (see [3]).

LEMMA 2. *Let Λ be a quasi-schurian algebra and δ a nontrivial path in kQ_Λ. If C is an oriented cycle then $\overline{C\delta} = \overline{\delta C} = 0$.*

Proof. Suppose that $\overline{C\delta} \neq 0$. Then we will prove that the set $\{\overline{\delta}, \overline{C\delta}\}$ is linearly independent over k. This gives a contradiction since Λ is quasi-schurian.

Let $a\overline{\delta} + b\overline{C\delta} = 0$ where a and b lie in k. If $a \neq 0$ then $(1 + ba^{-1}\overline{C})\overline{\delta} = 0$. But $ba^{-1}\overline{C}$ lies in the radical of Λ and so $1 + ba^{-1}\overline{C}$ is invertible in Λ. Thus $\overline{\delta} = 0$, a contradiction. So, a must be zero. This means that $b\overline{C\delta} = 0$ which also gives that $b = 0$. Then, the set $\{\overline{\delta}, \overline{C\delta}\}$ is linearly independent. □

3 MAIN RESULTS

Let Λ be a finite dimensional k-algebra. Recall that Λ is called weakly-symmetric if for any indecomposable projective Λ-module P we have that $soc(P) \simeq top(P)$. It can be proven (see [4]) that a weakly-symmetric algebra is self-injective. Moreover, symmetric implies weakly-symmetric. In case Λ is a quasi-schurian algebra, we give in this section an answer to the following questions.

1) When is Λ weakly-symmetric?.

2) When is Λ symmetric?.

The Supplement Property which was defined above for quasi-schurian algebras is very closely related with these questions, as we will see in this section.

THEOREM 3. *Let $\Lambda = kQ_\Lambda/I$ be a quasi-schurian algebra. Then the following conditions are equivalent*

I) Λ *is weakly-symmetric.*

II) Λ *satisfies the Supplement Property.*

III) For each non zero f in $Hom_\Lambda(P,Q)$ the induced morphism

$Hom_\Lambda(Q, f) : Hom_\Lambda(Q, P) \to End_\Lambda(Q)$ *is non zero, if P and Q are indecomposable non isomorphic projective Λ-modules.*

IV) Λ *satisfies the following conditions*

  a) If a minimal oriented cycle C is non zero, then $\overline{C(t)} \neq 0$ for each vertex t in the support $\underline{C}$ of C.

  b) Let $\{\underline{C_1}, \underline{C_2} \cdots, \underline{C_m}\}$ be the set of supports corresponding to the non zero oriented cycles. Then $Q_\Lambda = \cup_{i=1}^m \underline{C_i}$.

Before proving the theorem, we will need the following result.

LEMMA 4. *Let $\Lambda = kQ_\Lambda/I$ be a finite dimensional k-algebra, let i be a vertex in Q_Λ and γ a non trivial path in Q_Λ, nonzero in Λ. If $soc(P(i)) \simeq S(i)$ and $\overline{\gamma} \in soc(P(i))$, then γ is a cycle with origin at the vertex i.*

Proof. Assume that $soc(P(i)) \simeq S(i)$ and $\overline{\gamma}$ lies in $soc(P(i))$. Let $j = e(\gamma)$. Then $\overline{\gamma} \in I(j)$. But $k\overline{\gamma} = soc(P(i)) \simeq S(i)$, hence $k\overline{\gamma} \simeq S(i)$. But $\overline{\gamma}$ is in $I(j)$, then $k\overline{\gamma} = soc(I(j)) \simeq S(j)$. This means that $S(i) \simeq S(j)$ and hence $i = j$. □

REMARK: We recall that, if M is a Λ module then the socle of M is equal to the right annihilator of $rad(\Lambda)$ in M (see [4]). This property will be used in the next proof.

Proof. of Theorem 3:
$I) \Rightarrow II)$ Assume that Λ is weakly-symmetric. Let $\gamma = \alpha_r \alpha_{r-1} \cdots \alpha_1$ be a non zero path such that $o(\gamma) \neq e(\gamma)$. Therefore $\overline{\gamma} \notin soc(P(o(\gamma)))$: indeed, if this is not the case, then Lemma 4 would imply that $o(\gamma) = e(\gamma)$, a contradiction. Then there exists an arrow β such that $\beta\gamma$ is non zero. So, multiplying γ by the necessary number of arrows $\beta_1, \cdots, \beta_m$, we may assume that the non zero path $\delta = \beta_m \beta_{m-1} \cdots \beta_1 \gamma$ is an oriented cycle or $\overline{\delta}$ lies in the socle of $P(o(\delta))$. Hence the assertion is now a consequence of Lemma 2 and Lemma 4.

$II) \Rightarrow I)$ Assume that Λ satisfies the Supplement Property. Let i be a vertex in Q_Λ and γ a non zero path in Λ such that $\overline{\gamma} \in soc(P(i))$. By the Supplement Property, there exists a non zero minimal oriented cycle $\mathcal{C}$ containing the path γ and such that $o(\mathcal{C}) = i$. If $\gamma \neq \mathcal{C}$, then there is an arrow β in $\mathcal{C}$ such that $\overline{\beta\gamma} \neq 0$. Hence $\overline{\gamma}$ does not lie in $soc(P(i))$, giving a contradiction. Thus, $\gamma = \mathcal{C}$ and hence $soc(P(i)) = k\overline{\mathcal{C}}$. So, the socle of $P(i)$ is isomorphic to the simple $S(i)$.

$II) \Leftrightarrow III)$ $III)$ is just a restatement of $II)$.

$II) \Rightarrow IV)$
a): Let $\mathcal{C} = \alpha_n \alpha_{n-1} \cdots \alpha_2 \alpha_1$ be a non zero oriented cycle. Assume that $t = o(\alpha_i)$. Since $\overline{\mathcal{C}} \neq \overline{0}$ we have that the path $\gamma = \alpha_n \cdots \alpha_{i+1} \alpha_i$ is non zero. Then by the supplement property there is a path δ such that $\delta\gamma$ is a non zero minimal oriented cycle. Since the paths δ and $\alpha_{i-1} \cdots \alpha_1$ have the same starting and ending vertices we obtain that $\overline{\delta} = a\overline{\alpha_{i-1}} \cdots \overline{\alpha_1}$ where $a \in k - \{0\}$. Then $\overline{0} \neq \overline{\delta\gamma} = a\overline{\mathcal{C}(t)}$ and hence $\overline{0} \neq \overline{\mathcal{C}(t)}$.

b): Each arrow of Q_Λ is non zero in Λ. Hence by the Supplement Property we get that $Q_\Lambda = \cup_{i=1}^m \mathcal{C}_i$.

$IV) \Rightarrow II)$ Let γ be a non zero path. By b) and Lemma 2 we get that γ belongs to some non zero minimal oriented cycle C. Thus the Supplement Property holds since by a) we have that $\overline{C(o(\gamma))} \neq 0$. □

COROLLARY 5. *Let $\Lambda = kQ_\Lambda/I$ be a quasi-schurian weakly-symmetric algebra. Then the ordinary quiver Q_Λ is the union of all non zero minimal oriented cycles.*

The other main result in this section is the following theorem.

THEOREM 6. *Let $\Lambda = kQ_\Lambda/I$ be a quasi-schurian weakly-symmetric algebra. Let $\{\underline{\mathcal{C}_1}, \underline{\mathcal{C}_2} \cdots, \underline{\mathcal{C}_m}\}$ be the set of supports of the non zero minimal oriented cycles. The following statements are equivalent:*

I) Λ is a symmetric algebra.

II) There are non zero elements $a_1, \cdots, a_m$ in the field k such that, for each i and j with $(\underline{\mathcal{C}_i})_0 \cap (\underline{\mathcal{C}_j})_0 \neq \emptyset$ the following condition holds

$$\overline{\mathcal{C}_i(t)} = a_i a_j^{-1} \overline{\mathcal{C}_j(t)} \quad \forall t \in (\underline{\mathcal{C}_i})_0 \cap (\underline{\mathcal{C}_j})_0.$$

We will need the next lemma to give a proof of this theorem.

LEMMA 7. *Let $\Lambda = kQ_\Lambda/I$ be a symmetric k-algebra. Let $\varphi : \Lambda \to D(\Lambda)$ be an isomorphism of $\Lambda - \Lambda$ bimodules and $f = \varphi(1)$. Then the following conditions hold for every non zero minimal oriented cycle $\mathcal{C}$.*

a) *If $dim_k End_\Lambda(P(i)) = 2$ where $o(\mathcal{C}) = i$, then $f(\overline{\mathcal{C}}) \neq 0$.*

b) *$f(\overline{\mathcal{C}(j)}) = f(\overline{\mathcal{C}})$ for every $j \in (\underline{\mathcal{C}})_0$.*

Proof. b): Follows from β) in Proposition 1 since $\gamma_1\gamma_2 - \gamma_2\gamma_1 \in Ker f$ for every $\gamma_1, \gamma_2 \in \Lambda$.

a): By b) above it is sufficient to prove that $f(\overline{\mathcal{C}}) \neq 0$. Since $dim_k End_\Lambda(P(i)) = 2$ we get that $\{\overline{e_i}, \overline{\mathcal{C}}\}$ is a k-basis of $End_\Lambda(P(i))$ and $\overline{\mathcal{C}}^2 = 0$.

We know that $\overline{e_i}\overline{\mathcal{C}} \neq 0$. Then by Proposition 1 it follows that there exists $\lambda \in \Lambda$ such that $f(\overline{\mathcal{C}}\lambda\overline{e_i}) \neq 0$. In particular $0 \neq \lambda\overline{e_i} \in End_\Lambda(P_i)$, and we get that $\lambda\overline{e_i} = r\overline{e_i} + s\overline{\mathcal{C}}$ where $r, s \in k$. Then $\overline{\mathcal{C}}\lambda\overline{e_i} = r\overline{\mathcal{C}}\overline{e_i} + s\overline{\mathcal{C}}^2 = r\overline{\mathcal{C}}$ and this means that $f(\overline{\mathcal{C}}) \neq 0$ since $0 \neq f(\overline{\mathcal{C}}\lambda\overline{e_i}) = f(r\overline{\mathcal{C}})$. □

REMARK: Let $f : \Lambda \to k$ be as in Lemma 7, and $\mathcal{C}$ be a non zero minimal oriented cycle. It is clear by Lemma 7 that $f(\overline{\mathcal{C}(i)}) = f(\overline{\mathcal{C}(j)})$ for all vertices i, j in $\underline{\mathcal{C}}$. Hence f can be defined on the support $\underline{\mathcal{C}}$ as follows, fix a vertex j in $\underline{\mathcal{C}}$ and let $f(\underline{\mathcal{C}}) = f(\overline{\mathcal{C}(j)})$. In this way, we say that f is constant and non zero on $\underline{\mathcal{C}}$.

Proof. of Theorem 6:

I) $\Rightarrow$ II) : Assume that Λ is a symmetric algebra. Let $\varphi : \Lambda \to D(\Lambda)$ be an isomorphism of $\Lambda - \Lambda$ bimodules and $f = \varphi(1)$. To obtain the nonzero constants $a_1, \cdots, a_m$ we can use the above remark and define $a_i = f(\underline{\mathcal{C}_i})$.

II) $\Rightarrow$ I) : The idea of the proof is to construct a linear functional $f : \Lambda \to k$ such that the properties $\alpha'), \beta)$ in Proposition 1 hold. Let us start with the linear functional $F : kQ_\Lambda \to k$ defined on the basis of the paths in Q_Λ as follows: $F(\gamma) = a_i$ if there are i and t such that $\gamma = \mathcal{C}_i(t)$, and zero otherwise. Then II) implies that $\overline{\gamma} = F(\gamma)(F(\gamma'))^{-1}\overline{\gamma'}$, for nonzero cycles γ and γ' with the same origin. The next step is to check that $F : kQ_\Lambda \to k$ factors through the canonical epimorphism $\pi : kQ_\Lambda \to \Lambda$, that is, that $I \subseteq KerF$. Let $\gamma = \sum_{i=1}^n c_i\gamma_i \in I$ be a linear combination of paths γ_i starting at the vertex a and ending at the vertex b for $1 \leq i \leq n$. We may assume that $a = b$ and γ_i is a non zero oriented cycle for $i = 1, \cdots, n$. Since $\overline{\gamma_i} = (F(\gamma_i)/F(\gamma_1))\overline{\gamma_1}$ $i = 2, 3, \cdots, n$, we get that $0 = \overline{\gamma} = \sum_{i=1}^n c_i\overline{\gamma_i} = (\sum_{i=1}^n c_iF(\gamma_i)/F(\gamma_1))\overline{\gamma_1}$. But $\overline{\gamma_1} \neq 0$. So $\sum_{i=1}^n c_iF(\gamma_i)/F(\gamma_1) = 0$ and then $\gamma = \sum_{i=1}^n c_i(\gamma_i - (F(\gamma_i)/F(\gamma_1))\gamma_1)$, therefore $F(\gamma) = 0$. Hence there exists $f : \Lambda \to k$ such that $f = F\pi$.

We will prove that α') holds, that is $\lambda_1\Lambda \subseteq Kerf$ implies $\lambda_1 = 0$. Assume that $\lambda_1 = \sum_{j=1}^n c_j\overline{\gamma_j}$ be such that $\lambda_1\Lambda \subseteq Kerf$ where γ_j is a path in Q_Λ for $j = 1, 2, \cdots, n$. Observe that $\lambda_1 = \sum_{j=1}^n \lambda_1\overline{e_j}$, $\lambda_1\overline{e_i}\Lambda \subseteq \lambda_1\Lambda \subseteq Kerf$. Hence it is enough to prove α') only for each $\lambda_1\overline{e_i}$; that is, for all linear combination of paths starting at the vertex i. Then we may assume without loose of generality that $i = 1$ and $o(\gamma_j) = 1$ for $j = 1, 2, \cdots, n$.

Let $\{b_1, \cdots, b_r\}$ be the set of end points of the paths γ_j, for $j = 1, 2, \cdots, n$. Let $A_j = \{i : e(\gamma_i) = b_j\}$. Then we can write $\lambda_1 = \sum_{j=1}^r \sum_{i \in A_j} c_i\overline{\gamma_i}$. Let us prove that $\sum_{i \in A_1} c_i\overline{\gamma_i} = 0$. Assume that ν is a supplement to the paths $\{\gamma_i : i \in A_1\}$. This path exists since Λ is quasi-schurian and the Supplement Property holds. Fix an index i_1 in A_1, then $\overline{\gamma_j} = d_j\overline{\gamma_{i_1}}$ for some $d_j \in k$ and each $j \in A_1 - \{i_1\}$. Multiplying both sides of the above equality by $\overline{\nu}$ and applying f we get $d_j = f(\overline{\gamma_j\nu})/f(\overline{\gamma_{i_1}\nu})$. Now, by Lemma 2 and the fact that $f = F\pi$ we obtain $f(\overline{\gamma_i\nu}) = 0$ for all $i \in A_j$ with $j > 1$. Hence $0 = f(\lambda_1\overline{\nu}) = \sum_{i \in A_1} c_i f(\overline{\gamma_i\nu})$, and this implies that $\sum_{i \in A_1} c_i\overline{\gamma_i} = (\sum_{i \in A_1} c_i d_i)\overline{\gamma_{i_1}} = (\sum_{i \in A_1} c_i f(\overline{\gamma_i\nu}))\overline{\gamma_{i_1}}/f(\overline{\gamma_{i_1}\nu}) = 0$. We point out that the equality $\sum_{i \in A_j} c_i\overline{\gamma_i} = 0$ for $j = 2, 3, \cdots, r$ can be obtained in an analogous way. Hence $\lambda_1 = 0$, as we wanted.

We will prove that β) holds, that is $\lambda_1\lambda_2 - \lambda_2\lambda_1 \in Ker f$ for all $\lambda_1, \lambda_2 \in \Lambda$.

Let $\lambda_1 = \sum_i c_i\overline{\gamma_i}$ and $\lambda_2 = \sum_j d_j\overline{\delta_j}$, where γ_i, δ_j are paths for each i and j. Assume that $\overline{\gamma_i\delta_j} \neq 0$, hence $\gamma_i\delta_j$ lies in a non zero minimal oriented cycle $\mathcal{C}$ such that $o(\gamma_i\delta_j) = o(\mathcal{C})$. If $\gamma_i\delta_j = \mathcal{C}$ we obtain that the supports of $\gamma_i\delta_j$ and $\delta_j\gamma_i$ coincide. Hence $F(\gamma_i\delta_j) = F(\delta_j\gamma_i)$ and this implies that $f(\overline{\gamma_i\delta_j}) = f(\overline{\delta_j\gamma_i})$. In case $e(\gamma_i\delta_j) \neq e(\mathcal{C})$ we have $\overline{\delta_j\gamma_i} = 0$ and also $F(\gamma_i\delta_j) = 0$.

Therefore, $\overline{\gamma_i\delta_j} \neq 0$ implies that $f(\overline{\gamma_i\delta_j}) = f(\overline{\delta_j\gamma_i})$.

In the same way it can be proved that $\overline{\gamma_i\delta_j} = 0$ implies that $f(\overline{\gamma_i\delta_j}) = f(\overline{\delta_j\gamma_i})$. Hence the assertion follows. □

4 A COMBINATORIAL APPROACH TO THEOREM 6

Let $\Lambda = kQ_\Lambda/I$ be a weakly-symmetric and quasi-schurian k-algebra. We associate to Λ a graph $GS(\Lambda)$. The construction is as follows. Let $\{\underline{\mathcal{C}_1}, \underline{\mathcal{C}_2} \cdots, \underline{\mathcal{C}_m}\}$ be the set of supports of the non zero minimal oriented cycles. Then the set of vertices of $GS(\Lambda)$ is $\{1, 2, \cdots, m\}$ and the edges are determined as follows

a) If $m = 1$, the set of edges is empty.

b) If $m > 1$, there is only one edge with vertices $\{i, j\}$ in case $(\underline{\mathcal{C}_i})_0 \cap (\underline{\mathcal{C}_j})_0 \neq \emptyset$ and $i \neq j$.

It is not difficult to see that $GS(\Lambda)$ is a connected graph without loops and non parallel edges.

NOTATION: A chain C in $GS(\Lambda)$ joining the vertices v_1 and v_k is a sequence of vertices and edges $v_1 A_1 v_2 A_2 \cdots v_{k-1} A_{k-1} v_k$ where for each i the edge A_i has vertices v_i, v_{i+1}. We say that the length of C is $k - 1$. Let $B = w_1 B_1 w_2 B_2 \cdots w_{n-1} B_{n-1} w_n$ be another chain in $GS(\Lambda)$. We will say that the composition $A \circ B$ is defined if $v_k = w_1$ and we let $A \circ B$ be the chain

$$v_1 A_1 v_2 A_2 \cdots v_{k-1} A_{k-1} v_k B_1 w_2 B_2 \cdots w_{n-1} B_{n-1} w_n.$$

A chain $v_1 A_1 v_2 A_2 \cdots v_{k-1} A_{k-1} v_k$ is called reduced if $v_{i-1} \neq v_{i+1}$ for each $i = 2, 3, \cdots, k-1$. The set of all chains in $GS(\Lambda)$ is denoted by $Ch(GS(\Lambda))$. Usually we shall only be interested in reduced chains, and unless the contrary is explicitly stated, we shall assume that all chains under discussion are reduced.

A cycle C in $GS(\Lambda)$ is a chain of the form $v_1 A_1 v_2 A_2 \cdots v_{k-1} A_{k-1} v_1$. If the vertices $v_1, v_2, \cdots, v_{k-1}$ are all different, then the chain C is called a minimal cycle. We observe that a minimal cycle has at least three vertices.

Let C be the chain $v_1 A_1 v_2 A_2 \cdots v_{k-1} A_{k-1} v_k$ in $GS(\Lambda)$. We denote by $\underline{C}$ the support of C which is defined as the subgraph of $GS(\Lambda)$ with vertices $v_1, \cdots, v_k$ and edges $A_1, \cdots, A_{k-1}$. Given a minimal cycle $C = v_1 A_1 v_2 A_2 \cdots v_{k-1} A_{k-1} v_1$ in

$GS(\Lambda)$, we denote by $C(v_i)$ the cycle $v_i A_i v_{i+1} \cdots v_{k-1} A_{k-1} v_1 A_1 v_2 \cdots v_{i-1} A_{i-1} v_i$, for $1 \leq i \leq k-1$.

The set $\{\underline{C_1}, \underline{C_2} \cdots, \underline{C_m}\}$ of supports of the non zero minimal oriented cycles induces a family of nonzero constants $\lambda_{ij}(t) \in k$, such that $\overline{C_i(t)} = \lambda_{ij}(t) \overline{C_j(t)}$ for $t \in (\underline{C_i})_0 \cap (\underline{C_j})_0$. If the algebra Λ is symmetric we obtain by Theorem 6 that $\lambda_{ij} = \lambda_{ij}(t)$ $\forall t \in (\underline{C_i})_0 \cap (\underline{C_j})_0$. Thus, the non zero constants $\lambda_{ij}(t)$ do not depend on the common vertices $t \in (\underline{C_i})_0 \cap (\underline{C_j})_0$. Having this property as a motivation we will assume that the family of constants $\lambda_{ij}(t) \in k$, satisfies the following condition

$$\lambda_{ij} = \lambda_{ij}(t) \ \forall t \in (\underline{C_i})_0 \cap (\underline{C_j})_0.$$

Now, we can define a map $\phi_\Lambda : Ch(GS(\Lambda)) \to k$ in the following way

$$\phi_\Lambda(v_1 A_1 v_2 A_2 \cdots v_{k-1} A_{k-1} v_k) = \lambda_{v_1 v_2} \lambda_{v_2 v_3} \cdots \lambda_{v_{k-1} v_k}.$$

We point out that, if C_1 and C_2 are chains such that their composition is defined, then we have that $\phi_\Lambda(C_1 \circ C_2) = \phi_\Lambda(C_1) \phi_\Lambda(C_2)$.

Let D be the chain $v_1 A_1 v_2 A_2 \cdots v_{k-1} A_{k-1} v_k$. We denote by D^{-1} the chain $v_k A_{k-1} v_{k-1} \cdots v_2 A_1 v_1$. In this way $\phi_\Lambda(D \circ D^{-1}) = \phi_\Lambda(D^{-1} \circ D) = 1$ and hence $\phi_\Lambda(D^{-1}) = \phi_\Lambda(D)^{-1}$.

Let C be a minimal cycle in $GS(\Lambda)$ with support $\underline{C}$. It is clear that $\phi_\Lambda(C(v_i)) = \phi_\Lambda(C(v_j))$ for each v_i, v_j in $\underline{C}$. Hence ϕ_Λ can be defined on $\underline{C}$ as follows, fix a vertex v in $\underline{C}$ and let $\phi_\Lambda(\underline{C}) = \phi_\Lambda(C(v))$.

The existence of the non zero constants $a_1, a_2, \cdots, a_m$ which are required in Theorem 6 for Λ to be symmetric is very closely related with the structure of the graph $GS(\Lambda)$ and with the function $\phi_\Lambda : Ch(GS(\Lambda)) \to k$ as can be seen in the next theorem.

THEOREM 8. *Let Λ be a quasi-schurian weakly-symmetric k-algebra, and $\{\underline{C_1}, \underline{C_2} \cdots, \underline{C_m}\}$ be the set of supports of the non zero minimal oriented cycles. Suppose that the non zero constants $\lambda_{ij}(t)$ above defined do not depend on the common vertices $t \in (\underline{C_i})_0 \cap (\underline{C_j})_0$.*
Then

a) If the graph $GS(\Lambda)$ is a tree, then Λ is symmetric.

b) Suppose that $GS(\Lambda)$ is not a tree. Then Λ is symmetric if and only if the function $\phi_\Lambda : Ch(GS(\Lambda)) \to k$ satisfies $\phi_\Lambda(C) = 1$ for each minimal cycle C in $GS(\Lambda)$.

Proof. a): Assume that $GS(\Lambda)$ is a tree. Let us prove that in this case the required non zero constants always exist. Since $GS(\Lambda)$ is a tree we have that for each vertex $j \neq 1$ in the graph $GS(\Lambda)$ there exists only one nontrivial chain D_j in $GS(\Lambda)$ joining the vertex j with the vertex 1. Hence we can define $a_1, a_2, \cdots, a_m$ in the following way, let $a_1 = 1$ and $a_j = \phi_\Lambda(D_j)$ if $j \neq 1$. The next step is to prove that $\lambda_{ij} = a_i a_j^{-1}$ in case $(\underline{C_i})_0 \cap (\underline{C_j})_0 \neq \emptyset$ and $i \neq j$. Let A be the edge with vertices i and j. Since $GS(\Lambda)$ is a tree we have that $D_i = (iAj) \circ D_j$ and hence $a_i = \phi_\Lambda(D_i) = \phi_\Lambda((iAj) \circ D_j) = \phi_\Lambda(iAj) \phi_\Lambda(D_j) = \lambda_{ij} a_j$. Thus, $\lambda_{ij} = a_i a_j^{-1}$ as we wanted.

b):($\Rightarrow$) assume that Λ is symmetric, that is the non zero constants $a_1, a_2, \cdots, a_m$ of Theorem 6 exist.

Let $C = v_1 A_1 v_2 A_3 \cdots v_{k-1} A_{k-1} v_1$ be a minimal cycle in $GS(\Lambda)$. Then $\phi_\Lambda(C) = \lambda_{v_1 v_2} \lambda_{v_2 v_3} \cdots \lambda_{v_{k-1} v_1} = a_{v_1} a_{v_2}^{-1} a_{v_2} a_{v_3}^{-1} \cdots a_{v_{k-1}} a_{v_1}^{-1} = 1$.

($\Leftarrow$) Assume that $\phi_\Lambda(C) = 1$ for each minimal cycle C in $GS(\Lambda)$. We will need the next Lemma.

LEMMA: If B_j and D_j are two chains in $GS(\Lambda)$ joining the vertex j with the vertex 1 where $j \neq 1$. Then $\phi_\Lambda(B_j) = \phi_\Lambda(D_j)$.

Proof. If $B_j \circ D_j^{-1}$ is a minimal cycle, then by hypothesis we have $1 = \phi_\Lambda(B_j \circ D_j^{-1}) = \phi_\Lambda(B_j) \phi_\Lambda(D_j)^{-1}$. Hence $\phi_\Lambda(B_j) = \phi_\Lambda(D_j)$.

Assume that $B_j \circ D_j^{-1}$ is not a minimal cycle. Then it is not difficult to see that B_j and D_j have a decomposition $B_j = F_1 \circ F_2$, $D_j = G_1 \circ G_2$ in such way that $F_1 \circ G_1^{-1}$ is a minimal cycle in $GS(\Lambda)$. Hence $B_j \circ D_j^{-1} = F_1 \circ F_2 \circ G_2^{-1} \circ G_1^{-1}$ implies $\phi_\Lambda(B_j \circ D_j^{-1}) = \phi_\Lambda(F_1 \circ G_1^{-1}) \phi_\Lambda(F_2 \circ G_2^{-1})$. But we know by construction that $\phi_\Lambda(F_1 \circ G_1^{-1}) = 1$. Then $\phi_\Lambda(B_j \circ D_j^{-1}) = \phi_\Lambda(F_2 \circ G_2^{-1})$ but now, the length of the chain $F_2 \circ G_2^{-1}$ is smaller than the length of $B_j \circ D_j^{-1}$. Hence by induction we can obtain that $\phi_\Lambda(B_j \circ D_j^{-1}) = 1$ and conclude that $\phi_\Lambda(B_j) = \phi_\Lambda(D_j)$. $\square$

Now, using this lemma it is possible to define the required constants. Let D_j be a chain in $GS(\Lambda)$ joining the vertices j and 1 where $j \neq 1$. Then we define $a_1 = 1$ and $a_j = \phi_\Lambda(D_j)$ if $j \neq 1$. By the above lemma we have that the constant a_j is well defined. Now we will check that $\lambda_{ij} = a_i a_j^{-1}$ in case $(\underline{C}_i)_0 \cap (\underline{C}_j)_0 \neq \emptyset$ and $i \neq j$. Let A be the edge with vertices i and j, let D_j be a chain joining the vertex j with the vertex 1. Then the chain $B_i = (iAj) \circ D_j$ is a chain in $GS(\Lambda)$ joining the vertices i and 1. Hence $a_i = \phi_\Lambda(B_i) = \phi_\Lambda((iAj) \circ D_j) = \phi_\Lambda(iAj) \phi_\Lambda(D_j) = \lambda_{ij} a_j$. This implies that $\lambda_{ij} = a_i a_j^{-1}$. $\square$

From the above theorem we obtain the following corollaries.

Let Λ be a quasi-schurian and weakly-symmetric k-algebra, $\{\underline{C}_1, \cdots \underline{C}_m\}$ be the set of supports of the non zero minimal oriented cycles and $\lambda_{ij}(t) \in k$ be the family of non zero constants such that $\overline{C_i(t)} = \lambda_{ij}(t) \overline{C_j(t)}$ for $t \in (\underline{C}_i)_0 \cap (\underline{C}_j)_0$.

COROLLARY 9. *If the graph $GS(\Lambda)$ associated to the quasi-schurian weakly-symmetric k-algebra Λ is a tree, then the following conditions are equivalent*
 I) Λ is a symmetric algebra.
 II) $\lambda_{ij} = \lambda_{ij}(t)$ $\forall t \in (\underline{C}_i)_0 \cap (\underline{C}_j)_0$, $i \neq j$.

COROLLARY 10. *Suppose that the graph $GS(\Lambda)$ associated to the quasi-schurian and weakly-symmetric k-algebra Λ is not a tree. Then the following conditions are equivalent.*
 I) Λ is a symmetric algebra.
 II) $\lambda_{ij} = \lambda_{ij}(t)$ $\forall t \in (\underline{C}_i)_0 \cap (\underline{C}_j)_0$, $i \neq j$, and the function $\phi_\Lambda : Ch(GS(\Lambda)) \to k$ defined above satisfies $\phi_\Lambda(C) = 1$ for each minimal cycle C.

EXAMPLE: Let Λ be the factor algebra of the path algebra kQ for Q the quiver

Symmetric Quasi-Schurian Algebras

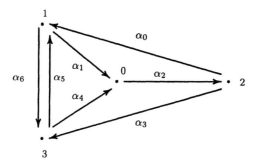

modulo the ideal $I = <\alpha_1\alpha_0 - a\alpha_4\alpha_3, \alpha_0\alpha_2\alpha_4, \alpha_3\alpha_2\alpha_1, \alpha_6\alpha_0, \alpha_6\alpha_5\alpha_6, \alpha_5\alpha_3, \alpha_1\alpha_5,$ $\alpha_4\alpha_6, \alpha_5\alpha_6\alpha_5, b\alpha_0\alpha_2\alpha_1 - \alpha_5\alpha_6, \alpha_2\alpha_1\alpha_0\alpha_2, \alpha_3\alpha_2\alpha_4 - c\alpha_6\alpha_5 >$ where $a, b, c \in k - \{0\}$.

Λ is quasi-schurian and moreover weakly-symmetric since this algebra satisfies the Supplement Property (see Theorem 3).

We will prove that Λ is symmetric if and only if $abc = 1$.

Let $C_1 = \alpha_1\alpha_0\alpha_2$, $C_2 = \alpha_4\alpha_3\alpha_2$ and $C_3 = \alpha_6\alpha_5$. Then $\{\underline{C_1}, \underline{C_2}, \underline{C_3}\}$ is the set of supports of the non zero minimal oriented cycles. Hence the graph $GS(\Lambda)$ is

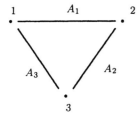

Let us now compute the family of non zero constants $\lambda_{ij}(t) \in k$, such that $\overline{C_i(t)} = \lambda_{ij}(t)\overline{C_j(t)}$ for $t \in (\underline{C_i})_0 \cap (\underline{C_j})_0$. In this case $C_1(0) = \alpha_1\alpha_0\alpha_2$, $C_2(0) = \alpha_4\alpha_3\alpha_2$, $C_1(2) = \alpha_2\alpha_1\alpha_0$, and $C_2(2) = \alpha_2\alpha_4\alpha_3$. Using the relations given in the ideal I we have $\overline{C_1(0)} = a\overline{C_2(0)}$, $\overline{C_1(2)} = a\overline{C_2(2)}$. Hence $\lambda_{12} = \lambda_{12}(0) = \lambda_{12}(2) = a$. In an analogous way we obtain that $\lambda_{13} = \lambda_{13}(1) = b^{-1}$, $\lambda_{23} = \lambda_{23}(3) = c$.

Consider the following minimal cycle C in $GS(\Lambda)$, $C = 1A_12A_23A_31$. Then $\phi_\Lambda(C) = \lambda_{12}\lambda_{23}\lambda_{31} = abc$, and $\{\underline{C}\}$ is the set of supports of the minimal cycles in $GS(\Lambda)$. So, by Corollary 10 we get Λ is symmetric if and only if $abc = 1$.

5 A CONNECXION WITH TRIVIAL EXTENSIONS OF ALGEBRAS

Let Λ be a quasi-schurian weakly-symmetric algebra, $\{\underline{C_1}, \cdots \underline{C_m}\}$ be the set of supports of the non zero minimal oriented cycles. In case it is possible to select exactly one arrow in each of the $\underline{C_i}$'s, we fix such a choice, and denote by $\mathcal{C}(\Lambda)$ the set consisting of the chosen arrows. Thus, $\mathcal{C}(\Lambda)$ is a set of arrows of Q_Λ such that $\mathcal{C}(\Lambda) \cap (\underline{C_i})_1$ has only one arrow for each $i = 1, 2, \cdots, m$. The ideal generated in Λ by $\mathcal{C}(\Lambda)$ will be denoted by $I_\mathcal{C}$. Moreover, $\mathcal{C}(\Lambda)$ induces a sub quiver Q_C of Q_Λ as follows $(Q_C)_0 = (Q_\Lambda)_0$ and $(Q_C)_1 = (Q_\Lambda)_1 - \mathcal{C}(\Lambda)$.

Let β be an arrow in Q_Λ. We will denote by $Suppl(\beta)$ the set of supplements of β. Thus, a path δ lies in $Suppl(\beta)$ if and only if $\delta\beta$ is a non zero minimal oriented cycle.

Our aim in this section is to give a proof and some consequences of the following result.

THEOREM 11. *Let Λ be a quasi-schurian symmetric algebra. If there exists a choice of arrows $\mathcal{C}(\Lambda)$ as above then Λ/I_C is a schurian algebra, and moreover $\Lambda \simeq T(\Lambda/I_C)$ where $T(\Lambda/I_C)$ is the trivial extension of Λ/I_C.*

In the proof of this theorem we will need the following lemmas.

LEMMA 12. *Let Λ be a quasi-schurian weakly-symmetric algebra. If there exists a choice of arrows $\mathcal{C}(\Lambda)$ as above, then Λ/I_C is a schurian algebra and Q_C is the ordinary quiver associated with Λ/I_C.*

Proof. Assume that $\Lambda = kQ_\Lambda/I$ where I is an admissible ideal. Let $\varphi : kQ_C \to \Lambda/I_C$ be defined as follows $\varphi(\delta) = \pi(\overline{\delta})$ where $\pi : \Lambda \to \Lambda/I_C$ is the canonical epimorphism. We get that $rad(\Lambda/I_C) = rad\Lambda/I_C$ and $\Lambda/rad\Lambda \simeq (\Lambda/I_C)/rad(\Lambda/I_C)$ since $I_C \subseteq rad\Lambda$ and π is an epimorphism. Hence Λ/I_C is basic and $\{\varphi(i) : i \in (Q_C)_0\}$ is a complete family of orthogonal primitive idempotents of Λ/I_C. So, to obtain that $Q_C = Q_{\Lambda/I_C}$ it is enough to prove that $\{\varphi(\alpha) : \alpha \in (Q_C)_1\}$ is a k-basis of $rad(\Lambda/I_C)/rad^2(\Lambda/I_C)$. First, observe that $\alpha \notin \mathcal{C}(\Lambda)$ implies that $\varphi(\alpha)$ is non zero and also does not lie in $rad^2(\Lambda/I_C)$. Therefore $\{\varphi(\alpha) : \alpha \in (Q_C)_1\}$ is a k-basis of $rad(\Lambda/I_C)/rad^2(\Lambda/I_C)$ since $rad(\Lambda/I_C) = rad\Lambda/I_C$.

Finally let us prove that Λ/I_C is schurian. Using the k-module isomorphisms $\pi(\overline{e_j})(\Lambda/I_C)\pi(\overline{e_i}) \simeq \overline{e_j}\Lambda\overline{e_i}/\overline{e_j}I_C\overline{e_i}$ for all i,j. It is enough to prove that $dim_k \overline{e_i}I_C\overline{e_i} = 1$ for each i, since Λ is quasi-schurian. But this follows easily because each non zero oriented cycle contains an arrow from $\mathcal{C}(\Lambda)$. □

DEFINITION: Let Γ be a finite dimensional k-algebra. We say that $x \in \Gamma$ is maximal in case $x \neq 0$ and $xw = wx = 0$ for all $w \in rad\Gamma$.

REMARK Let $\Gamma = Q_\Gamma/I$ with I an admissible ideal. Let δ be a non zero path in Q_Γ. Then δ is maximal in Γ if and only if $\overline{\delta\alpha} = \overline{\alpha\delta} = 0$ for all arrows $\alpha \in Q_\Gamma$.

LEMMA 13. *Let $\Lambda = Q_\Lambda/I$ be a quasi-schurian weakly-symmetric algebra with I an admissible ideal. If there exists a choice of arrows $\mathcal{C}(\Lambda)$ as above and Q_Λ has no loops then the next statements hold, where $\pi : \Lambda \to \Lambda/I_C$ is the canonical epimorphism.*

a) $\overline{e_{o(\beta)}}I_C\overline{e_{e(\beta)}} = 0$ for all $\beta \in \mathcal{C}(\Lambda)$. Therefore $\pi(\overline{\gamma}) \neq 0$ for any supplement γ of an arrow β in $\mathcal{C}(\Lambda)$.

b) Let γ be a path in Q_Λ. Then $\pi(\overline{\gamma})$ is maximal if and only if there exists β in $\mathcal{C}(\Lambda)$ such that $\gamma \in Suppl(\beta)$.

c) Let $\mathcal{C}(\Lambda) = \{\beta_1, \beta_2, \cdots, \beta_r\}$. Then the set $\{\pi(\overline{\delta_i}) : \delta_i \in Suppl(\beta_i), 1 \leq i \leq r\}$ is a k-basis of the vector space generated by all the maximal paths in Λ/I_C.

Proof. a): Suppose that γ is a non zero path and $\overline{\gamma} \in \overline{e_{o(\beta)}}I_C\overline{e_{e(\beta)}}$ for some $\beta \in \mathcal{C}(\Lambda)$. Let δ be a supplement of γ, which exists since Λ is weakly-symmetric (see Theorem 3). Now, we may assume that γ contains an arrow β' of $\mathcal{C}(\Lambda)$ because $\overline{\gamma} \in I_C$ and Λ is quasi-schurian. But β and δ have the same starting and ending vertices. Hence

$\delta = \beta$ and the non zero oriented cycle $\beta\gamma$ has two arrows in $\mathcal{C}(\Lambda)$, a contradiction. So, $\overline{\gamma} = 0$ and therefore $\overline{e_{o(\beta)}} I_\mathcal{C} \overline{e_{e(\beta)}} = 0$.

b): Let γ be a path in Q_Λ. Assume that $\pi(\overline{\gamma})$ is maximal. So, $\overline{\gamma} \neq 0$. Let δ be a supplement of γ. Then $\overline{\delta}$ contains an arrow β in $\mathcal{C}(\Lambda)$ since $\overline{\gamma} \notin I_\mathcal{C}$ and $\delta\gamma$ is a non zero minimal oriented cycle. Now we will prove that $\delta = \beta$ using the fact that $\pi(\gamma)$ is maximal. Considering a decomposition of δ as $\gamma_1 \beta \gamma_2$ we obtain by a) that $0 \neq \pi(\overline{\gamma_2 \gamma \gamma_1}) = \pi(\overline{\gamma_2}) \pi(\overline{\gamma}) \pi(\overline{\gamma_1})$ and therefore γ_1, γ_2 are trivial paths by the maximality of $\pi(\overline{\gamma})$. Hence $\delta = \beta$ and this means that $\gamma \in Suppl(\beta)$.

Assume now that $\gamma \in Suppl(\beta)$ with β in $\mathcal{C}(\Lambda)$, and let us prove that $\pi(\overline{\gamma})$ is maximal. From a) we obtain that $\pi(\overline{\gamma}) \neq 0$. Let α be an arrow in Q_Λ such that $\pi(\overline{\alpha})\pi(\overline{\gamma}) \neq 0$. So $\overline{\alpha\gamma} \neq 0$ and therefore there exists a supplement μ of $\alpha\gamma$. But β and $\mu\alpha$ have the same starting an ending vertices. Then there is c in $k - \{0\}$ such that $\overline{\beta} = c\overline{\mu\alpha}$ since Λ is quasi-schurian, a contradiction because I is an admissible ideal. Hence $\pi(\overline{\alpha})\pi(\overline{\gamma}) = 0$ for all arrow $\alpha \in Q_\Lambda$. In an analogous way we obtain that $\pi(\overline{\gamma})\pi(\overline{\alpha}) = 0$ for all arrow $\alpha \in Q_\Lambda$.

c): The set $\{\pi(\overline{\delta_i}) \ 1 \leq i \leq r\}$ generates all the maximal paths since b) holds and $\Lambda/I_\mathcal{C}$ is schurian by Lemma 12. Let $\sum_{i=1}^{r} a_i \pi(\overline{\delta_i}) = 0$ with $a_i \in k$. Then $\sum_{i=1}^{r} a_i \overline{\delta_i} \in I_\mathcal{C}$, and using a) we obtain $a_j \overline{\delta_j} = \overline{e_{o(\beta_j)}} (\sum_{i=1}^{r} a_i \overline{\delta_i}) \overline{e_{e(\beta_i)}} = 0$ since Λ is quasi-schurian and β_i, β_j do not have the same starting and ending vertices for $i \neq j$. So, $a_j = 0$ for $j = 1, 2, \cdots, r$. Therefore $\{\pi(\overline{\delta_i}) \ 1 \leq i \leq r\}$ is a linearly independent set and hence a k-basis since we knew that it generates all the maximal paths. $\square$

LEMMA 14. *Let $\Lambda = kQ_\Lambda/I$ be a quasi-schurian weakly-symmetric algebra with I an admissible ideal. If Q_Λ has a loop then $\Lambda \simeq k[x]/ < x^2 >$.*

Proof. Let n be the number of vertices of Q_Λ. If $n = 1$ then Q_Λ has only one loop since Λ is quasi-schurian and I is an admissible ideal. Therefore in this case $\Lambda \simeq k[x]/ < x^2 >$ since Λ is quasi-schurian. Assume $n > 1$ and let α be a loop. As before we have only one loop at the vertex $o(\alpha)$. Let β be another arrow starting at $o(\alpha)$ and let δ be a supplement of β (see Theorem 3). Hence there exists $c \in k - \{0\}$ such that $\overline{\alpha} = c\overline{\delta\beta}$ since Λ is quasi-schurian, a contradiction because I is admissible. So, n has to be 1 and in this case we have already proved the lemma. $\square$

Before giving a proof of Theorem 11, let us recall from [3] (see also in [7],[8]) the description of the ordinary quiver and relations of the trivial extension $T(A)$ of a schurian algebra A.

Let $A = kQ_A/I$ be a schurian algebra with I an admissible ideal, and $p_1, p_2, \cdots, p_t$ be paths in Q_A such that $\{\overline{p_1}, \overline{p_2}, \cdots, \overline{p_t}\}$ is a k-basis of the vector space generated by all the maximal paths in A. Then the vertices of $Q_{T(A)}$ are the vertices of Q_A and $(Q_{T(A)})_1 = (Q_A)_1 \cup \{\beta_{p_1}, \beta_{p_2}, \cdots, \beta_{p_t}\}$, where β_{p_i} is an arrow starting at $e(p_i)$, ending at $o(p_i)$ and not belonging to Q_A for $i = 1, 2, \cdots, t$. We observe that all arrows of $Q_{T(A)}$ are in oriented cycles. An oriented cycle C in $Q_{T(A)}$ is called elementary if there exists a vertex j in C with $C(j) = q\beta_{p_i}$ for some $i = 1, 2, \cdots, t$ and some path q maximal in A with the same starting and ending vertices as p_i. We can describe now the relations of $T(A)$ given in [3].

THEOREM 15. *([3]) Let $A = kQ_A/I$ be a schurian algebra with I an admissible ideal. Let $I_{T(A)}$ be the ideal of $kQ_{T(A)}$ generated by the following relations:*

i) The composition of $n+1$ arrows in an elementary oriented cycle of lenght n.

ii) The composition of arrows not belonging to a same elementary oriented cycle.

iii) The elements $q - bq'$ where q, q' are paths in $Q_{T(A)}$ having the same ending and starting vertices, and such that one of the following conditions holds.

a) $\overline{q} = b\overline{q'}$ with $b \in k - \{0\}$ and q, q' paths in Q_A.

b) There is a path ν in Q_A such that $\nu q = \alpha_{r-1} \cdots \alpha_2 \alpha_1 \beta_{p_i} \alpha_n \cdots \alpha_{r+1} \alpha_r$ and $\nu q' = \alpha'_{s-1} \cdots \alpha'_2 \alpha_1 \beta_{p_j} \alpha'_m \cdots \alpha'_{s+1} \alpha'_s$ are elementary cycles. Then b is defined by $b = a_1/a_2$ for non zero $a_1, a_2 \in k$ with $\overline{\alpha_n \cdots \alpha_1} = a_1 \overline{p_i}$ and $\overline{\alpha'_m \cdots \alpha'_1} = a_2 \overline{p_j}$.

Then the ideal $I_{T(A)}$ is admissible and $T(A) \simeq kQ_{T(A)}/I_{T(A)}$.

Now recall the well known isomorphism of $T(A) - T(A)$ bimodules $\psi : T(A) \to D(T(A))$ given by $\psi(x_1, f_1)(x_2, f_2) = f_1(x_2) + f_2(x_1)$, and consider the linear map $F = \psi(1, 0) : T(A) \to k$. This functional is handy to describe the constansts of *iii)* in the above theorem. We start by giving an explicit description of it. Let $\{\overline{p_1}, \overline{p_2}, \cdots, \overline{p_t}, \overline{p_{t+1}}, \overline{p_{t+2}}, \cdots, \overline{p_s}\}$ be a basis of A, extending the chosen basis $\{\overline{p_1}, \overline{p_2}, \cdots, \overline{p_t}\}$ of the vector space generated by all the maximal paths in A, we will denote by $\{\overline{p_1}*, \overline{p_2}*, \cdots, \overline{p_s}*\}$ the basis of $D(A)$ dual to $\{\overline{p_1}, \overline{p_2}, \cdots, \overline{p_s}\}$. Let $\phi : kQ_{T(A)} \to T(A)$ be the map defined by $\phi(\alpha) = (\overline{\alpha}, 0)$ for $\alpha \in (Q_A)_1$ and $\phi(\beta_{p_i}) = (0, \overline{p_i}*)$. It is not difficult to see that the induced linear functional $f : kQ_{T(A)} \to k$ defined by the composition $F\phi$ satisfies the following condition: f is constant and non zero on the support $\underline{C}$ of each minimal oriented cycle C non zero in $T(A)$ (see Lemma 7 and its remark). Moreover, for the elementary oriented cycle $q\beta_{p_i}$ we have that $\overline{q} = f(q\beta_{p_i})\overline{p_i}$. Therefore the condition *iii)* in Theorem 15 can be changed by

iii)' The elements $\delta_1 - f(\delta_1\nu)(f(\delta_2\nu))^{-1}\delta_2$, where δ_1, δ_2 are paths in $Q_{T(A)}$ having the same ending and starting vertices and such that there exists a path ν with $\delta_1\nu$ and $\delta_2\nu$ elementary oriented cycles.

This last condition will be used in the proof of Theorem 11.

Proof. of Theorem 11: Let $\Lambda = Q_\Lambda/I$ where I is an admissible ideal and denote by $\overline{\gamma}$ the congruence class $\gamma + I$ in Λ. If Q_Λ has a loop we get by Lemma 14 that $\Lambda/I_C \simeq k$ and hence $\Lambda \simeq T(\Lambda/I_C)$.

Assume that Q_Λ has no loops and let $\mathcal{C}(\Lambda) = \{\beta_1, \beta_2, \cdots, \beta_t\}$. Let $A = \Lambda/I_C$, then by Lemma 12 we have that A is schurian and its ordinary quiver is Q_C. Now, for each $i = 1, 2, \cdots, t$ we choose a path p_i in $Supp(\beta_i)$. So, By Lemma 13 the set $\{\pi(\overline{p_i}) \; 1 \leq i \leq t\}$ is a k-basis of the vector space generated by all the maximal paths in A where $\pi : \Lambda \to A = \Lambda/I_C$ is the canonical epimorphism. By Theorem 15 we have that $Q_{T(A)} = Q_A \cup \mathcal{C}(\Lambda) = Q_\Lambda$. Hence $\psi : kQ_{T(A)} \to \Lambda$ where $\psi(\gamma) = \overline{\gamma}$ is an epimorphism of k-algebras. So, we have to check that $Ker\psi = I_{T(A)}$ to obtain $T(A) \simeq \Lambda$. We will need the following Lemmas:

LEMMA (A): *Let C be an oriented minimal cycle in $Q_{T(A)}$. Then C is non zero in $T(A)$ if and only if it is non zero in Λ.*

Proof. ($\Rightarrow$) : Assume C is non zero in $T(A)$. Then by Theorem 15 there is an elementary oriented cycle $q\beta_i$ such that $C(j) = q\beta_i$ for some vertex j in C. Therefore

$C(j)$ is non zero in Λ and by Lemma 7, C is non zero in Λ.

($\Leftarrow$) : Suppose that C is non zero in Λ. Then by the definition of $\mathcal{C}(\Lambda)$ we get that C contains an arrow from $\mathcal{C}(\Lambda)$. So, C is an elementary oriented cycle and therefore it is non zero in $T(A)$. □

Let $\{\underline{\mathcal{C}}_1, \cdots \underline{\mathcal{C}}_m\}$ be the set of supports of the non zero minimal oriented cycles in Λ. Since Λ is symmetric we have by Lemma 7 (see also its remark) a linear functional $\varphi : \Lambda \to k$ non zero on the supports $\underline{\mathcal{C}}_i$ for all $i = 1, 2, \cdots, m$. Making a change of variables $\beta_i \mapsto a_i \beta_i$ with adequate $a_i \in k - \{0\}$ for all $i = 1, 2, \cdots, t$ we may assume that $\varphi(\overline{p_i \beta_i}) = 1$ for all $i = 1, 2, \cdots, t$. Now, this functional satisfies the following property.

LEMMA (B): *Let q be a path in Q_A such that $\pi(\overline{q}) = c_q \pi(\overline{p_i})$ with $c_q \in k - \{0\}$ and some $i = 1, 2, \cdots, t$. Then $c_q = \varphi(\overline{q \beta_i})$.*

Proof. Since $\overline{q} - c_q \overline{p_i} \in Ker\pi = I_C$ we get by Lemma 13 a) that $\overline{q} = c_q \overline{p_i}$. Therefore $c_q = \varphi(\overline{q \beta_i})$ since $\varphi(\overline{p_i \beta_i}) = 1$. □

REMARK: Let C be a minimal oriented cycle non zero in $T(A)$. Then this lemma give us that $\varphi(\overline{C}) = f(C)$ where $f : kQ_{T(A)} \to k$ is the linear functional defined above.

Let us prove that $Ker\psi \supseteq I_{T(A)}$: By Lemma(A) and Lemma 2 we obtain that the relations i) and ii) in Theorem 15 are zero in Λ. Using Lemma(B) and its remark we obtain that the above relations $iii)'$ are zero in Λ because it is quasi-schurian. So, $Ker\psi \supseteq I_{T(A)}$.

Finally, we will check that $Ker\psi \subseteq I_{T(A)}$: Let $\gamma \in Ker\psi$. Then γ is zero in A and therefore γ is zero in $T(A)$ since A is a sub algebra of $T(A)$. So, $\gamma \in I_{T(A)}$ because $kQ_{T(A)}/I_{T(A)} \simeq T(A)$. Hence $Ker\psi \subseteq I_{T(A)}$.

Now it is easy to obtain the main result of this section.

THEOREM 16. *Let Λ be basic connected finite dimensional k-algebra. The following statements are equivalent*

1) There exists a schurian basic triangular algebra Λ' such that $\Lambda \simeq T(\Lambda')$.

2) Λ is symmetric quasi-schurian, and we can choose a set $\mathcal{C}(\Lambda)$ such that the quiver Q_C has non oriented cycles.

If these conditions hold, then $\Lambda' \simeq \Lambda/I_C$ where I_C is the ideal generated by $\mathcal{C}(\Lambda)$ in Λ.

Proof. Follows from Theorems 11 and 15. □

Another application of Theorem 11 is the following result.

THEOREM 17. *Let Q be a quiver without oriented cycles, and Λ an iterated tilted algebra of type Q. If $\Gamma = T(\Lambda)$ is quasi-schurian then, for each choice $\mathcal{C}(\Gamma)$ as above and such that the quiver Q_C has non oriented cycles, Γ/I_C is an iterated tilted algebra of type Q.*

Proof. It follows immediately from Theorem 11 and the next lemma. □

LEMMA 18. *Let Q be a quiver without oriented cycles, and Λ an iterated tilted algebra of type Q. Let Λ' be a basic finite dimensional k-algebra.*

If $T(\Lambda) \simeq T(\Lambda')$ and Λ' has finite global dimension then Λ' is iterated tilted of type Q.

Proof. The proof is based on known results about derived categories and repetitive algebras (see [2],[5] and [6]). Since $T(\Lambda) \simeq T(\Lambda')$ we get that the repetitive algebra $\hat{\Lambda}$ of Λ is isomorphic to the repetitive algebra $\hat{\Lambda}'$ of Λ'. In particular, we obtain that the triangulated category $\underline{mod}\hat{\Lambda}$ is triangle equivalent to $\underline{mod}\hat{\Lambda}'$. Since Λ and Λ' have finite global dimension we have the diagram

$$D^b(\Lambda) \xrightarrow{\sim} \underline{mod}\hat{\Lambda} \xrightarrow{\sim} \underline{mod}\hat{\Lambda}' \xleftarrow{\sim} D^b(\Lambda')$$

where $\xrightarrow{\sim}$ denotes a triangle equivalence. Thus $D^b(\Lambda)$ is triangle equivalent to $D^b(\Lambda')$, and therefore Λ' is an iterated tilted algebra of type Q (see [5] or [6]). □

We get now a useful approach to obtain iterated tilted algebras of a given tree class, generalizing an analogous result proven in [3] for Dynkin quivers.

COROLLARY 19. *Let Q be an oriented tree and $\Gamma = T(kQ)$. For each choice $\mathcal{C}(\Gamma)$ as above with $Q_\mathcal{C}$ without oriented cycles we have that $\Gamma/I_\mathcal{C}$ is an iterated tilted algebra of type Q.*

Proof. It follows immediately from Theorem 17. □

EXAMPLE: Let Q be the following oriented tree

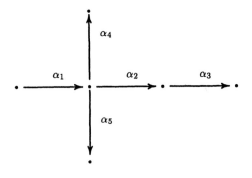

and $\Gamma = T(kQ)$ be the trivial extension of kQ. Considering the maximal paths $p_1 = \alpha_3\alpha_2\alpha_1$, $p_2 = \alpha_4\alpha_1$, and $p_3 = \alpha_5\alpha_1$ we obtain by Theorem 15 that Q_Γ is

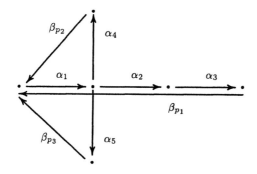

Let $\mathcal{C}(\Gamma) = \{\alpha_3, \alpha_4, \alpha_5\}$, and $\Lambda = \Gamma/I_\mathcal{C}$. By Lemma 12 we get that Q_Λ is

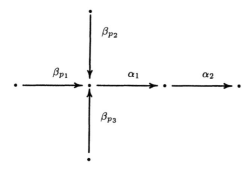

So, by Corollary 19 we have that $\Lambda \simeq kQ_\Lambda / <\alpha_2\alpha_1\beta_{p_2}, \alpha_2\alpha_1\beta_{p_3}>$ is an iterated tilted algebra of type Q.

ACKNOWLEDGMENTS

I would like to thank Prof. María Inés Platzeck for her many, very helpful comments and suggestions, and for a careful reading of this paper. Finally, I also thank the referee for the comments about the paper.

REFERENCES

[1] I.Assem,D.Happel,O.Roldán. "Representation-finite trivial extension algebras". J.Pure Appl. Algebra 33(1984). 235-242.

[2] D.Hughes,J.Wachbusch. "Trivial extensions of tilted algebras" Proc. London Math. Soc. 46(3)(1983) 347-364.

[3] E.Fernández. Ph.D. Thesis "Extensiones triviales y álgebras inclinadas iteradas", 1999.

[4] F.W.Anderson, K.R.Fuller. "Ring and categories of modules". Second edition. Springer-Verlag 1992.

[5] I. Assem, "Tilting Theory"- an introduction. Topics in algebra. Banach Center Publ., vol 26, part 1 (1990).

[6] D. Happel "Triangulated Categories in the representation Theory of finite dimensional algebras". Cambridge Unit Press (1988).

[7] E. Fernández, M.I. Platzeck "Presentations of trivial extensions and a theorem of S. Brenner". Preprint (2000).

[8] H. Asashiba "The derived Equivalence Classification of Representation-Finite Selfinjective Algebras". Journal of Algebra 214, 182-221 (1999).

On lattices at the ends of connected components of the Auslander-Reiten quiver

ALFREDO JONES Centro de Matemática, Facultad de Ciencias, Iguá 4225, Montevideo, Uruguay, E-mail: ajones@cmat.edu.uy

ABSTRACT Let R be a complete discrete valuation ring with radical $R\pi$ and residue field $R/\pi R$ of characteristic p dividing the order of a finite group G. We show that a virtually irreducible RG-lattice L with exponent π^a lies at the end of its Auslander Reiten component if and only if $L/\pi^{a-1}L$ is indecomposable.

1 INTRODUCTION

Let G be a finite group, p a prime that divides $|G|$, the order of G, and R a complete rank one valuation ring of characteristic zero with maximal ideal $R\pi$ such that p is the rational prime with $Rp \subseteq R\pi$. We consider RG-lattices, that is finitely generated RG-modules that are free as R-modules. In [2] it was shown that an absolutely irreducible RG-lattice which is indecomposable module π lies at the end of a connected component of the stable Auslander Reiten quiver. In this note we extend this result giving a necessary and sufficient condition for a virtually irreducible lattice to lie at the end of its component of the stable Auslander Reiten quiver. Absolutely irreducible lattices, as well as absolutely indecomposable lattices L with rank, rk L, prime to p, are special cases of virtually irreducible lattices. If RG is of infinite type there exist infinitely many indecomposables of rank prime to p, therefore if these are absolutely indecomposable, there exist infinitely many virtually irreducible lattices.

2 DEFINITIONS AND NOTATIONS

For any indecomposable non projective RG-lattice L, the almost split sequence $\mathcal{A}(L)$ of L is of the form

$$\mathcal{A}(L): 0 \longrightarrow \Omega L \longrightarrow \mathrm{M}(L) \longrightarrow L \longrightarrow 0$$

where Ω is the Heller operator. The lattice L lies at the end of its component of the stable Auslander Reiten quiver when the projective free part of $\mathrm{M}(L)$ is indecomposable. Set

$$\underline{\mathrm{Hom}}(L,L) = \frac{\mathrm{Hom}_{RG}(L,L)}{\mathrm{Proj}\,\mathrm{Hom}_{RG}(L,L)}$$

where $\mathrm{Proj}\,\mathrm{Hom}_{RG}(L,L)$ is the submodule of those homorphisms which factor through a projective lattice. Then, as shown in [4], $\underline{\mathrm{Hom}}(L,L)$ has a simple socle as a module over itself and $\mathcal{A}(L)$ is a pull-back of a projective cover of L along a generator of this socle.

The exponent of a lattice L, as defined in [1], is $\exp L = \pi^a$ if $R\pi^a$ is the annihilator of the torsion module $\underline{\mathrm{Hom}}(L,L)$, thus $\exp L$ is the least power of π such that multiplication of L by $\exp L$ factors through a projective lattice. If $R|G| = R\pi^n$ then $0 \leq a \leq n$, where $a = 0$ for L projective.

A lattice L with $\exp L = \pi^a$ is virtually irreducible if it is absolutely indecomposable and

$$\mathrm{soc}\,\underline{\mathrm{Hom}}(L,L) = \underline{\mathrm{Hom}}(L,L)\pi^{a-1}$$

This condition is equivalent to the following: if $\exp \mathrm{M}(L) = \pi^b$ then $b < a$. We remark that if L is virtually irreducible then $\exp L$ is a power of π because then $R(\exp L)(\mathrm{rk}\,L) = R|G|$. For these and other results on virtually irreducible lattices we refer to [1], [4] and [5].

3 THE THEOREM

Let v denote the p-adic valuation of R.

LEMMA. *If L is virtually irreducible then for every lattice X which is a direct summand of $\mathrm{M}(L)$, $v(\mathrm{rk}\,X) > v(\mathrm{rk}\,L)$.*

Proof. Let $\exp L = \pi^a$ and $R|G| = \pi^n$, then since L is virtually irreducible,

$$v(\mathrm{rk}\,L) = v(|G|) - v(\exp L) = n - a$$

On the other hand, if Tr is the trace function, from [1, Proposition 4.2] we know that for any RG-lattice X

$$R(\exp X)\,\mathrm{Tr}\big(\mathrm{Hom}_{RG}(X,X)\big) = R|G|$$

Thus if $\exp X = \pi^b$

$$R(\mathrm{rk}\,X) \subseteq \mathrm{Tr}\big(\mathrm{Hom}_{RG}(X,X)\big) = R\pi^{n-b}$$

Therefore $v(\mathrm{rk}\,X) \geq n - b$, but $b < a$. $\square$

THEOREM. *If L is a virtually irreducible lattice with $\exp L = \pi^a$, $a \geq 2$, then the projective free part of $\mathrm{M}(L)$ is indecomposable if and only if $\overline{L} = L/\pi^{a-1}L$ is indecomposable.*

Proof. The condition is clearly necessary because since $\pi^{a-1}\,\mathrm{id}_L \in \mathrm{soc}\,\underline{\mathrm{Hom}}(L,L)$, $\mathrm{M}(L)$ is the kernel of a projective cover of the module $\overline{L}$ ([1, Theorem 2.4]).

Assume now $\overline{L}$ indecomposable. As is well known the Auslander Reiten sequence of L decomposes module π^{a-1}, so $\overline{\mathrm{M}(L)} \cong \overline{L} \oplus \Omega\overline{L}$. Therefore if $\mathrm{M}(L)$ decomposes it must have an indecomposable direct summmand X such that $\overline{X} \cong \overline{L}$. But then $\mathrm{rk}\,X = \mathrm{rk}\,L$, and this is a contradiction with the lemma. □

COROLARY. *If a virtually irreducible lattice L is indecomposable module π then the projective free part of $\mathrm{M}(L)$ is indecomposable.*

Proof. It suffices to remark that if $\exp L = \pi$ then the projective cover of L gives the Auslander Reiten sequence for L so the result also holds in this case. □

REFERENCES

[1] J. F. Carlson, A. Jones. *An exponential property of lattices over group rings.* J. London Math. Soc. **39** (1989), 467–479.

[2] A. Jones, S. Kawata, G. O. Michler. *On exponents and Auslander Reiten components of irreducible lattices.* Archiv der Mathematik, to appear.

[3] R. Knörr. *Virtually irreducible lattices.* Proc. London Math. Soc. **59** (1989), 99–132.

[4] K. W. Roggenkamp. *The construction of almost split sequences for integral group rings and orders.* Comm. Algebra **5** (1977), 1363–1373.

[5] J.Thévenaz. *Duality in G-algebras.* Math. Z. **200** (1988), 47–85.

Factorisations of morphisms for wild hereditary algebras

OTTO KERNER Mathematisches Institut, Heinrich-Heine-Universität, Universitätsstraße 1, D-40225 Düsseldorf, Germany, E-mail: kerner@mx.cs.uni-duesseldorf.de

ABSTRACT Let H be a connected wild hereditary path-algebra. It will be shown that morphisms between modules, contained in the different parts of the module category H-mod have strong factorisation properties.

If A is a finite dimensional connected tame hereditary algebra, X a preprojective, respectively Y a preinjective, module and $\mathcal{T}$ a regular component, that is a regular tube in the Auslander-Reiten quiver $\Gamma(A)$ of A, then each homomorphism $f: X \to Y$ factorises through add $\mathcal{T}$, the additive closure of $\mathcal{T}$. It is the aim of the paper, to show that much stronger factorisation properties hold, if H is wild hereditary.

Let $H = k\mathcal{Q}$ be a connected wild hereditary path-algebra, over some field k. This means that $\mathcal{Q}$ is a finite connected quiver without oriented cycles which is neither of Dynkin nor of Euclidean type. Since H is hereditary the Auslander-Reiten translations $\tau = \tau_H \cong \operatorname{D}\operatorname{Ext}^1_H(-, H)$, respectively $\tau^- = \tau_H^- \cong \operatorname{Ext}^1_H(\operatorname{D}-, H)$, are left exact, respectively right exact, functors, where $\operatorname{D} = \operatorname{Hom}_k(-, k)$. Recall that an H-module X is called *preprojective*, respectively *preinjective*, if $\tau^m X$, respectively $\tau^{-m} X$, is zero for $m \gg 0$. A module X is called *regular*, if $\tau^m(\tau^{-m} X) \cong X$, for all integers m.

We say that a morphism $f: X \to Y$ *factorises* through a module M, if there exist morphisms $f_1: X \to M$ and $f_2: M \to Y$ such that $f = f_1 f_2$. The main result of the paper is:

THEOREM. *Let $H = k\mathcal{Q}$ be a finite dimensional connected wild hereditary algebra, X_1 a preprojective, X_2 a regular and X_3 a preinjective module. If $R \neq 0$ is regular, then one has.*
 (a) *Each homomorphism $f: X_1 \to X_2$ factorises through $\tau^{-m} R$ for $m \gg 0$.*
 (b) *Each homomorphism $g: X_2 \to X_3$ factorises through $\tau^m R$ for $m \gg 0$.*

(c) *Each homomorphism $h : X_1 \to X_3$ factorises through $\tau^m R$ for $|m| \gg 0$.*

For the proof of the theorem, a result of Lukas [9] is essential. It says that for any two nonzero regular H-modules X and Y there exist monos $X \to \tau^m Y$, respectively epis $\tau^{-m} X \to Y$, for $m \gg 0$. F. Lukas used infinite dimensional H-modules for his proof. A proof of this result, without infinite dimensional modules, was sketched in [8, 6.5]. For the convenience of the reader, this proof will be presented in section 1.

It was shown in [8, 6.4] that for any two nonzero preprojective, respectively preinjective, modules X and Y, there exist monos $X \to \tau^{-m} Y$, respectively epis $\tau^m X \to Y$, for $m \gg 0$. These results on the existence of monos, respectively epis, can be extended to the case, that the modules X and Y are in essentially different parts of the category H-mod of finite dimensional left H-modules.

COROLLARY. *Let $X_1 \neq 0$ be a preprojective, $X_2 \neq 0$ a regular and $X_3 \neq 0$ a preinjective module. Then one has.*
  (a) *There exists a mono $X_1 \to \tau^m X_2$, for $|m| \gg 0$.*
  (b) *There exist monos $X_i \to \tau^m X_3$, $(i = 1, 2)$ for $m \gg 0$.*
  (a') *There exists an epi $\tau^m X_2 \to X_3$, for $|m| \gg 0$.*
  (b') *There exist epis $\tau^{-m} X_1 \to X_i$, $(i = 2, 3)$ for $m \gg 0$.*

Since $H = kQ$ is a path-algebra, the category H-mod is equivalent to the category of finite dimensional k-linear representations of Q, and we will not distinguish between these categories. Morphisms will be written opposite to the scalars. For general results on the representation theory of finite dimensional algebras I refer to [1, 11], for standard results on wild hereditary algebras one may consult [5].

1 MONOS AND EPIS BETWEEN REGULAR MODULES

1.1 Let H be a connected wild hereditary algebra and X, Y be nonzero regular modules. It was shown by Baer [2] that $\mathrm{Hom}_H(X, \tau^m Y) \neq 0$ for $m \gg 0$. On the other hand $\mathrm{Hom}_H(\tau^m X, Y) = 0$ for $m \gg 0$ [3].

Denote by $\langle -, - \rangle : \mathbb{Z}^n \times \mathbb{Z}^n \to \mathbb{Z}$ the homological bilinear form, see [11]. Then we have $\langle \underline{\dim} X, \underline{\dim} \tau^m Y \rangle = \dim \mathrm{Hom}_H(X, \tau^m Y) - \dim \mathrm{Ext}_H^1(X, \tau^m Y)$. Consequently, for $m \gg 0$, we get $\langle \underline{\dim} X, \underline{\dim} \tau^m Y \rangle = \dim \mathrm{Hom}_H(X, \tau^m Y)$, since $\mathrm{Ext}_H^1(X, \tau^m Y) \cong \mathrm{D}\,\mathrm{Hom}_H(\tau^m Y, \tau X) = 0$ for $m \gg 0$. It follows from the spectral properties of the Coxeter transformation, that $\langle \underline{\dim} X, \underline{\dim} \tau^m Y \rangle$ grows exponentially in m, see for example [10]. This implies the well known

LEMMA. $\dim \mathrm{Hom}_H(X, \tau^m Y) \gg 0$ *for $m \gg 0$.*

1.2 The main result of this section is the following.

PROPOSITION. *Let $H = kQ$ be connected wild hereditary and let X, Y be nonzero regular H-modules. Then there exists a mono $f : X \to \tau^m Y$, respectively an epi $g : \tau^{-m} X \to Y$, for $m \gg 0$.*

This result first was shown by Lukas [9, 2.3] using infinite dimensional H-modules. The proof given here has already been indicated in [8]. By duality it is enough to show the first part. The proof is based on the following two lemmas.

1.3 LEMMA. *Let X, Y be nonzero regular H-modules. Then X is cogenerated by $\tau^m Y$, respectively generated by $\tau^{-m} Y$, for $m \gg 0$.*

For a proof see [9, 2.2] or [5, 10.7].

1.4 Call an indecomposable regular H-module E *additively elementary*, respectively *elementary*, if each short exact sequence $0 \to U \to E^r \to V \to 0$ with U, V regular and $r \geq 1$, respectively $r = 1$, splits, see [7, 6]. Since the Auslander-Reiten translation τ defines an equivalence on the category H-reg of regular H-modules, E is (additively) elementary, if and only if so is $\tau^m E$, for any integer m. Call a linear map $f : X \to Y$ *right minimal*, if no indecomposable direct summand of X is in the kernel $\operatorname{Ker} f$ of f, see [1].

LEMMA. *For an indecomposable regular H-module E there are equivalent.*

(a) *E is additively elementary.*

(b) *Let R be regular and $f : E^r \to R$ right minimal. Then $\operatorname{Ker} f$ is preprojective.*

(c) *Let R be regular and $f : E^r \to R$ right minimal. Then $\tau^m f : \tau^m E^r \to \tau^m R$ is injective for $m \gg 0$.*

Proof. The implications (c)$\Rightarrow$(b)$\Rightarrow$(a) are clear, the implication (a)$\Rightarrow$(c) follows from [7, 1.2]. $\square$

1.5 If H is connected wild hereditary and M is indecomposable preprojective or preinjective, then $\operatorname{Ext}_H^1(M, M) = 0$ and $\operatorname{Hom}_H(M, M) = k$, hence $q_H(\underline{\dim} M) = \langle \underline{\dim} M, \underline{\dim} M \rangle = 1$. Consequently an indecomposable module X with $q_H(\underline{\dim} X) < 1$ is regular.

If $H = kQ$ and Q has two vertices $1, 2$ and $r \geq 3$ arrows $\alpha_1, \ldots, \alpha_r$ from 1 to 2, then $q_H((x, y)) = (x - y)^2 - (r - 2)xy$. Let for example E be the indecomposable representation with $\underline{\dim} E = (1, 1)$ and linear maps $E(\alpha_1) = id : k \to k$ and $E(\alpha_i) = 0$ for $i > 1$. Then E is regular, but in addition it is additively elementary. Indeed, let I be the ideal of H, generated by $\alpha_2, \ldots, \alpha_r$ and $\mathcal{K}$ be the full subcategory of H-mod, consisting of modules annihilated by I. Since $\mathcal{K}$ is closed under submodules and factors, contains E and is isomorphic to H'-mod, where H' is the path-algebra of the Dynkin quiver A_2, it immediately follows that E is additively elementary in H-mod. If E' is any indecomposable regular H-module with $\underline{\dim} E' = (1, 1)$, then there exists an automorphism σ of H, such that $E' = {}_\sigma E$. Consequently E' also is additively elementary.

1.6 The proof of 1.2 now will be given. It is done in two steps.

(a) For each connected wild hereditary algebra $H = kQ$ there exist additively elementary modules. This was shown in [4], if the quiver Q has $n > 2$ vertices. If Q has two vertices, the module E considered in 1.5 is additively elementary.

Take an additively elementary H-module E. Then there exists an integer m_0 such that for all integers m with $m \geq m_0$ the following hold:

(i) X is cogenerated by $\tau^m E$, see 1.3.

Let $g_m : X \to \tau^m E^r$ be a mono, for some $r \geq 1$.

(ii) If R is regular and $h : \tau^m E \to R$ is nonzero, then h is injective, see [7, 1.3] or 1.4.

(b) Take $s \geq 0$, such that $t = \dim \operatorname{Hom}_H(E, \tau^s Y) \geq r$, see 1.1, and let $f_1, \ldots f_t$ be a basis of $\operatorname{Hom}_H(E, \tau^s Y)$. Then $f' = (f_1, \ldots, f_t)^t : E^t \to \tau^s Y$ is right minimal. Choose $m_1 \geq m_0$, such that

(α) $\tau^{m_1} f' : \tau^{m_1} E^t \to \tau^{m_1+s} Y$ is injective, see 1.4.

(β) $\operatorname{Hom}_H(\tau^{m_0} E, \tau^{m_1} E) \neq 0$ [2].

Let $h : \tau^{m_0} E \to \tau^{m_1} E$ be nonzero, hence injective, by (ii). Then $h \oplus \cdots \oplus h : \tau^{m_0} E^r \to \tau^{m_1} E^r$ is injective, too. Since $r \leq t$, there exists a mono $e : \tau^{m_1} E^r \to \tau^{m_1} E^t$.

The map $g_{m_0}(h \oplus \cdots \oplus h) e(\tau^{m_1} f') : X \to \tau^{m_1+s} Y$ is injective. □

2 PROOF OF THE THEOREM

2.1 Besides proposition 1.2, the proof of the theorem, respectively the corollary, is based on the following lemma.

LEMMA. *If $0 \neq I$ is* preinjective, respectively $0 \neq P$ is preprojective, then

(a) $\tau^m I$ contains a regular nonzero submodule, respectively $\tau^{-m} P$ contains a regular nonzero factor module, for $m \gg 0$.

(b) There exists a short exact sequence $0 \to R_1 \to R_2 \to I \to 0$, respectively $0 \to P \to R_1 \to R_2 \to 0$, with R_i regular.

Proof. By duality, it suffices to show the first parts.

(a) Let $X \neq 0$ be a regular module. There exists an $m_0 \geq 0$, such that for all $m \geq m_0$ the following hold (see for example [5]):

(i) $\operatorname{Hom}_H(X, \tau^m I) \neq 0$.

(ii) $\dim \tau^m I > \dim X$.

(iii) If Y is indecomposable preinjective with $\dim Y < \dim X$, then $\operatorname{Hom}_H(Y, \tau^m I) = 0$.

Let $f : X \to \tau^m I$ be a nonzero morphism. f is not surjective, by (ii). Let R be the image of f. By (iii) R has no nonzero preinjective direct summand, hence it is regular.

(b) Consider first the case where the quiver $\mathcal{Q}$ has at least 3 vertices. Let T be a regular H-tilting module [12]. Then $\operatorname{Ext}_H^1(T, I) = 0$, so I is generated by T. Let $f_1, \ldots, f_r$ be a basis of $\operatorname{Hom}_H(T, I)$ and K be the kernel of the surjection $f = (f_1, \ldots, f_r)^t : T^r \to I$. Application of $\operatorname{Hom}_H(T, -)$ to the short exact sequence $0 \to K \to T^r \to I \to 0$ shows $\operatorname{Ext}_H^1(T, K) = 0$. Consequently K is generated by T, hence it has no nonzero preprojective direct summand. Since it also is a submodule of the regular module T^r, it is regular, too.

If $\mathcal{Q}$ has two vertices, let $S(2)$ be the simple projective module and $S(1)$ be the simple injective module. It is enough, to show the assertion for I indecomposable preinjective. Denote by $E(i)$ the injective hulls of $S(i)$.

Consider first the preinjective modules $\tau_H^i E(1)$: Let H' be the Kronecker-algebra and $S'(1)$ simple injective in H'-mod. Consider in H'-mod the nonsplit short exact sequence
$$\eta: \ 0 \to M \to E' \to S'(1) \to 0$$
where E' is the injective hull of the simple module $S'(2)$ in H'-mod and M is indecomposable regular with $\underline{\dim} M = (1,1)$. Since H' is a factor algebra of H, there exists a full exact embedding H'-mod $\to H$-mod and η can be considered as short exact sequence in H-mod, by this embedding. In H-mod the modules M and E' are regular, since $q_H(\underline{\dim} M) = 2 - r < 0$ and $q_H(\underline{\dim} E') = 5 - 2r < 0$, see 1.5. Application of τ_H^i then gives
$$\eta': \ 0 \to \tau_H^i M \to \tau_H^i E' \to \tau_H^i E(1) \to 0$$
and $\tau_H^i M$, respectively $\tau_H^i E'$, are regular.

Consider now $\tau_H^i E(2)$: If $f_m: \tau^m E(1) \to \tau^{m-1} E(2)$, for $m \geq 1$ is an irreducible map, then f_m is surjective with kernel K_m. It follows for example from [13] that K_m is a brick with $\dim \operatorname{Ext}_H^1(K_m, K_m) = r - 1$, hence it is indecomposable regular. Consider the following commutative and exact diagram in H-mod

$$\begin{array}{ccccccccc}
& & & & & & 0 & & \\
& & & & & & \downarrow & & \\
& & 0 & & & & K_m & & \\
& & \downarrow & & & & \downarrow & & \\
0 & \to & \tau_H^m M & \to & \tau_H^m E' & \to & \tau_H^m E(1) & \to & 0 \\
& & \downarrow & & \| & & \downarrow f_m & & \\
0 & \to & K^{(m)} & \to & \tau_H^m E' & \to & \tau_H^{m-1} E(2) & \to & 0 \\
& & \downarrow & & & & \downarrow & & \\
& & K_m & & & & 0 & & \\
& & \downarrow & & & & & & \\
& & 0 & & & & & &
\end{array}$$

Since the category of regular H-modules is closed under extensions, $K^{(m)}$ is regular.

Therefore the second row of the diagram, respectively η', show assertion (b), if Q has 2 vertices.

2.2 The proof of (a) now will be given, (b) follows from duality. The proof is divided into 3 steps.

(A) Let $P \neq 0$ be projective (possibly decomposable), X_2, respectively $R \neq 0$, be regular and $s \geq 0$. Then each homomorphism $f: \tau^{-s} P \to X_2$ factorises through $\tau^{-m} R$, for $m \gg 0$.

Indeed, consider $\tau^s f: P \to \tau^s X_2$. By 1.1 there exists an epi $g_2: \tau^{-m} R \to \tau^s X_2$ for $m \gg 0$. Since P is projective, $\tau^s f$ factorises through g_2, that is $\tau^s f = g_1 g_2$, where $g_1: P \to \tau^{-m} R$. Application of τ^{-s} gives $f = \tau^{-s}(\tau^s f) = (\tau^{-s} g_1)(\tau^{-s} g_2)$.

(B) Let $X_1 = \oplus_{i=1}^t \tau^{-s_i} P_i$ with $P_i \neq 0$ projective and let $f = (f_1, \ldots f_t)^t: X_1 \to X_2$, where $f_i: \tau^{-s_i} P_i \to X_2$. Then f factorises through $\tau^{-m} R^t$, for $m \gg 0$:

By (A) there exists an m_0, such that for all $m \geq m_0$ the maps f_i factorise through $\tau^{-m} R$. Let $f_i = g_1^i g_2^i$ be such a factorisation. Define $g_1 = g_1^1 \oplus \cdots \oplus g_1^t: X_1 \to \tau^{-m} R^t$ and $g_2 = (g_2^1, \ldots, g_2^t)^t: \tau^{-m} R^t \to X_2$. Then $f = g_1 g_2$ is a factorisation through $\tau^{-m} R^t$.

(C) By 1.2 there exists an epi $h : \tau^{-r}R \to R^t$ for some $r > 0$. Let K be the kernel of h. Then $K = K_0 \oplus K_1$, with K_0 preprojective and K_1 regular. Take an integer $m_0 \geq 0$, such that for all $m \geq m_0$ and all preprojective modules P, $\text{Ext}^1_H(X_1, \tau^{-m}P) = 0$.

Application of τ^{-m} to the short exact sequence $0 \to K \longrightarrow \tau^{-r}R \xrightarrow{h} R^t \to 0$ gives

$$0 \to \tau^{-m}K \longrightarrow \tau^{-m-r}R \xrightarrow{\tau^{-m}h} \tau^{-m}R^t \to 0$$

We get $\text{Ext}^1_H(X_1, \tau^{-m}K_1) = 0$, since $\tau^{-m}K_1$ is regular and $\text{Ext}^1_H(X_1, \tau^{-m}K_0) = 0$ for $m \geq m_0$, by the choice of m_0.

Therefore, the map $(X_1, \tau^{-m}h) : \text{Hom}_H(X_1, \tau^{-m-r}R) \to \text{Hom}_H(X_1, \tau^{-m}R^t)$ is surjective, for $m \geq m_0$. Combining (B) and (C) gives the assertion (a).

Finally, (c) will be shown. Using (a), we will prove that $h : X_1 \to X_3$ factorises through $\tau^{-m}R$ for $m \gg 0$. The other part of the statement (c) is dual.

By 2.1 there exists a short exact sequence $0 \to R_1 \longrightarrow R_2 \xrightarrow{g} X_3 \to 0$, with R_i regular. Since $\text{Ext}^1_H(X_1, R_1) = 0$, the map $(X_1, g) : \text{Hom}_H(X_1, R_2) \to \text{Hom}_H(X_1, X_3)$ is surjective, that is $h = fg$ with $f : X_1 \to R_2$. By (a) the map f factorises through $\tau^{-m}R$ for $m \gg 0$ and so does h. $\square$

2.3 REMARK. The proof tells a little bit more, than in the theorem was stated. For example, given X_2, X_3 and R as in the theorem. Let $s \leq t$ be nonnegative integers. Then there exists an integer $m(s,t)$ such that for any preprojective module $\oplus_{i=1}^t \tau^{-s_i} P_i$ with P_i projective and $s_1 < s_2 \cdots s_t \leq s$, each homomorphism $f : \oplus_{i=1}^t \tau^{-s_i} P_i \to X_i$ ($i = 2, 3$) factorises through $\tau^{-m}R$ for $m \geq m(s,t)$.

3 PROOF OF THE COROLLARY

(a) Let $0 \to X_1 \xrightarrow{f} R_1 \longrightarrow R_2 \to 0$ be a short exact sequence with R_i regular, see 2.1(b). By 1.2 there exists a mono $h : R_1 \to \tau^m X_2$, for $m \gg 0$, hence $fh : X_1 \to \tau^m X_2$ is injective. By the theorem, f factorises through $\tau^{-m} X_2$ for $m \gg 0$, that is $f = f_1 f_2$ with $f_1 : X_1 \to \tau^{-m} X_2$. Since f is injective, so is f_1.

(b) By 2.1(a), for $s \gg 0$ there exists a mono $e : R \to \tau^s X_3$, for some regular module $0 \neq R$.

By (a) there exists a mono $g_1 : X_1 \to \tau^m R$, for $m \gg 0$, by 1.2 there exists a mono $g_2 : X_2 \to \tau^m R$, for $m \gg 0$. Since τ is a left exact functor, $\tau^m e$ is injective, for $m \geq 0$. Consequently the maps $g_i(\tau^m e) : X_i \to \tau^{m+s} X_3$ are injective, for $m \gg 0$.

(a') and (b') are shown dually.

REFERENCES

[1] M. AUSLANDER, I. REITEN AND S. SMALØ. *Representation theory of artin algebras.* Cambridge Studies in Advanced Mathematics, 1994

[2] D. BAER. *Homological properties of wild hereditary algebras.* In: V. Dlab, P. Gabriel and G. Michler (eds.) Representation theory I, Springer Lect. Notes in Math. **1177** (1986), 1–12.

[3] O. KERNER. *Tilting wild algebras.* J. London Math. Soc. **39** (1989), 29–47.

[4] O. KERNER. *Elementary stones.* Comm. Algebra **22** (1994), 1797–1806.

[5] O. KERNER. *Representations of wild quivers.* In: R. Bautista, R. Martinez Villa and J. A. de la Peña (eds), Representation theory of algebras and related topics, CMS Conf. Proc. **19** (1996), 65–107.

[6] O. KERNER. *Minimal approximations, orbital elementary modules and orbit algebras of regular modules.* J. Algebra **217** (1999), 528–554

[7] O. KERNER AND F. LUKAS. *Elementary modules.* Math. Z. **223** (1996), 421–434.

[8] O. KERNER AND M. TAKANE. *Mono orbits, epi orbits and elementary vertices of representation infinite quivers.* Comm. Algebra **25** (1997), 51–77.

[9] F. LUKAS. *Infinite dimensional modules over wild hereditary algebras.* J. London Math. Soc. **44** (1991) 401–419.

[10] J. A. DE LA PEÑA AND M. TAKANE. *Spectral properties of Coxeter transformations.* Arch. Math. **55** (1990), 120–134.

[11] C. M. RINGEL. *Tame algebras and integral quadratic forms.* Lecture Notes in Math. **1099**, Springer, Berlin, 1984.

[12] C. M. RINGEL. *The regular components of the Auslander-Reiten quiver of a tilted algebra.* Chinese Ann. Math. Ser. B **9** (1988), 1–18.

[13] L. UNGER. *On wild tilted algebras which are squids* Arch. Math. **55** (1990), 542–550.

A note on concealed-canonical Artin algebras

DIRK KUSSIN Fachbereich 17 Mathematik, Universität Paderborn, D–33095 Paderborn Germany, Email: dirk@math.uni-paderborn.de

ZYGMUNT POGORZAŁY Faculty of Mathematics and Informatics, Nicholas Copernicus University, ul. Chopina 12/18, 87-100 Toruń Poland,
Email: zypo@mat.uni.torun.pl

ABSTRACT In this article some omnipresence condition is given which assures that a derived-canonical algebra is already concealed-canonical. The proof exploits the theory of coherent sheaves over exceptional curves.

1   INTRODUCTION

Throughout this article let k be an arbitrary field, and A be a finite dimensional k-algebra. We shall use the term module for a finitely generated right A-module. The category of (finitely generated right) A-modules is denoted by $\mathrm{mod}(A)$. Moreover, the derived category of bounded complexes of A-modules (see [4]) will be denoted by $\mathrm{D}^b(A)$. We call A *derived-canonical*, if there is a canonical algebra Λ (in the sense of Ringel/Crawley-Boevey [16]) such that $\mathrm{D}^b(A) \simeq \mathrm{D}^b(\Lambda)$ as triangulated categories. If moreover Λ is of tubular type, then we call A *derived-tubular*. Note that a derived-canonical algebra is connected since its derived category is. The Grothendieck group of $\mathrm{mod}(A)$ will be denoted by $\mathrm{K}_0(A)$, the Coxeter transformation on $\mathrm{K}_0(A)$ by Φ.

Recall from [16] that for a canonical algebra Λ the module category $\mathrm{mod}(\Lambda)$ is trisected into $\mathrm{mod}_+(\Lambda) \vee \mathrm{mod}_0(\Lambda) \vee \mathrm{mod}_-(\Lambda)$, where $\mathrm{mod}_0(\Lambda)$ is a stable separating tubular family, and there are no non-zero morphisms going from right to left. Recall from [11] that a k-algebra A is called *concealed-canonical* (*almost concealed-canonical*, resp.), if for some canonical algebra Λ there exists a tilting module lying in $\mathrm{mod}_+(\Lambda)$ (in $\mathrm{mod}_+(\Lambda) \vee \mathrm{mod}_0(\Lambda)$, resp.) and whose endomorphism algebra is isomorphic to A. If additionally Λ is of tubular type, then we call A a *tubular* algebra. Concealed-canonical algebras (in particular: tubular and canonical algebras) were studied by several authors (see for example [6, 9, 11, 13, 14, 16, 17], also [1, 2] and [5, 10, 15]).

It is well-known that the class of concealed-canonical algebras is not closed under derived equivalence. The aim of this note is to present a condition under which it follows that a derived-canonical algebra is concealed-canonical. The essential property will be the existence of some omnipresent indecomposable module. The notion of omnipresence was also successfully used in a similar context in [14, 17]. Recall that an A-module M is called *omnipresent*, if each simple A-module occurs as a composition factor of M. Moreover, an Auslander-Reiten component is called *regular*, if it contains neither a projective nor an injective module, and it is called *semi-regular*, if it does not contain at the same time a projective and an injective module.

The main result of this note is the following

THEOREM. *Let A be a finite dimensional k-algebra over a field k. Then the following conditions (1) and (2) are equivalent*

(1) (a) A is derived-canonical, and

   (b) there is an omnipresent indecomposable $M \in \mathrm{mod}(A)$, such that

   (i) the class $[M] \in \mathrm{K}_0(A)$ has finite Φ-period.

   (ii) M lies in some regular Auslander-Reiten component in $\mathrm{mod}(A)$.

(2) A is concealed-canonical.

REMARKS. (1) As the proof of the theorem will show, condition (b) can be replaced by the following condition:

(b') There is a (finite) family of indecomposables $M_i \in \mathrm{mod}(A)$ ($i \in I$) such that their direct sum is omnipresent, and such that all M_i ($i \in I$) lie in regular components in $\mathrm{mod}(A)$ and in the *same* tubular family in $\mathrm{D}^b(A)$.

(2) The *almost* concealed-canonical algebra A over an algebraically closed field, which is given as path algebra of the quiver

$$1 \underset{y}{\overset{x}{\rightrightarrows}} 2 \xrightarrow{z} 3$$

with relation $zx = 0$, shows, that in condition (ii) regularity cannot be replaced by semi-regularity. Namely, A can be realized as endomorphism algebra of a tilting sheaf over the weighted projective line of weight type $(1,2)$ (see [3]). The indecomposable projective A-module $P(3)$ is omnipresent, lying in a semi-regular tube of A.

If we restrict to the tubular case we have a stronger result.

COROLLARY. *Let A be a finite dimensional k-algebra over a field k. Then the following conditions are equivalent*

(1) A is derived-tubular, and there is an omnipresent indecomposable $M \in \mathrm{mod}(A)$ lying in some semi-regular Auslander-Reiten component in $\mathrm{mod}(A)$.

(2) A is tubular.

REMARK. (3) Let k be algebraically closed and A be the poset algebra given by the quiver

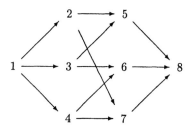

with all 6 possible commutativity relations. Then A is derived-canonical (of tubular type $(3,3,3)$), but not tubular (see [12]). The indecomposable projective injective A-module $P(8) = I(1)$ is omnipresent lying in a component in $\mathrm{mod}(A)$ which is not semi-regular. Thus, semi-regularity of the component in the corollary is indispensable.

Note, that in the theorem and in the corollary the implication (2) $\Longrightarrow$ (1) is trivial. In the proof of our result we shall use the coherent sheaves technique approach to the representation theory [3, 7]. This approach makes our proof rather simple. The following characterization of concealed-canonical algebras from [9] is of great importance for our proof: A is concealed-canonical if and only if there exists an exceptional curve $\mathbb{X}$ (see [7]) – that is, a weighted projective line if k is algebraically closed – and a torsion-free tilting object in the category $\mathrm{coh}(\mathbb{X})$ of coherent sheaves whose endomorphism algebra is isomorphic to A.

2 THE DERIVED CATEGORY OF A CANONICAL ALGEBRA

Let Λ be a canonical k-algebra over the field k (compare [16]). By [16] $\mathrm{mod}(\Lambda)$ contains a stable separating tubular family $\mathrm{mod}_0(\Lambda)$, which is a coproduct of uniserial connected length categories $\mathcal{U}_x$ (called stable tubes). By the construction of [9] there is a small k-category $\mathcal{H}$, which is abelian, hereditary (that is, $\mathrm{Ext}^i_{\mathcal{H}}(-,-) = 0$ for all $i \geq 2$), noetherian, locally-finite (that is, all Hom and Ext^1 spaces are of finite dimension over k), containing no non-zero projective object and admitting a torsion-free tilting object with endomorphism algebra isomorphic to Λ. Each indecomposable object in $\mathcal{H}$ is either in $\mathcal{H}_0$, the full subcategory of objects of finite length (so-called torsion objects), or in $\mathcal{H}_+$, the full subcategory formed by the torsion-free objects, which do not contain any non-zero torsion subobject. The relation $\mathrm{Hom}_{\mathcal{H}}(\mathcal{H}_0, \mathcal{H}_+) = 0$ holds. Moreover, $\mathcal{H}_0 = \mathrm{mod}_0(\Lambda)$.

There is an auto-equivalence $\tau : \mathcal{H} \longrightarrow \mathcal{H}$, called *Auslander-Reiten translation*, such that Serre duality holds naturally in $X, Y \in \mathcal{H}$:

$$\mathrm{Ext}^1_{\mathcal{H}}(X, Y) \simeq \mathrm{D}\,\mathrm{Hom}_{\mathcal{H}}(Y, \tau X),$$

where D denotes the duality $\mathrm{Hom}_k(-,k)$. Moreover, $\mathcal{H}$ admits almost split sequences, and for indecomposable end term X in such a sequence the starting term is given by τX (see [9, Thm. 6.1]).

The category $\mathcal{H}$ is also denoted by $\mathrm{coh}(\mathbb{X})$, and $\mathbb{X}$ equipped with $\mathrm{coh}(\mathbb{X})$ is called *exceptional curve* [7]. By tilting theory the categories $\mathrm{coh}(\mathbb{X})$ and $\mathrm{mod}(\Lambda)$ are derived-equivalent, $\mathrm{D}^b(\mathbb{X}) = \mathrm{D}^b(\Lambda)$, in particular also have isomorphic Grothendieck groups: $\mathrm{K}_0(\mathbb{X}) = \mathrm{K}_0(\Lambda)$. For each object X in $\mathcal{H}$ denote by $[X]$ the class in $\mathrm{K}_0(\mathbb{X})$. We then have $[\tau X] = \Phi[X]$. Since $\mathcal{H}$ is hereditary, we have

$$\mathcal{D} := \mathrm{D}^b(\mathbb{X}) = \mathrm{add}\left(\bigcup_{n \in \mathbb{Z}} \mathcal{H}[n]\right),$$

where the $\mathcal{H}[n]$ are (disjoint) copies of $\mathcal{H}$; for each $X \in \mathcal{H}$ the copy in $\mathcal{H}[n]$ is denoted by $X[n]$. Each indecomposable object in $\mathcal{D}$ is of the form $X[n]$ for some (indecomposable) $X \in \mathcal{H}$ and some $n \in \mathbb{Z}$. For all $X, Y \in \mathcal{H}$ and all $m, n \in \mathbb{Z}$ we have

$$\mathrm{Hom}_\mathcal{D}(X[m], Y[n]) = \mathrm{Ext}_\mathcal{H}^{n-m}(X, Y);$$

in particular, if $m > n$ or $n > m+1$, then $\mathrm{Hom}_\mathcal{D}(X[m], Y[n]) = 0$.

The Auslander-Reiten translation τ extends canonically to an auto-equivalence $\tau : \mathcal{D} \longrightarrow \mathcal{D}$ (which we denote by the same symbol).

3 PROOF OF THE RESULTS

Assume that condition (1) from the theorem holds, and that $\mathrm{D}^b(A) = \mathrm{D}^b(\Lambda)$, where Λ is canonical, and let $\mathbb{X}$ and $\mathcal{H}$ be as above. The proof has three steps:

First step: The omnipresent indecomposable $M \in \mathrm{mod}(A)$ lies in $\mathcal{H}_0[n]$ for some $n \in \mathbb{Z}$. Without loss of generality, we assume $n = 0$.

Second step: Realize A as (endomorphism algebra of) a tilting complex T in $\mathcal{D}$. By omnipresence, we immediately see that $T \in \mathcal{H}_0[-1] \cup \mathcal{H}$.

Third step: We have to show, that (using regularity) actually $T \in \mathcal{H}_+$, that is, A can be realized as (endomorphism algebra of) a torsion-free tilting object in A and hence is concealed-canonical (see [9]).

The *second* step is clear. For the *first*: We assume $M \in \mathcal{H}$. For non-tubular $\mathbb{X}$ and for non-zero $M \in \mathcal{H}_+$ it follows as in [8, Prop. 4.5], that $[M]$ has no finite Φ-period. Thus, $M \in \mathcal{H}_0$, and M lies in a stable tube $\mathcal{T}$ of finite rank. Observe, that in the tubular case, M lies in a stable tube in any case (since $\mathrm{ind}\,\mathcal{H}$ consists entirely of stable tubes, compare [6]), not necessarily in $\mathcal{H}_0$, but after a possible change of the chosen separating tubular family $\mathrm{mod}_0(\Lambda)$ (and thus changing $\mathcal{H}$, compare [6, Prop. 7]) we can assume $M \in \mathcal{H}_0$.

It remains to prove the *third* step. We assume more generally, that M lies in a semi-regular component $\mathcal{C}$ of A. Then $\mathcal{C}$ contains either no projective or no injective A-module.

Case 1. $\mathcal{C}$ contains no projective. Let P be an indecomposable direct summand of the tilting complex T, which is an indecomposable projective $A = \mathrm{End}(T)$-module. Assume that $P \in \mathcal{H}_0$. By omnipresence, $\mathrm{Hom}_A(P, M) \neq 0$, and by orthogonality of the stable tubes, P also lies in the tube $\mathcal{T}$. By assumption, P and M lie in different Auslander-Reiten components of A, therefore $\mathrm{Rad}_A^\infty(P, M) \neq 0$, and then also $\operatorname{Rad}_{\mathcal{D}}^\infty(P, M) \neq 0$, which gives a contradiction since P and M lie in the same stable tube $\mathcal{T}$, which is standard ([15]). Therefore, no indecomposable summand of T lies in $\mathcal{H}_0$, hence $T \in \mathcal{H}_0[-1] \cup \mathcal{H}_+$ and therefore A is dual to an almost concealed-canonical algebra.

Case 2. The component $\mathcal{C}$ contains no injective. Assume moreover, that there is an indecomposable projective A-module P lying in $\mathcal{H}_0[-1]$. Then consider the corresponding injective A-module $I = \tau P[1]$. By omnipresence, $\operatorname{Hom}_A(M, I) \neq 0$, and by proceeding as above we see that $T \in \mathcal{H}_+ \cup \mathcal{H}_0$, and thus A is almost concealed-canonical.

Now by [11], if $\mathcal{C}$ is regular, or if Λ is of tubular type, it follows, that A is concealed-canonical. This proves the theorem and the corollary.

ACKNOWLEDGEMENT

This note was written during a stay of the second named author at the University of Paderborn. He would like to express his gratitude to Helmut Lenzing for his hospitality. He also acknowledges a partial support of the Polish Scientific Grant KBN 2 PO3A 012 14. Both authors would like to thank Helmut Lenzing for stimulating discussions on the subject.

REFERENCES

[1] M. Barot, *Representation-finite derived tubular algebras*, Arch. Math. (Basel) **74** (2000), no. 2, 89–94.

[2] M. Barot and J. A. de la Peña, *Derived tubular strongly simply connected algebras*, Proc. Amer. Math. Soc. **127** (1999), no. 3, 647–655.

[3] W. Geigle and H. Lenzing, *A class of weighted projective curves arising in representation theory of finite dimensional algebras*, Singularities, Representation of Algebras and Vector Bundles (Lambrecht 1985) (Berlin-Heidelberg-New York), Lecture Notes in Math., vol. 1273, Springer-Verlag, 1987, pp. 265–297.

[4] D. Happel, *Triangulated categories in the representation theory of finite dimensional algebras*, London Math. Soc. Lecture Note Series, no. 119, Cambridge University Press, 1988.

[5] D. Happel and C. M. Ringel, *The derived category of a tubular algebra*, Representation Theory I. Finite Dimensional Algebras (Ottawa 1984) (Berlin-Heidelberg-New York), Lecture Notes in Math., vol. 1177, Springer-Verlag, 1986, pp. 156–180.

[6] D. Kussin, *Non-isomorphic derived-equivalent tubular curves and their associated tubular algebras*, J. Algebra **226** (2000), 436–450.

[7] H. Lenzing, *Representations of finite dimensional algebras and singularity theory*, Trends in ring theory. Proceedings of a conference at Miskolc, Hungary, July 15–20, 1996 (V. Dlab et al., ed.), CMS Conf. Proc., vol. 22, Amer. Math. Soc., Providence, R. I., 1998, pp. 71–97.

[8] H. Lenzing and J. A. de la Peña, *Wild canonical algebras*, Math. Z. **224** (1997), 403–425.

[9] _____ , *Concealed-canonical algebras and separating tubular families*, Proc. London Math. Soc. **78** (1999), no. 3, 513–540.

[10] H. Lenzing and H. Meltzer, *Sheaves on a weighted projective line of genus one, and representations of a tubular algebra*, Representations of Algebras (Ottawa 1992) (V. Dlab and H. Lenzing, eds.), CMS Conf. Proc., vol. 14, Amer. Math. Soc., Providence, R. I., 1993, pp. 313–337.

[11] _____ , *Tilting sheaves and concealed-canonical algebras*, Representation Theory of Algebras (Cocoyoc, 1994) (R. Bautista, R. Martínez-Villa, and J. A. de la Peña, eds.), CMS Conf. Proc., vol. 18, Amer. Math. Soc., Providence, R. I., 1996, pp. 455–473.

[12] H. Lenzing and I. Reiten, *Additive functions for quivers with relations*, Colloq. Math. **82** (1999), no. 1, 85–103.

[13] H. Meltzer, *Auslander-Reiten components for concealed-canonical algebras*, Colloq. Math. **71** (1996), no. 2, 183–202.

[14] I. Reiten and A. Skowroński, *Sincere stable tubes*, Preprint 99–011, Bielefeld, 1999.

[15] C. M. Ringel, *Tame algebras and integral quadratic forms*, Lecture Notes in Math., vol. 1099, Springer-Verlag, Berlin-Heidelberg-New York, 1984.

[16] _____ , *The canonical algebras*, Topics in Algebra, Banach Center Publ., no. 26, 1990, with an appendix by William Crawley-Boevey, pp. 407–432.

[17] A. Skowroński, *On omnipresent tubular families of modules*, Representation Theory of Algebras (Cocoyoc, 1994) (R. Bautista, R. Martínez-Villa, and J. A. de la Peña, eds.), CMS Conf. Proc., vol. 18, Amer. Math. Soc., Providence, R. I., 1996, pp. 641–657.

Koszul algebras and the Gorenstein condition

ROBERTO MARTINEZ-VILLA[1] Instituto de Matemáticas, Universidad Nacional Autónoma de México, México 04510, D.F., México. e-mail: mvilla@matem.unam.mx

ABSTRACT Non commutative versions of regular algebras appear naturally in representation theory as the Yoneda algebras of selfinjective Koszul algebras, they have been studied in [4], [10], [11]. Here we extend these notions by considering algebras such that some of the simple satisfy the Gorenstein condition [1], [9], [12]. When they are Koszul, we study the corresponding Yoneda algebras, examples of such algebras will be the Auslander algebra, the preprojective algebra and selfinjective algebras of radical cube zero of infinite representation type. We will prove that by taking tensor products we can construct new algebras satisfying the Gorenstein condition.

1 NOTATION AND KNOWN RESULTS

We will consider graded quiver algebras over an algebraically closed field K, this is: positively graded K-algebras $\Lambda = \bigoplus_{i \geq 0} \Lambda_i$ such that $\Lambda_0 = K \times K \cdots \times K$, where K is a field and for all i we have $\dim_K \Lambda_i < \infty$ and for all i, j there is an equality $\Lambda_i \Lambda_j = \Lambda_{i+j}$.

We know [6] such algebras are isomorphic to algebras of the form KQ/I, where Q is a finite quiver and I is an homogeneous ideal of KQ in the grading given by path length and I is contained in J^2 where J is the ideal generated by the arrows.

Given a Z-graded module $M = \{M_i\}_{i \in Z}$ we denote by $M[n]$ the nth-shift defined by $M[n]_i = M_{n+i}$.

We consider the category $l.f.\text{mod}_\Lambda$ of locally finite Z-graded modules $M = \{M_i\}_{i \in Z}$, such that $\dim_K M_i < \infty$.

---

[1] Part of this paper was written during my visit to Universidad de Murcia on December 1998 and part during my visit to Northeastern University on July 1999. I thank both Manolo Saorin and Alex Martzinkovsky for their kind hospitality, for exchanging ideas and for their encouragement, to the mentioned universities for funding.

We know by [6], there exists a duality $D : l.f.\text{mod}_\Lambda \to l.f.\text{mod}_{\Lambda^{op}}$ given by $D(M)_j = Hom_K(M_{-j}, K)$.

The category of graded Λ-modules and degree zero maps, $Hom_\Lambda(M,N)_0$, will be denoted $Gr\,Mod_\Lambda$, and by Mod_Λ the category of graded modules and maps $Hom_\Lambda(M,N) = \bigoplus_{i \in Z} Hom_\Lambda(M,N)_i$, with $Hom_\Lambda(M,N)_n$ the degree n maps.

We have isomorphisms: $Hom_\Lambda(M[-n], N)_0 \cong Hom_\Lambda(M, N[n])_0 \cong Hom_\Lambda(M,N)_n$.

In a similar way the k extensions of degree zero, $Ext^k_\Lambda(-,?)_0$ are defined as the derived functors of $Hom_\Lambda(-,?)_0$. We define $Ext^k_\Lambda(M,N)_n = Ext^k_\Lambda(M, N[n])_0$ and $Ext^k_\Lambda(M,N)$ is the graded vector space: $Ext^k_\Lambda(M,N) = \bigoplus_{n \geq 0} Ext^k_\Lambda(M,N)_n$.

We recall the following definitions and results concerning Koszul algebras [3], [6], [7]:

DEFINITION. Let $\Lambda = KQ/I$ be a graded quiver algebra. We say that a graded Λ-module M is Koszul if M has a graded projective resolution: $\cdots \to P_n[-n] \to P_{n-1}[-n+1] \to \cdots P_1[-1] \to P_0[0] \to M \to 0$, with each $P_j[-j]$ finitely generated with all generators in degree j.

We say that Λ is Koszul if all graded simple are Koszul.

THEOREM 1.1. *Let $\Lambda = KQ/I$ be a Koszul algebra. Then the following statements are true:*

i) *The algebra Λ is quadratic, this is: I is generated by linear combinations of paths of length 2.*

ii) *Let $V_2 = (KQ)_2$ be the vector space generated by all paths of length 2 and $\langle,\rangle : V_2 \times V_2 \to K$ the bilinear form defined by $\langle \alpha\beta, \beta'\alpha' \rangle = 1$ if $\alpha = \alpha'$ and $\beta = \beta'$ and 0 otherwise. Let L_2 be the orthogonal $I_2^\perp$ of the vector space $I_2 = I \cap (KQ)_2$.*

Denote by KQ^{op} the quiver algebra of the opposite quiver and by L the ideal generated by L_2.

Then the Yoneda algebra $\Gamma = \bigoplus_{k \geq 0} Ext^k_\Lambda(\Lambda_0, \Lambda_0)$ is Koszul and $\Gamma \cong KQ^{op}/L$.

iii) *Let K_Λ and $K_{\Gamma^{op}}$ be the full subcategories of $Gr\,Mod_\Lambda$ and $Gr\,Mod_{\Gamma^{op}}$ consisting of Koszul Λ and Γ^{op}-modules, respectively. Then the functor $F(M) = \bigoplus_{k \geq 0} Ext^k_\Lambda(M, \Lambda_0)$ is a duality from K_Λ to $K_{\Gamma^{op}}$ satisfying $F(J^k M) = \Omega^k F(M)$, where Ω^k denotes the kth syzygy.*

2 KOSZUL DUALITY AND EXTENSIONS

All the algebras in this section will be Koszul, we will study the relations between the extension groups of two Koszul modules and the corresponding extension groups under Koszul duality. The main result is contained in the following:

THEOREM 2.1. *Let $\Lambda = KQ/I$ be a Koszul algebra and M and N two Koszul modules. Then for any pair of integers k and l, with $k \geq 0$, the following two statements are true:*

Koszul Algebras and the Gorenstein Condition

i) If $Ext_\Lambda^k(M, N[l])_0 \neq 0$, then $k \geq -l$.

ii) If $k \geq -l$, then there exists a vector space isomorphism:

$$Ext_\Lambda^k(M, N[l])_0 \cong Ext_{\Gamma^{op}}^{k+l}(F(N)[l], F(M))_0.$$

Proof. i) Assume $Ext_\Lambda^k(M, N[l])_0 \neq 0$.
We have an exact sequence:

$$0 \to \Omega^k(M) \to P_{k-1}[-k+1] \to \Omega^{k-1}(M) \to 0$$

with $P_{k-1}[-k+1]$ the projective cover of $\Omega^{k-1}(M)$.
Then we have an exact sequence:

$$0 \to Hom_\Lambda(\Omega^{k-1}(M), N[l])_0 \to Hom_\Lambda(P_{k-1}[-k+1], N[l])_0 \to$$
$$\to Hom_\Lambda(\Omega^k(M), N[l])_0 \to Ext_\Lambda^k(M, N[l])_0 \to 0$$

Since $Ext_\Lambda^k(M, N[l])_0 \neq 0$, also $Hom_\Lambda(\Omega^k(M), N[l])_0 \neq 0$.
The module $\Omega^k(M)$ is generated in degree k and $N[l]$ in degree $-l$. It follows $k \geq -l$.

ii) We consider first the case $k = 0$.
There is a short exact sequence: $0 \to J^l N[l] \to N[l] \to N[l]/J^l N[l] \to 0$.
It induces an exact sequence:

$$0 \to Hom_\Lambda(M, J^l N[l])_0 \to Hom_\Lambda(M, N[l])_0 \to Hom_\Lambda(M, N[l]/J^l N[l])_0$$

Since $Hom_\Lambda(M, N[l]/J^l N[l])_0 = 0$ the vector spaces $Hom_\Lambda(M, J^l N[l])_0$ and $Hom_\Lambda(M, N[l])_0$ are isomorphic.
The modules M and $J^l N[l]$ are both generated in degree zero and Koszul. It follows there exists isomorphisms:

* $\quad Hom_\Lambda(M, N[l])_0 \cong Hom_\Lambda(M, J^l N[l])_0 \cong Hom_{\Gamma^{op}}(\Omega^l F(N)[l], F(M))_0.$

As before, we have an exact sequence: $0 \to \Omega^l F(N)[l] \to P_{l-1}[1] \to \Omega^{l-1} F(N)[l] \to 0$, with $P_{l-1}[1]$ the projective cover of $\Omega^{l-1} F(N)[l]$.
Hence there exists an exact sequence:

$$0 \to Hom_{\Gamma^{op}}(\Omega^{l-1} F(N[l]), F(M))_0 \to Hom_{\Gamma^{op}}(P_{l-1}[1], F(M))_0 \to$$
$$\to Hom_{\Gamma^{op}}(\Omega^l F(N)[l], F(M))_0 \to Ext_{\Gamma^{op}}^l(F(N[l]), F(M))_0 \to 0$$

Since $P_{l-1}[1]$ is generated in degree -1 and $F(M)$ in degree zero the maps: $Hom_{\Gamma^{op}}(P_{l-1}[1], F(M))_0 = 0$ and there exists an isomorphism:

** $\quad Hom_{\Gamma^{op}}(\Omega^l F(N)[l], F(M))_0 \cong Ext_{\Gamma^{op}}^l(F(N)[l], F(M))_0,$

by using * and ** we get the result for $k = 0$.
Consider the case $k \geq 1$ and $k = -l$.
The exact sequence $0 \to \Omega^k M \to P_{k-1}[-k+1] \to \Omega^{k-1} M \to 0$ induces an exact sequence:

$$0 \to Hom_\Lambda(\Omega^{k-1} M, N[l])_0 \to Hom_\Lambda(P_{k-1}[-k+1], N[l])_0 \to$$

$$\to Hom_\Lambda(\Omega^k M, N[l])_0 \to Ext^k_\Lambda(M, N[l])_0 \to 0$$

The module $P_{k-1}[-k+1]$ is projective generated in degree $k-1$ and $N[l]$ is generated in degree $k = -l$. As above, $Hom_\Lambda(P_{k-1}[-k+1], N[l])_0 = 0$ and $Hom_\Lambda(\Omega^k M, N[l])_0 \cong Ext^k_\Lambda(M, N[-k])_0$. Since both $\Omega^k M$ and $N[-k]$ generated in degree k the duality F induces an isomorphism: $Hom_\Lambda(\Omega^k M, N[-k])_0 \cong Hom_{\Gamma^{op}}(F(N)[-k], J^k F(M))_0$.

The exact sequence: $0 \to J^k F(M) \to F(M) \to F(M)/J^k F(M) \to 0$ induces an exact sequence:

$$0 \to Hom_{\Gamma^{op}}(F(N)[-k], J^k F(M))_0 \to Hom_{\Gamma^{op}}(F(N)[-k], F(M))_0 \to$$

$$\to Hom_{\Gamma^{op}}(F(N)[-k], F(M)/J^k F(M))_0.$$

It is clear $Hom_{\Gamma^{op}}(F(N)[-k], F(M)/J^k F(M))_0 = 0$.
Then it follows $Ext^k_\Lambda(M, N[-k])_0 \cong Hom_{\Gamma^{op}}(F(N)[-k], F(M))_0$.
It remains to consider the case $k \geq 1$ and $k > -l$.
Assume $k = 1$ and $l \geq 0$.
The exact sequence: $0 \to J^l N[l] \to N[l] \to N[l]/J^l N[l] \to 0$ induces an exact sequence: $0 \to Hom_\Lambda(M, J^l N[l])_0 \to Hom_\Lambda(M, N[l])_0 \to Hom_\Lambda(M, N[l]/J^l N[l])_0$.
Since $Hom_\Lambda(M, N[l]/J^l N[l])_0 = 0$ there is an isomorphism:
$*\quad Hom_\Lambda(M, J^l N[l])_0 \cong Hom_\Lambda(M, N[l])_0$.
Applying the duality F we obtain an isomorphism:
$Hom_\Lambda(M, J^l N[l])_0 \cong Hom_{\Gamma^{op}}(\Omega^l F(N)[l], F(M))_0$.
There is an isomorphism:
$**\quad Hom_{\Gamma^{op}}(\Omega^l F(N)[l], F(M))_0 \cong Ext^1_{\Gamma^{op}}(F(N)[l], F(M))_0$.
Using $*$ and $**$ we obtain the result for $k = 1$.
Assume $k > 1$ and $k > -l$.
The exact sequence: $0 \to J^l N[l] \to N[l] \to N[l]/J^l N[l] \to 0$ induces an exact sequence:

$$\to Ext^{k-1}_\Lambda(M, N[l]/J^{l+k-1} N[l])_0 \to Ext^k_\Lambda(M, J^{l+k-1} N[l])_0 \to$$

$$\to Ext^k_\Lambda(M, N[l])_0 \to Ext^k_\Lambda(M, N[l]/J^{l+k-1} N[l])_0 \to$$

The exact sequences: $0 \to \Omega^{k-1} M \to P_{k-2}[-k+2] \to \Omega^{k-2} M \to 0$ and $0 \to \Omega^k M \to P_{k-1}[-k+1] \to \Omega^{k-1} M \to 0$, with $P_{k-2}[-k+2]$, $P_{k-1}[-k+1]$ projectives generated in degrees $k-2$ and $k-1$, respectively, induce exact sequences:
$0 \to Hom_\Lambda(\Omega^{k-2} M, N[l]/J^{l+k-1} N[l])_0 \to Hom_\Lambda(P_{k-2}[-k+2], N[l]/J^{l+k-1} N[l])_0$
$\to Hom_\Lambda(\Omega^{k-1} M, N[l]/J^{l+k-1} N[l])_0 \to Ext^{k-1}_\Lambda(M, N[l]/J^{l+k-1} N[l])_0 \to 0$ and
$0 \to Hom_\Lambda(\Omega^{k-1} M, N[l]/J^{l+k-1} N[l])_0 \to Hom_\Lambda(P_{k-1}[-k+1], N[l]/J^{l+k-1} N[l])_0$
$\to Hom_\Lambda(\Omega^k M, N[l]/J^{l+k-1} N[l])_0 \to Ext^k_\Lambda(M, N[l]/J^{l+k-1} N[l])_0 \to 0$.
Since $Hom_\Lambda(\Omega^{k-1} M, N[l]/J^{l+k-1} N[l])_0 = 0$ and there is an equality: $Hom_\Lambda(\Omega^k M, N[l]/J^{l+k-1} N[l])_0 = 0$, it follows $Ext^{k-1}_\Lambda(M, N[l]/J^{l+k-1} N[l])_0 = 0$ and $Ext^k_\Lambda(M, N[l]/J^{l+k-1} N[l])_0 = 0$.
Therefore: $Ext^k_\Lambda(M, J^{l+k-1} N[l])_0 \cong Ext^k_\Lambda(M, N[l])_0$.
Now $\Omega^{k-1} M$ and $J^{l+k-1} N[l]$ are both generated in degree $k-1$.
Hence, we have an isomorphism: $x \to Fx$.
$Ext^1_\Lambda(\Omega^{k-1} M, J^{l+k-1} N[l])_0 \cong Ext^1_{\Gamma^{op}}(F(J^{l+k-1} N[l]), F(\Omega^{k-1} M))_0$ given as follows:

Koszul Algebras and the Gorenstein Condition

If $x \in Ext_\Lambda^1(\Omega^{k-1}M, J^{l+k-1}N[l])_0$ is the exact sequence:

$$x: 0 \to J^{l+k-1}N[l] \to E \to \Omega^{k-1}M \to 0,$$

then Fx is defined as the exact sequence:

$$0 \to F(\Omega^{k-1}M) \to F(E) \to F(J^{k+l-1}N[l]) \to 0.$$

We have isomorphisms:

$$Ext_\Lambda^k(M, N[l])_0 \cong Ext_\Lambda^k(M, J^{l+k-1}N[l])_0 \cong Ext_\Lambda^1(\Omega^{k-1}(M), J^{l+k-1}N[l])_0 \cong$$

$$\cong Ext_{\Gamma^{op}}^1(F(J^{l+k-1}N[l]), F(\Omega^{k-1}M))_0 \cong Ext_{\Gamma^{op}}^1(\Omega^{l+k-1}F(N)[l], J^{k-1}F(M))_0.$$

By an argument similar to the given above, $Ext_{\Gamma^{op}}^{k+l}(F(N)[l], J^{k-1}F(M))_0 \cong Ext_{\Gamma^{op}}^{k+l}(F(N)[l], F(M))_0$. □

3 GORENSTEIN RINGS

A ring Λ will be called Gorenstein [1], [2] if it has finite injective dimension both as left and right module. The rings considered may not be noetherian but we will consider modules with minimal projective resolutions consisting of finitely generated projectives, in case Λ is graded; the modules, the maps and the extension groups will be graded.

PROPOSITION 3.1. *Let Λ be a Gorenstein ring, let M be a module with minimal projective resolution consisting of finitely generated modules, assume there exists an integer n such that $Ext_\Lambda^k(M, \Lambda) = 0$ for $k \neq n$. Then $Ext_\Lambda^n(M, \Lambda)$ satisfies the following conditions:*

$Ext_{\Lambda^{op}}^i(Ext_\Lambda^n(M, \Lambda), \Lambda^{op}) = 0$ *for* $i \neq n$.

$Ext_{\Lambda^{op}}^n(Ext_\Lambda^n(M, \Lambda), \Lambda^{op}) \cong M$.

In case Λ is graded the isomorphism is as graded modules.

Proof. Assume M has finite projective dimension. Then $pdM = n$.

Let $n = 0$. Then M is projective and M^* is also projective, hence, $Ext_{\Lambda^{op}}^k(M^*, \Lambda^{op}) = 0$ for k different from zero and $M^{**} \cong M$.

Assume $pdM < \infty$ and $n > 0$.

Let $0 \to P_n \to P_{n-1} \to \cdots \to P_0 \to M \to 0$ be the minimal projective resolution. Dualizing with respect to the ring we obtain a complex: $*) \ 0 \to P_0^* \to P_1^* \to \cdots \to P_n^* \to 0$.

By hypothesis, it is exact except at the index n where the homology is $Ext_\Lambda^n(M, \Lambda)$, then the sequence: $0 \to P_0^* \to P_1^* \to \cdots \to P_n^* \to Ext_\Lambda^n(M, \Lambda) \to 0$ is a minimal projective resolution of $Ext_\Lambda^n(M, \Lambda)$.

Dualizing again and using the fact that the complex: $0 \to P_n^{**} \to P_{n-1}^{**} \to \cdots \to P_0^{**} \to 0$ is exact except at zero, where the homology is M, we obtain:

$Ext_{\Lambda^{op}}^i(Ext_\Lambda^n(M, \Lambda), \Lambda^{op}) = 0$ for $i \neq n$ and $Ext_{\Lambda^{op}}^n(Ext_\Lambda^n(M, \Lambda), \Lambda^{op}) \cong M$.

Assume M has infinite projective dimension and $n = 0$.

Let $\ldots \to P_j \to P_{j-1} \to \ldots P_1 \to P_0 \to M \to 0$ be the minimal projective resolution of M. Dualizing we obtain an exact sequence:

$$0 \to M^* \to P_0^* \to P_1^* \to \ldots P_{k+1}^* \to P_{k+2}^* \to Y \to 0$$

Suppose $id_\Lambda \Lambda = k$. Then for $i > 0$ we have isomorphisms:

$$Ext^i_{\Lambda^{op}}(M^*, \Lambda^{op}) \cong Ext^i_{\Lambda^{op}}(\Omega^{k+2}(Y), \Lambda^{op}) \cong Ext^{k+i+2}_{\Lambda^{op}}(Y, \Lambda^{op}) = 0.$$

The exact sequence: $0 \to M^* \to P_0^* \to \Omega^{k+1}(Y) \to 0$ induces an exact sequence:

$$0 \to (\Omega^{k+1}(Y))^* \to P_0^{**} \to M^{**} \to Ext^1_{\Lambda^{op}}(\Omega^{k+1}(Y), \Lambda^{op}) \to 0.$$

Since $Ext^1_{\Lambda^{op}}(\Omega^{k+1}(Y), \Lambda^{op}) \cong Ext^{k+2}_{\Lambda^{op}}(Y, \Lambda^{op}) = 0$, the sequence $0 \to (\Omega^{k+1}(Y))^* \to P_0^{**} \to M^{**} \to 0$ is exact.

The sequence $0 \to \Omega^{k+1}(Y) \to P_1^* \to \Omega^k(Y) \to 0$ induces an exact sequence:

$$0 \to (\Omega^k(Y))^* \to P_1^{**} \to (\Omega^{k+1}(Y))^* \to Ext^1_{\Lambda^{op}}(\Omega^k(Y), \Lambda^{op}) \to 0.$$

Since $Ext^1_{\Lambda^{op}}(\Omega^k(Y), \Lambda^{op}) \cong Ext^{k+1}_{\Lambda^{op}}(Y, \Lambda^{op}) = 0$, the sequence:
$0 \to (\Omega^k(Y))^* \to P_1^{**} \to (\Omega^{k+1}(Y))^* \to 0$ is exact.
We have proved the sequence: $P_1^{**} \to P_0^{**} \to M^{**} \to 0$ is exact.
It follows $M \cong M^{**}$ and $Ext^i_{\Lambda^{op}}(M^*, \Lambda^{op}) = 0$ for i different from zero.
Assume $pdM = \infty$ and $n > 0$.
Let $\cdots \to P_n \to P_{n-1} \to \cdots \to P_0 \to M \to 0$ be the minimal projective resolution of M. We obtain by dualizing the complex:

$$0 \to M^* \to P_0^* \to P_1^* \to \cdots \to P_{n-1}^* \xrightarrow{f_n^*} P_n^* \xrightarrow{f_{n+1}^*} P_{n+1}^* \to \cdots$$

which is exact except at the index n where $Ext^n_\Lambda(M, \Lambda) \cong Ker f_{n+1}^* / Im f_n^*$.
Let $C = P_n^* / Im f_n^*$ and $X = Im f_{n+1}^* = Ker f_{n+2}^* = P_n^* / Ker f_{n+1}^*$.
We have an exact sequence: $*)\ 0 \to Ext^n_\Lambda(M, \Lambda) \to C \to X \to 0$.
Consider the exact sequence:

$$0 \to X \to P_{n+1}^* \to P_{n+2}^* \to \cdots \to P_{n+k}^* \to Y \to 0.$$

Then $\Omega^k Y \cong X$.
Suppose $id_\Lambda \Lambda = k$. Then for $i \geq 1$ we have isomorphisms:

$$Ext^i_\Lambda(X, \Lambda) \cong Ext^i_\Lambda(\Omega^k Y, \Lambda) \cong Ext^{i+k}_\Lambda(Y, \Lambda) = 0.$$

From the exact sequence: $*)$ we obtain the exact sequence:

$$Ext^i_{\Lambda^{op}}(X, \Lambda^{op}) \to Ext^i_{\Lambda^{op}}(C, \Lambda^{op}) \to Ext^i_{\Lambda^{op}}(Ext^n_\Lambda(M, \Lambda), \Lambda^{op}) \to Ext^{i+1}_{\Lambda^{op}}(X, \Lambda^{op}) \to$$

Since $Ext^i_\Lambda(X, \Lambda) \cong Ext^{i+1}_\Lambda(X, \Lambda) = 0$ for $i \geq 1$, it follows the sequence: $0 \to X^* \to C^* \to Ext^n_\Lambda(M, \Lambda)^* \to 0$ is exact and $Ext^i_{\Lambda^{op}}(C, \Lambda^{op}) \cong Ext^i_{\Lambda^{op}}(Ext^n_\Lambda(M, \Lambda), \Lambda^{op})$ for $i \geq 1$.
Since $M^* = 0$ and the sequence: $0 \to P_0^* \to P_1^* \to \cdots \to P_{n-1}^* \to P_n^* \to C \to 0$ is a minimal projective resolution of C the projective dimension of C is n.

Koszul Algebras and the Gorenstein Condition

From the fact that the complexes:

$$0 \to P_n \to P_{n-1} \to \cdots \to P_0 \to 0 \text{ and } 0 \to P_n^{**} \to P_{n-1}^{**} \to \cdots \to P_0^{**} \to 0$$

are isomorphic it follows $Hom_{\Lambda^{op}}(C, \Lambda^{op}) \cong \Omega^{n+1}M$, the module $Ext^i_{\Lambda^{op}}(C, \Lambda^{op}) = 0$ for $i \neq n$ and $Ext^n_{\Lambda^{op}}(C, \Lambda^{op}) \cong M$.

Therefore: $Ext^i_{\Lambda^{op}}(Ext^n_\Lambda(M, \Lambda), \Lambda^{op}) = 0$ for $i \neq 0$ and $i \neq n$ and $Ext^n_{\Lambda^{op}}(Ext^n_\Lambda(M, \Lambda), \Lambda^{op}) \cong M$.

Consider the following exact diagram:

$$
\begin{array}{ccccccc}
 & & 0 & & & & \\
 & & \downarrow & & & & \\
P^*_{n-1} & \to & \text{Ker} f^*_{n+1} & \to & Ext^n_\Lambda(M, \Lambda) & \to & 0 \\
{\scriptstyle id}\downarrow & & \downarrow & & \downarrow & & \\
P^*_{n-1} & \xrightarrow{f^*_n} & P^*_n & \xrightarrow{f^*_{n+1}} & P^*_{n+1} & & \\
 & & \downarrow & \nearrow & & & \\
 & & \text{Im} f^*_{n+1} & & & & \\
 & & \downarrow & & & & \\
 & & 0 & & & &
\end{array}
$$

Dualizing we obtain the diagram:

$$
\begin{array}{ccccccc}
 & & & & 0 & & \\
 & & & & \downarrow & & \\
 & & & & (\text{Im} f^*_{n+1})^* & & \\
 & & & \nearrow & \downarrow & & \\
 & & P^{**}_{n+1} & \xrightarrow{f^{**}_{n+1}} & P^{**}_n & \xrightarrow{f^*_n} & P^{**}_{n-1} \\
 & & \downarrow t & & \downarrow p & & \downarrow id \\
0 & \to & Ext^n_\Lambda(M, \Lambda)^* & \xrightarrow{s} & (\text{Ker} f^*_{n+1})^* & \to & P^{**}_{n-1}
\end{array}
$$

with exact rows and exact middle column.

Hence; $pf^{**}_{n+1} = 0 = st$ implies $t = 0$.

By five's lemma, t is an epimorphism.

Therefore: $Ext^n_\Lambda(M, \Lambda)^* = 0$. □

PROPOSITION 3.2. *Let Λ be any ring and let M be a finitely presented indecomposable Λ-module with minimal projective presentation $P_1 \to P_0 \to M \to 0$ such that $Ext^1_\Lambda(M, \Lambda) = Ext^2_\Lambda(M, \Lambda) = 0$. Then the following statements are true:*

i) $tr M \cong \Omega^2(M)^*$.

ii) $\Omega^2(M)$ *is reflexive.*

iii) $\Omega^2(M)$, $\Omega(M)$ *are indecomposable.*

Proof. By hypothesis we have exact sequences:

$$0 \to M^* \to P_0^* \to \Omega(M)^* \to 0 \text{ and } 0 \to \Omega(M)^* \to P_1^* \to \Omega^2(M)^* \to 0.$$

Gluing the two sequences we obtain an exact sequence:
*) $0 \to M^* \to P_0^* \to P_1^* \to \Omega^2(M)^* \to 0.$
It follows $trM \cong \Omega^2(M)^*$.
Dualizing the sequence *) we obtain a commutative exact diagram:

$$\begin{array}{ccccccc} 0 \to & \Omega^2(M) & \to & P_1 & \to & P_0 \\ & \downarrow & & \downarrow \cong & & \downarrow \cong \\ 0 \to & \Omega^2(M)^{**} & \to & P_1^{**} & \to & P_0^{**} \end{array}$$

Hence: $\Omega^2(M) \cong \Omega^2(M)^{**}$.
From the fact trM is indecomposable follows $\Omega^2(M)^*$ is indecomposable.
Assume $\Omega^2(M) = X \oplus Y$ with $X \neq 0 \neq Y$.
Dualizing we obtain $\Omega^2(M)^* = X^* \oplus Y^*$.
Since $\Omega^2(M) \subseteq P_1$ the modules X^* and Y^* are not zero. Therefore: $\Omega^2(M)^*$ decomposes, which can not hapen because $\Omega^2(M)^* = trM$ and trM is indecomposable if M is.
We have proved $\Omega^2(M)$ is indecomposable.
Assume now $\Omega(M) = X \oplus Y$ with $X \neq 0 \neq Y$. Then $\Omega^2(M) = \Omega(X) \oplus \Omega(Y)$, therefore either X or Y is projective. Suppose $\Omega(M) = X \oplus Q$ with Q projective. Taking push outs, the projection map $q : \Omega(M) \to Q$ induces a commutative exact diagram:

$$\begin{array}{ccccccccc} 0 \to & \Omega(M) & \stackrel{i}{\to} & P_0 & \stackrel{p}{\to} & M & \to 0 \\ & q \downarrow & & \downarrow & & \downarrow id & \\ 0 \to & Q & \to & W & \to & M & \to 0 \\ & \downarrow & & \downarrow & & & \\ & 0 & & 0 & & & \end{array}$$

Since $Ext^1_\Lambda(M, \Lambda) = 0$ the projection map q factors through P_0, there exists a map $t : P_0 \to Q$ with $ti = 1$. Let $s : Q \to \Omega(M)$ be a map such that $qs = 1$. Then $tsj = 1$ and t is a split epimorphism, contradicting the fact $\Omega(M) \subseteq rad\, P_0$.
We have proved $\Omega(M)$ is indecomposable. □

COROLLARY. Assume M has a minimal projective resolution with all projectives finitely generated and $Ext^i_\Lambda(M, \Lambda) = 0$ for $1 \leq i \leq n-1$. Then $\Omega^i(M)$ is reflexive for all $2 \leq i \leq n-1$, if M is indecomposable, then for $1 \leq i \leq n-1$ the module $\Omega^i(M)$ is indecomposable.

4 GORENSTEIN KOSZUL ALGEBRAS

In this section we study graded Gorenstein algebras Λ which are in addition Koszul. Using the duality, we will describe relations between the graded Λ simple satisfying the Gorenstein condition and the indecomposable projectives over the Yoneda algebra Γ satisfying the dual property.

Koszul Algebras and the Gorenstein Condition

PROPOSITION 4.1. [10] *Let Λ be a Koszul algebra and $\Gamma = \bigoplus_{i \geq 0} Ext^i_\Lambda(\Lambda_0, \Lambda_0)$ its Yoneda algebra. Then for any graded simple Λ-module S the conditions 1) and 2) are equivalent:*

1) *The simple S satisfies the following conditions:*
   i) $pdS = n$.
   ii) $Ext^i_\Lambda(S, \Lambda) = 0$ for $i \neq n$.
   iii) $Ext^n_\Lambda(S, \Lambda) = S'$ is a graded simple Λ^{op}-module.

2) *The module $F(S) = \bigoplus_{i \geq 0} Ext^i_\Lambda(S, \Lambda_0)$ is projective injective of finite length.*

Proof. First we show 1 implies 2.

Let $0 \to P_n \to P_{n-1} \to \cdots \to P_0 \to S \to 0$ be the minimal projective resolution of S, the simple S' has a minimal projective resolution: $0 \to P_0^* \to P_1^* \to \cdots \to P_{n-1}^* \to P_n^* \to S' \to 0$.

Dualizing with respect to the field we obtain a minimal injective coresolution of $D(S')$:
$$0 \to D(S') \to D(P_n^*) \to D(P_{n-1}^*) \to \cdots \to D(P_0^*) \to 0.$$

Let $I_j = D(P_{n-j}^*)$ be the jth-injective in the coresolution. Then there is a chain of isomorphisms:
$$\begin{aligned} soc\, I_j &= soc\, \Omega^{-j} D(S') = D(D\Omega^{-j}D(S')/JD\Omega^{-j}D(S')) \cong \\ &\cong D(\Omega^j S'/(\Omega^j S')J) \cong D(P_{n-j}^*/P_{n-j}^*J) \cong \\ &\cong P_{n-j}/JP_{n-j} \cong \Omega^{n-j}S/J\Omega^{n-j}S. \end{aligned}$$

Let P be the projective $P = \bigoplus_{i \geq 0} Ext^i_\Lambda(S, \Lambda_0)$, and $I = D\left(\bigoplus_{i \geq 0} Ext^i_{\Lambda^{op}}(S', \Lambda_0)\right)$ is injective.

We have the following isomorphisms:
$$\begin{aligned} soc^{k+1}I/soc^k I &\cong D(J^k D(I)/J^{k+1}D(I)) \cong D(Ext^k_{\Lambda^{op}}(S', \Lambda^{op})) \cong \\ &\cong D(Hom_{\Lambda_0^{op}}(\Omega^k S'/(\Omega^k S')J, \Lambda_0^{op})) \cong \\ &\cong D(soc\, \Omega^{-k}D(S')) \cong D(soc\, D(P_{n-k}^*)) \cong \\ &\cong P_{n-k}^*/P_{n-k}^* J \cong D(P_{n-k}/JP_{n-k}) \cong \\ &\cong Hom_{\Lambda_0}(\Omega^{n-k}S/J\Omega^{n-k}S, \Lambda_0) \cong \\ &\cong Ext^{n-k}_\Lambda(S, \Lambda_0) \cong PJ^{n-k}/PJ^{n-k+1}. \end{aligned}$$

In particular, $soc\, I \cong PJ^n$.

Let T be a simple in the socle of P. Then there exists an integer $1 \leq k \leq n-1$ such that $T \subseteq PJ^k$ and T is not contained in PJ^{k+1}. Hence; $PJ^k = T \oplus X$.

Applying Koszul duality we have isomorphisms:
$$\bigoplus_{s \geq 0} Ext^s_{\Gamma^{op}}(J^k P, \Gamma_0) \cong F^{-1}(PJ^k) \cong \Omega^k F^{-1}(P) \cong \Omega^k(S) \cong F^{-1}(T) \oplus F^{-1}(X).$$

Set $Q \cong F^{-1}(T)$ and $F^{-1}(X) \cong Y$.

We have an exact commutative diagram:

$$\begin{array}{ccccccccc} 0 \to & Q \oplus Y & \stackrel{i}{\to} & P_{k-1} & \to & \Omega^{k-1}(S) & \to 0 \\ & p \downarrow & & \downarrow & & \downarrow id & \\ x : 0 \to & Q & \to & W & \to & \Omega^{k-1}(S) & \to 0 \\ & \downarrow & & \downarrow & & & \\ & 0 & & 0 & & & \end{array}$$

But $x \in Ext^1_\Lambda(\Omega^{k-1}S, Q) \cong Ext^k_\Lambda(S, Q) = 0$ implies the projection p factors through P_{k-1}, contradicting $Q \subseteq JP_{k-1}$.

We have proved $soc\, P \cong J^n P$, hence $P \subseteq I$.

Since P and I have the same composition factors, $P \cong I$, which shows 1 implies 2.

Assume $F(S) = P$ is projective injective of finite length and $Ext^k_\Lambda(S, \Lambda) \neq 0$ for some $k \geq 1$.

We have an exact sequence: $0 \to Q \to W \to \Omega^{k-1}(S) \to 0$ and an induced commutative exact diagram:

$$\begin{array}{ccccccc} 0 \to & \Omega^k(S) & \to & P_{k-1} & \to & \Omega^{k-1}(S) & \to 0 \\ & \downarrow f & & \downarrow & & \downarrow id & \\ 0 \to & Q & \to & W & \to & \Omega^{k-1}(S) & \to 0 \end{array}$$

If $Im\, f$ is not contained in JQ, then Q is a summand of $\Omega^k(S)$, this is: $\Omega^k(S) \cong Q \oplus X$ and we have isomorphisms: $F(\Omega^k(S)) \cong J^k F(S) \cong F(Q) \oplus F(X)$.

The module $F(S)$ is an indecomposable projective injective of length n, this implies $k = n$ and $X = 0$.

If $Im\, f \subseteq JQ$, then there exists an integer $t \geq 1$ such that $Im\, f \subseteq J^t Q$ and $Im\, f$ is not contained in $J^{t+1}Q$.

We have a commutative exact diagram:

$$\begin{array}{ccccccc} 0 \to & \Omega^k(S) & \to & P_{k-1} & \to & \Omega^{k-1}(S) & \to 0 \\ & \downarrow & & \downarrow & & \downarrow id & \\ 0 \to & J^{t-1}Q & \to & Z & \to & \Omega^{k-1}(S) & \to 0 \end{array}$$

with $0 \to J^{t-1}Q \to Z \to \Omega^{k-1}(S) \to 0$ non split and $J^{t-1}Q, Z, \Omega^{k-1}(S)$ generated in degree $k - 1$.

Applying F we obtain a non split exact sequence:

$*)\ 0 \to J^{k-1}F(S) \stackrel{i}{\to} F(Z) \to \Omega^{t-1}F(Q) \to 0$.

Since $F(S)$ is projective injective, the inclusion map: $j : J^{k-1}F(S) \to F(S)$ extends to $F(Z)$. This is: there exists a degree zero map $\overline{j} : F(Z) \to F(S)$ with $\overline{j}i = j$, but $Im\, \overline{j} \subseteq J^{k-1}F(S)$, contradicting the sequence $*)$ does not split.

Now assume $Hom_\Lambda(S, \Lambda) \neq 0$, then there exists an indecomposable projective Q such that $S \subseteq J^t Q$ and S is not contained in $J^{t+1}Q$. As above, $J^t Q \cong S \oplus X$. Applying F, there is an isomorphism: $\Omega^t F(Q) \cong F(S) \oplus F(X)$, with $F(S)$ projective injective.

Let $0 \to \Omega^t F(Q) \to P_{t-1} \to \Omega^{t-1}F(Q) \to 0$ be exact with P_{t-1} the projective cover of $\Omega^{t-1}F(Q)$. Since $F(S)$ is injective the inclusion $F(S) \to P_{t-1}$ splits, contradicting $F(S) \subseteq JP_{t-1}$.

Koszul Algebras and the Gorenstein Condition

Therefore: $Hom_\Lambda(S,\Lambda) = 0$.

It remains to prove that $Ext_\Lambda^n(S,\Lambda) \cong M$ is simple.

Consider the minimal projective resolution of S: $0 \to P_n \to P_{n-1} \to \cdots \to P_0 \to S \to 0$. Dualizing we obtain an exact sequence: $0 \to P_0^* \to P_1^* \to \cdots \to P_{n-1}^* \to P_n^* \to M \to 0$.

Since P_j^* is generated in degree $-j$ and P_{j-1}^* is generated in degree $-j+1$ the module M is Koszul up to shifting. Applying Koszul duality F^o in the opposite ring to M we obtain isomorphisms: $J^t F^o(M) \cong F^o(\Omega^t(M))$, hence; there is a chain of isomorphisms:

$$\begin{aligned}
J^t F^o(M)/J^{t+1} F^o(M) &\cong \bigoplus_{k\geq 0} Ext_{\Lambda^{op}}^k(\Omega^t(M),\Lambda_o)/\bigoplus_{k\geq 0} Ext_{\Lambda^{op}}^k(\Omega^{t+1}(M),\Lambda_o) \cong \\
&\cong \bigoplus_{k\geq 0} Ext_{\Lambda^{op}}^{k+t}(M,\Lambda_o)/\bigoplus_{k\geq 0} Ext_{\Lambda^{op}}^{k+t+1}(M,\Lambda_o) \cong \\
&\cong Ext_{\Lambda^{op}}^t(M,\Lambda_o) \cong \\
&\cong Hom_{\Lambda^{op}}(\Omega^t(M)/\Omega^t(M)J,\Lambda_o) \cong \\
&\cong Hom_{\Lambda^{op}}(P_{n-t}^*/P_{n-t}^*J,\Lambda_o) \cong P_{n-t}/JP_{n-t}.
\end{aligned}$$

In the other hand we have isomorphisms:

$$\begin{aligned}
soc^{t+1} DF(S)/soc^t DF(S) &\cong D(J^t F(S)/J^{t+1} F(S)) \cong D(Ext_\Lambda^t(S,\Lambda_o)) \cong \\
&\cong D(Hom_\Lambda(\Omega^t(S)/J\Omega^t(S),\Lambda_o)) \cong \\
&\cong \Omega^t(S)/J\Omega^t(S) \cong P_t/JP_t.
\end{aligned}$$

We have proved $soc^{n-t+1} DF(S)/soc^{n-t} DF(S) \cong J^t F(M)/J^{t+1} F(M)$.

In particular, $J^n F(M) \cong soc\, DF(S)$.

Suppose there exists a simple $T \subseteq J^k F(M)$ and T is not contained in $J^{k+1} F(M)$. Then $J^k F(M) \cong T \oplus X$ and $\Omega^k(M) \cong F^{-1}(T) \oplus F^{-1}(X)$, thus $F^{-1}(T)$ is a projective summand of $\Omega^k(M)$, but $Ext_\Lambda^k(M,\Lambda) = 0$ for all $k \neq n$. It follows $k = n$ and $soc\, F(M) \cong J^n F(M)$.

From the fact $DF(S)$ is injective it follows $F(M)$ is isomorphic to a submodule of $DF(S)$, but $F(M)$ and $DF(S)$ have the same length, hence; $F(M) \cong DF(S)$.

Therefore: $F(M)$ is projective, hence M is simple. □

DEFINITION. Let $\Lambda = KQ/I$ be a graded quiver algebra. A graded simple S generated in degree zero satisfies the Gorenstein condition if there exists a non negative integer n_s, called the depth of S, such that:

i) $Ext_\Lambda^k(S,\Lambda) = 0$ for $k \neq n_s$.

ii) There exists a graded Λ^{op} simple S' generated in degree zero and an integer l such that $Ext_\Lambda^{n_s}(S,\Lambda) \cong S'[-l]$, as graded Λ^{op}-modules.

We will denote by $\mathcal{G}(\Lambda)$ the set of isomorphism classes of graded simple satisfying the Gorenstein condition.

PROPOSITION 4.2. *Let $\Lambda = KQ/I$ be a Gorenstein graded quiver algebra, this is: $id_\Lambda \Lambda < \infty$ and $id\Lambda_\Lambda < \infty$, such that all graded Λ and Λ^{op} simple have projective resolutions consisting of finitely generated projectives. Then there exists a bijection $\varphi : \mathcal{G}(\Lambda) \to \mathcal{G}(\Lambda^{op})$ between the sets of graded Λ and Λ^{op} simple satisfying the Gorenstein condition.*

Proof. Let S be in $\mathcal{G}(\Lambda)$ and define $\varphi(S) = S'$, where S' is a graded Λ^{op} simple such that $Ext_\Lambda^{n_s}(S, \Lambda) \cong S'[-l]$. We must prove that $S' \in \mathcal{G}(\Lambda^{op})$.

We know by proposition 3.1, $Ext_{\Lambda^{op}}^k(S'[-l], \Lambda^{op}) = 0$ for $k \neq n_s$. Hence; $Ext_{\Lambda^{op}}^k(S', \Lambda^{op}) = 0$ for $k \neq n_s$.

We have isomorphisms: $Ext_{\Lambda^{op}}^{n_s}(Ext_\Lambda^{n_s}(S, \Lambda)[l], \Lambda^{op})) \cong S[-l] \cong Ext_{\Lambda^{op}}^{n_s}(S', \Lambda^{op})$.

Defining in a similar way a map $\varphi' : \mathcal{G}(\Lambda^{op}) \to \mathcal{G}(\Lambda)$, it is clear that φ and φ' are inverse maps. □

DEFINITION. Let $\Lambda = KQ/I$ be a graded quiver algebra, a graded indecomposable projective Q satisfies the co-Gorenstein condition if there exists a non negative integer n_Q, called the co-depth, such that:

i) The module $Ext_\Lambda^k(\Lambda_0, Q) = 0$ for $k \neq n_Q$.

ii) There exists a graded Λ^{op} simple S' generated in degree zero and an integer l such that $Ext_\Lambda^{n_Q}(\Lambda_0, Q) \cong S'[-l]$, as graded Λ^{op}-modules.

We denote by $\mathcal{CG}(\Lambda)$ the set of indecomposable projectives satisfying the co-Gorenstein condition.

THEOREM 4.3. *Let Λ be a Koszul K-algebra and Γ its Yoneda algebra. Then the Koszul duality $F(M) = \underset{k \geq 0}{\oplus} Ext_\Lambda^k(M, \Lambda)$ induces a bijection $\phi : \mathcal{G}(\Lambda) \to \mathcal{CG}(\Gamma^{op})$ given by $\phi(S) = F(S)$.*

Proof. Let S be in $\mathcal{G}(\Lambda)$. By hypothesis, given S there exist two integers n and l such that $Ext_\Lambda^k(S, \Lambda[m])_0 = 0$ if $k \neq n$ or $m \neq l$ and $Ext_\Lambda^n(S, \Lambda[l])_0 = S'[-l]$ is a Λ^{op}-simple concentrated in degree l.

By theorem 2.1, for $k \geq -m$ there is an isomorphism of K-vector spaces: $Ext_\Lambda^k(S, \Lambda[m])_0 \cong Ext_{\Gamma^{op}}^{k+m}(F(\Lambda[m]), F(S))_0$, where $F(\Lambda) \cong \Gamma_0^{op}$ and $F(S) = Q$ is an indecomposable projective.

Hence; $Ext_{\Gamma^{op}}^{k+m}(\Gamma_0^{op}[m], Q)_0 = 0$ for $k + m \geq 0$, unless $k = n$ and $m = l$.

We have proved $Ext_{\Gamma^{op}}^t(\Gamma_0^{op}, Q) = 0$ for $t \neq n + l$ and $Ext_{\Gamma^{op}}^{n+l}(\Gamma_0^{op}, Q) \cong Ext_{\Gamma^{op}}^{n+l}(\Gamma_0^{op}[l], Q)_0 \cong Ext_\Lambda^n(S, \Lambda[l])_0 \cong S'[l]$.

Therefore: $\dim_K Ext_{\Gamma^{op}}^{n+l}(\Gamma_0^{op}, Q) = 1 = \dim S'[l]$.

It follows $Ext_{\Gamma^{op}}^{n+l}(\Gamma_0^{op}, Q)$ is simple as $\Gamma^{op}K$-module, hence $Ext_{\Gamma^{op}}^{n+l}(\Gamma_0^{op}, Q) \cong T[l]$, with T a Γ simple.

It was proved above $\phi(S) = F(S) = Q \in \mathcal{CG}(\Gamma^{op})$.

The converse is proved using again theorem 2.1.

Let Q be an indecomposable projective Γ^{op}-module such that there exist integers n and l with $Ext_{\Gamma^{op}}^k(\Gamma_0^{op}, Q) = 0$ for $k \neq n$ and $Ext_{\Gamma^{op}}^n(\Gamma_0^{op}, Q) \cong T[l]$, with T a simple.

Then we have $Ext_{\Gamma^{op}}^k(\Gamma_0^{op}[m], Q)_0 = 0$ if $k \neq n$ or $m \neq l$ and $Ext_{\Gamma^{op}}^n(\Gamma_0^{op}[l], Q) \cong T[l]$.

Let S be a graded simple such that $F(S) \cong Q$ and let t be a non negative integer such that $Ext_\Lambda^t(S, \Lambda) \neq 0$. Then there exists an integer m with $t + m \geq 0$ such that $Ext_\Lambda^t(S, \Lambda[m])_0 \neq 0$.

By theorem 2.1,

$$Ext_\Lambda^t(S, \Lambda[m])_0 \cong Ext_{\Gamma^{op}}^{t+m}(F(\Lambda)[m], F(S))_0 \cong Ext_{\Gamma^{op}}^{t+m}(\Gamma_0^{op}[m], Q)_0 \neq 0.$$

Koszul Algebras and the Gorenstein Condition

It follows $t + m = n$ and $m = l$.
Hence; $Ext_\Lambda^t(S, \Lambda[m])_0 = 0$ unless $k = n - l$ and $m = l$.
We have the following vector space isomorphisms:

$$Ext_\Lambda^{n-l}(S, \Lambda) = Ext_\Lambda^{n-l}(S, \Lambda[l])_0 \cong T[l].$$

Therefore: $\dim_K Ext_\Lambda^{n-l}(S, \Lambda) = 1$ and $Ext_\Lambda^{n-l}(S, \Lambda)$ is a simple concentrated in degree l.
We have proved $Ext_\Lambda^{n-l}(S, \Lambda) \cong S'[-l]$. □

One example of the situation considered above is the standard Auslander algebra. We know by [6] that a simple S over an Auslander algebra Λ satisfies the Gorenstein condition if and only if S has projective dimension 2. If Γ is the Yoneda algebra of Λ and F is the Koszul duality, then $F(S)$ is projective injective if and only if S has projective dimension 2. It is clear that the conditions of the proposition are satisfied.

PROPOSITION 4.4. *Let $\Lambda = KQ/I$ be a Gorenstein algebra such that all graded Λ and Λ^{op} simple have finite projective resolutions and let $\mathcal{G}(\Lambda)$ be the set of graded simple satisfying the Gorenstein condition. Then for any simple $S_j \in \mathcal{G}(\Lambda)$ there exists an indecomposable projective $Q_{\sigma(j)}$ and an integer l such that $Ext_\Lambda^{ns_j}(S_j, Q_{\sigma(j)}[l])_0 \neq 0$ and σ is an injective function from $\mathcal{G}(\Lambda)$ to the graded indecomposable projective Λ-modules.*

Proof. We have the following isomorphisms:

$$Ext_\Lambda^n(S_j, \Lambda) \cong \bigoplus_{k \in \mathcal{Z}} Ext_\Lambda^n(S_j, \Lambda)_k \cong S'_j$$

with S'_j generated in degree l.
Since $\dim_K S'_j = 1$ there exist isomorphisms:

$$Ext_\Lambda^n(S_j, \Lambda) = Ext_\Lambda^n(S_j, \Lambda)_l \cong S'_j \cong \bigoplus_{k=1}^m Ext_\Lambda^n(S_j, Q_k[l])_0.$$

It follows there exists some integer $\sigma(j)$ such that $Ext_\Lambda^n(S_j, Q_{\sigma(j)}[l])_0 \neq 0$ and

$$Ext_\Lambda^n(S_j, Q_k[l])_0 = 0 \text{ for } k \neq \sigma(j).$$

Then there are isomorphisms:

$$Ext_\Lambda^{ns_j}(S_j, Q_{\sigma(j)})_l \cong \bigoplus_{k=1}^m Ext_\Lambda^n(S_j, Q_{\sigma(j)})_m \cong Ext_\Lambda^n(S_j, Q_{\sigma(j)}) \neq 0.$$

We will show now σ is injective:
Assume for $S_k \in \mathcal{G}(\Lambda)$ there exist an isomorphism $Q_{\sigma(j)} \cong Q_{\sigma(k)}$.
Then both $Ext_\Lambda^{ns_j}(S_j, Q_{\sigma(j)})$ and $Ext_\Lambda^{ns_k}(S_k, Q_{\sigma(k)})$ are different from zero.
There exists natural isomorphisms:

$$Ext_\Lambda^{ns_j}(S_j, Q_{\sigma(j)}) \cong Ext_\Lambda^{ns_j}(S_j, \Lambda) \otimes_\Lambda Q_{\sigma(j)} \cong S'_j \otimes_\Lambda Q_{\sigma(j)} \text{ and}$$

$$Ext_\Lambda^{ns_k}(S_k, Q_{\sigma(k)}) \cong Ext_\Lambda^{ns_k}(S_k, \Lambda) \otimes_\Lambda Q_{\sigma(k)} \cong S'_k \otimes_\Lambda Q_{\sigma(k)}.$$

Dualizing we have:

$$D(S'_j \otimes_\Lambda Q_{\sigma(j)}) \cong Hom_\Lambda(S'_j, D(Q_{\sigma(j)})) \text{ and}$$

$$D(S'_k \otimes_\Lambda Q_{\sigma(k)}) \cong Hom_\Lambda(S'_k, D(Q_{\sigma(k)})).$$

Hence; $S'_j \cong soc\, D(Q_{\sigma(j)})$ and $S'_k \cong soc\, D(Q_{\sigma(k)})$. Since $Q_{\sigma(j)} \cong Q_{\sigma(k)}$, then $S'_j \cong S'_k$.

By proposition 3.1 $S_j \cong S_k$ and σ is injective. □

DEFINITION. We say that an (graded) algebra is weakly Gorenstein if there exists an integer $n > 0$ such that for all (graded) Λ-modules M and all (graded) Λ^{op} modules N of finite length and all integers $k > n$ we have $Ext^k_\Lambda(M, \Lambda) = 0 = Ext^k_{\Lambda^{op}}(N, \Lambda^{op})$.

THEOREM 4.5. *Let Λ be a Gorenstein Koszul algebra such that all graded simple satisfy the Gorenstein condition and let Γ be the Yoneda algebra of Λ. Then all graded Γ and Γ^{op} simple satisfy the Gorenstein condition, in particular Γ is weakly Gorenstein.*

Proof. By proposition 4.4, for each graded simple S_j there exists a unique indecomposable projective $Q_{\sigma(j)}$ and an integer l such that $Ext^{ns_j}_\Lambda(S_j, Q_{\sigma(j)}[l])_0 \neq 0$ hence; $Ext^{ns_j}_\Lambda(S_j, Q_{\sigma(j)}) \cong Ext^{ns_j+l}_{\Gamma^{op}}(F(\Lambda)[l], F(S_j))_0 \neq 0$.

There are isomorphisms:

$$Ext^{ns_j}_\Lambda(S_j, \Lambda[l])_0 \cong Ext^{ns_j}_\Lambda(S_j, Q_{\sigma(j)}[l])_0 \cong Ext^{ns_j+l}_{\Gamma^{op}}(F(Q_{\sigma(j)})[l], F(S_j))_0.$$

Set $T_{\sigma(j)} = F(Q_{\sigma(J)})$ and $F(S_j) = P$.

Assume there exists an indecomposable projective Γ^{op}-module P' and integers k and m such that $Ext^m_{\Gamma^{op}}(T_{\sigma(j)}[k], P')_0 \neq 0$.

There is an isomorphism: $Ext^m_{\Gamma^{op}}(T_{\sigma(j)}[k], P')_0 \cong Ext^{m-k}_\Lambda(F^{-1}(P'), Q_{\sigma(j)}[k])_0$.

By proposition 4.4, $Ext^{m-k}_\Lambda(F^{-1}(P'), Q_{\sigma(j)}[k])_0 \neq 0$ and $Ext^{ns_j}_\Lambda(S_j, Q_{\sigma(j)}[l])_0 \neq 0$ imply $n_{S_j} = m - k$, $F^{-1}(P') = S_j$ and $k = l$.

We have proved for each Γ^{op}-simple $T_{\sigma(j)}$ there exists an integer m such that $\dim_K Ext^m_{\Gamma^{op}}(T_{\sigma(j)}, \Gamma^{op}) = 1$ and $Ext^m_{\Gamma^{op}}(T_{\sigma(j)}, \Gamma^{op}) = 0$ for $k \neq m$.

It follows $Ext^m_{\Gamma^{op}}(T_{\sigma(j)}, \Gamma^{op})$ is a Γ-simple.

If we consider Λ^{op} instead of Λ we obtain in a similar way that all graded Γ-modules satisfy the Gorenstein condition. □

5 TENSOR PRODUCT OF ALGEBRAS AND THE GORENSTEIN CONDITION

In this section we will construct examples of algebras such that all graded simple satisfy the Gorenstein condition, we know that for selfinjective and generalized Auslander regular algebras (see [4], [10], [11] and section 6, below) all simple satisfy this condition, we will prove that given two algebras Λ_1, Λ_2 such that all the graded simple satisfy the Gorenstein condition the tensor product $\Lambda_1 \otimes \Lambda_2$ has the same property, hence the tensor product of a selfinjective and a generalized Auslander regular algebra will be an example of an infinite dimensional algebra of infinite global dimension such that all graded simple satisfy the Gorenstein condition.

We will start by recalling Künneth relations, refereeing to [5] for the proof.

PROPOSITION 5.1. *Let Λ be a ring, A a left Λ complex and C a right complex, $H(A)$, $H(C)$, $H(A \otimes_\Lambda C)$, the homology of the complexes A, C, $A \otimes_\Lambda C$, respectively. Then there exists an exact sequence with α a degree zero map and β a degree one map:*

$$0 \to H(A) \otimes H(C) \xrightarrow{\alpha} H(A \otimes C) \xrightarrow{\beta} Tor_1^\Lambda(H(A), H(C)) \to 0$$

Explicitly, for any integer n an exact sequence:

$$0 \to \sum_{p+q=n} H(A)_p \otimes H_q(C) \to H_n(A \otimes C) \to \sum_{p+q=n-1} Tor_1^\Lambda(H_p(A), H_q(C)) \to 0$$

LEMMA 5.2. *Let R, T be two K–algebras over a field, M, X left R-modules and N, Y right T-modules with M, N finitely presented. Then there exists a natural isomorphism:*

$$\psi : Hom_R(M, X) \underset{K}{\otimes} Hom_T(N, Y) \to Hom_{R \otimes T}(M \otimes N, X \otimes Y)$$

given by $\psi(f \otimes g)(m \otimes n) = f(m) \otimes g(n)$.

Proof. If $R = M$ and $N = T$, then ψ is an isomorphism, since is the composition of the natural isomorphisms: $Hom_R(R, X) \underset{K}{\otimes} Hom_T(T, Y) \cong X \otimes Y \cong Hom_{R \otimes T}(R \otimes T, X \otimes Y)$.

Assume $N = T$ and let $R^m \to R^n \to M \to 0$ be a presentation of M, it induces exact sequences:

$$R^m \otimes T \to R^n \otimes T \to M \otimes T \to 0,$$

$$0 \to Hom_{R \otimes T}(M \otimes T, X \otimes Y) \to Hom_{R \otimes T}(R^n \otimes T, X \otimes Y) \to$$
$$\to Hom_{R \otimes T}(R^m \otimes T, X \otimes Y)$$

$$0 \to Hom_R(M, X) \to Hom_R(R^n, X) \to Hom_R(R^m, X)$$

$$0 \to Hom_R(M, X) \otimes Hom_T(T, Y) \to Hom_R(R^n, X) \otimes Hom_T(T, Y) \to$$
$$\to Hom_R(R^m, X) \otimes Hom_T(T, Y)$$

The maps:

$$\psi : Hom_R(R^n, X) \otimes Hom_T(T, Y) \to Hom_{R \otimes T}(R^n \otimes T, X \otimes Y)$$

$$\psi : Hom_R(R^m, X) \otimes Hom_T(T, Y) \to Hom_{R \otimes T}(R^m \otimes T, X \otimes Y)$$

are isomorphisms, since the first ψ is a composition of the following isomorphisms:

$$Hom_R(R^n, X) \otimes Hom_T(T, Y) \cong (\underset{n}{\oplus} Hom_{R \otimes T}(R, X) \otimes Hom_T(T, Y)) \cong$$
$$\oplus_n (Hom_R(R, X) \otimes Hom_T(T, Y)) \cong \underset{n}{\oplus} Hom_{R \otimes T}(R \otimes T, X \otimes Y) \cong$$
$$Hom_{R \otimes T}(\underset{n}{\oplus} R \otimes T, X \otimes Y) \cong Hom_{R \otimes T}((\underset{n}{\oplus} R) \otimes T, X \otimes Y).$$

Similarly, for the second ψ.
Hence they induce an isomorphism:

$$\psi: Hom_R(M,X) \otimes Hom_T(T,Y) \to Hom_{R \otimes T}(M \otimes T, X \otimes Y).$$

Assume now N has a presentation: $T^k \to T^l \to N \to 0$, it induces the following exact sequences:

$$M \otimes T^k \to M \otimes T^l \to M \otimes N \to 0,$$

$$0 \to Hom_{R \otimes T}(M \otimes N, X \otimes Y) \to Hom_{R \otimes T}(M \otimes T^l, X \otimes Y) \to$$
$$\to Hom_{R \otimes T}(M \otimes T^k, X \otimes Y)$$

$$0 \to Hom_T(N,Y) \to Hom_T(T^l,Y) \to Hom_T(T^k,Y)$$

$$0 \to Hom_R(M,X) \otimes Hom_T(T,Y) \to Hom_R(M,X) \otimes Hom_T(T^l,Y) \to$$
$$\to Hom_R(M,X) \otimes Hom_T(T^k,Y)$$

The natural isomorphisms:

$$\psi: Hom_R(M,X) \otimes Hom_T(T^l,Y) \to Hom_{R \otimes T}(M \otimes T^l, X \otimes Y)$$

$$\psi: Hom_R(M,X) \otimes Hom_T(T^k,Y) \to Hom_{R \otimes T}(M \otimes T^k, X \otimes Y)$$

Induce an isomorphism:

$$\psi: Hom_R(M,X) \otimes Hom_T(N,Y) \to Hom_{R \otimes T}(M \otimes N, X \otimes Y),$$

as claimed. □

PROPOSITION 5.3. *Let R,T be two algebras over a field K and M,X left R-modules, N,Y right T-modules. Assume M,N have projective resolutions consisting of finitely generated modules. Then for all $n \geq 0$ there exists a natural isomorphism:*

$$Ext^n_{R \otimes T}(M \otimes N, X \otimes Y) \cong \bigoplus_{i+j=n} Ext^i_R(M,X) \otimes Ext^j_T(N,Y).$$

Proof. Consider projective resolutions of M and N as R and T-modules, respectively, and assume all projectives in the resolution are finitely generated:

$*) \quad \cdots \to P_k \overset{f_{k-1}}{\to} P_{k-1} \to \cdots P_1 \overset{f_1}{\to} P_0 \overset{f_0}{\to} M \to 0$

$**) \quad \cdots \to Q_t \overset{g_{t-1}}{\to} Q_{t-1} \to \cdots Q_1 \overset{g_1}{\to} Q_0 \overset{g_0}{\to} N \to 0.$

Since K is semisimple, it follows by Künneth formulas that the following sequence is a projective resolution of $M \otimes N$ as $R \otimes T$-modules:

$***) \quad \cdots \sum_{i+j=n} P_i \otimes Q_j \to \sum_{i+j=n-1} P_i \otimes Q_j \to \cdots P_1 \otimes Q_0 \oplus P_0 \otimes Q_1$
$\to P_0 \otimes Q_0 \to M \otimes N \to 0$

Applying the functors $Hom_R(-,X)$, $Hom_T(-,Y)$ and $Hom_{R \otimes T}(-, X \otimes Y)$ to the sequences: $*), **), ***)$, respectively, we obtain complexes:

Koszul Algebras and the Gorenstein Condition

$\diamond$) $0 \to Hom_R(P_0, X) \to Hom_R(P_1, X) \to \cdots Hom_R(P_n, X) \to \cdots$

$\diamond\diamond$) $0 \to Hom_T(Q_0, X) \to Hom_T(Q_1, X) \to \cdots Hom_T(Q_m, X) \to \cdots$

$\diamond\diamond\diamond$) $0 \to Hom_{R \otimes T}(P_0 \otimes Q_0, X \otimes Y) \to$

$\to Hom_{R \otimes T}(P_1 \otimes Q_0 \oplus P_0 \otimes Q_1, X \otimes Y) \to \cdots$

$\to Hom_{R \otimes T}\left(\sum_{i+j=n} P_i \otimes Q_j, X \otimes Y\right) \to \cdots$

We have natural isomorphisms:

$$Hom_{R \otimes T}\left(\sum_{i+j=n} P_i \otimes Q_j, X \otimes Y\right) \cong \sum_{i+j=n} Hom_{R \otimes T}(P_i \otimes Q_j, X \otimes Y) \cong$$
$$\cong Hom_R(P_i, X) \otimes Hom_T(Q_j, Y).$$

Hence; $\diamond\diamond\diamond$) is isomorphic to the tensor product $\diamond\diamond\diamond\diamond$) of the complexes $\diamond$) and $\diamond\diamond$).

Where $\diamond\diamond\diamond\diamond$) is :

$0 \to Hom_R(P_0, X) \otimes Hom_T(Q_0, Y) \to Hom_R(P_0, X) \otimes Hom_T(Q_1, Y)$

$\oplus Hom_R(P_1, X) \otimes Hom_T(Q_0, Y) \to$

$\to \cdots \sum_{i+j=n} Hom_R(P_i, X) \otimes Hom_T(Q_j, Y) \to \cdots$

Let A be the complex $\diamond$) and C the complex $\diamond\diamond$). Then $A \otimes C$ is isomorphic to the complex $\diamond\diamond\diamond$).

We have isomorphisms: $H_p(A) \cong Ext_R^p(M, X)$, $H_q(C) \cong Ext_T^q(N, Y)$ and $H_n(A \otimes C) \cong Ext_{R \otimes T}^n(M \otimes N, X \otimes Y)$.

The result follows by Künneth relations. $\square$

COROLLARY. Under the conditions of the proposition for each integer $n \geq 0$ there exists a natural isomorphism of $(R \otimes T)^{op}$-modules:

$$Ext_{R \otimes T}^n(M \otimes N, R \otimes T) \cong \bigoplus_{i+j=n} Ext_R^i(M, R) \otimes Ext_T^j(N, T).$$

Proof. It is clear that the map:

$$\psi : Hom_R(M, R) \underset{K}{\otimes} Hom_T(N, T) \to Hom_{R \otimes T}(M \otimes N, R \otimes T)$$

given by $\psi(f \otimes g)(m \otimes n) = f(m) \otimes g(n)$ is an isomorphism of $(R \otimes T)^{op}$-modules. From this it follows that the isomorphism of extension groups given in the proof of the proposition is an isomorphism of $(R \otimes T)^{op}$-modules. $\square$

PROPOSITION 5.4. *Let R, T be two algebras over an algebraically closed field K of small injective dimension $si(R)$ and $si(T)$, respectively. Then $R \otimes T$ has small injective dimension and $si(R \otimes T) = si(R) + si(T)$.*

Proof. Since K is algebraically closed, all $R \otimes T$-simples are of the form $S = X \otimes Y$, with X an R-simple an Y a T-simple.

Set $n = si(R)$ and $m = si(T)$.

Then for any integer $k > n + m$: $Ext^k_{R \otimes T}(X \otimes Y, R \otimes T) \cong \underset{i+j=k}{\oplus} Ext^i_R(X, R) \otimes Ext^j_T(Y, T) = 0$.

Let M be an R-module of finite length with $Ext^n_R(M, R) \neq 0$ and N a T-module of finite length with $Ext^m_T(N, T) \neq 0$. Then $Ext^{n+m}_{R \otimes T}(M \otimes N, R \otimes T) \cong Ext^n_R(M, R) \otimes Ext^m_T(N, T) \neq 0$.

We have proved $is(R \otimes T) = n + m$. □

THEOREM 5.5. *Let R and T be two graded quiver algebras over an algebraically closed field K, let S_i and S_j be R and T graded simple, respectively, satisfying the Gorenstein condition. This is: there exists non negative integers n_i and n_j such that: $Ext^k_R(S_i, R) = 0$ for all $k \neq n_i$ and $Ext^l_T(S_j, T) = 0$ for all $l \neq n_j$ and $Ext^{n_i}_R(S_i, R) = S'_i[l]$ is a graded R^{op}-module and $Ext^{n_j}_T(S_j, T) = S'_j[m]$ is a graded T^{op}-simple.*

Then $S_i \otimes S_j$ is a $R \otimes T$-simple satisfying the Gorenstein condition.

Proof. It follows from the isomorphisms:

$$Ext^k_{R \otimes T}(S_i \otimes S_j, R \otimes T) \cong \underset{s+t=k}{\oplus} Ext^s_R(S_i, R) \otimes Ext^t_T(S_j, T) = 0$$

if $s \neq n_i$ or $t \neq n_j$ and

$$Ext^{n_i+n_j}_{R \otimes T}(S_i \otimes S_j, R \otimes T) \cong Ext^{n_i}_R(S_i, R) \otimes Ext^{n_j}_T(S_j, T) \cong$$
$$S'_i[l] \otimes S'_j[m] \cong S'_i \otimes S'_j[l + m].$$

□

COROLLARY. *Let R, T be two graded quiver algebras over an algebraically closed field K such that all graded simple satisfy the Gorenstein condition. Then all graded $R \otimes T$-simple satisfy the Gorenstein condition.*

We end this section with an example:

Let $A = T_n(K)$ be the triangular $n \times n$ matrix ring. Then all non projective simple satisfy the Gorenstein condition and all have depth 1, the unique projective simple has depth zero, but since its dual with respect to the ring is not simple, it does not satisfy the Gorenstein condition. We do not known examples of algebras such that all simple modules satisfy the Gorenstein condition and some of them have different depth.

6 APPENDIX: SELFINJECTIVE KOSZUL ALGEBRAS

In [10], selfinjective Koszul algebras and their Yoneda algebras were studied, in particular the following generalization of a theorem by Bondal-Politshchuk and P. S. Smith was proved. (see also proposition 4.1).

THEOREM 6.1. *Let $\Lambda = KQ/I$ be an indecomposable Koszul algebra then the following two conditions are equivalent:*

1) *The Yoneda algebra of Λ is a selfinjective algebra with radical r such that $r^n \neq 0$ and $r^{n+1} = 0$, with $n \geq 2$.*

2) *The algebra Λ satisfies the following conditions:*

   i) *There is an integer $n \geq 2$ such that all graded simples have projective dimension n.*

   ii) *Given a graded simple S for all $k \neq n$, we have: $Ext^k_\Lambda(S, \Lambda) = 0$.*

   iii) *There exists a bijection between the graded Λ and Λ^{op}-simple given by $Ext^n_\Lambda(-, \Lambda)$.*

The algebras satisfying the conditions of the theorem were called generalized Auslander regular.

DEFINITION. A graded quiver algebra Λ has small global dimension n if $n = \sup \{pdM | M$ is of finite length$\}$.

The aim of this section is to prove the following:

THEOREM 6.2. *Let Λ be a noetherian generalized Auslander regular Koszul algebra of global dimension n. Let $0 \to \Lambda \to E_0 \to E_1 \to E_2 \to \cdots \to E_n \to 0$ be the minimal injective coresolution of Λ. Then $E_n \cong D(\Lambda)[n]$.*

(See [1] for related results).
We will use freely the results and definitions from [9], [10].

PROPOSITION 6.3. *Let Λ be a generalized Auslander regular Koszul algebra of small global dimension n. If M is a graded torsion free module of finite projective dimension such that all projectives in the minimal projective resolution are finitely generated, then $pdM < n$.*

Proof. We know by [9], there exists a Koszul submodule N of M such that $L = M/N$ is of finite length. Since N is a Koszul torsion free module, it follows from [9], that $pdN < n$. Hence; $pdL = n$ and $pdM \leq n$.

Let $0 \to P_n \to \cdots \to P_{n-1} \to P_0 \to M \to 0$ be the minimal projective resolution of M.

The exact sequence $0 \to N \to L \to M \to 0$ induces an exact sequence:

$$Ext^n_\Lambda(L, \Lambda) \to Ext^n_\Lambda(M, \Lambda) \to Ext^n_\Lambda(N, \Lambda).$$

Since $Ext^n_\Lambda(N, \Lambda) = 0$ and $Ext^n_\Lambda(L, \Lambda)$ is of finite length as Λ^{op}-module, $Ext^n_\Lambda(M, \Lambda)$ has finite length as Λ^{op}-module. Therefore: $pdExt^n_\Lambda(M, \Lambda) = n$.

Let $0 \to Q^*_0 \to Q^*_1 \to \cdots \to Q^*_n \to Ext^n_\Lambda(M, \Lambda) \to 0$ be a minimal projective resolution of $Ext^n_\Lambda(M, \Lambda)$. Since Λ^{op} is generalized Auslander regular and $Ext^n_\Lambda(M, \Lambda)$ is of finite length, $Ext^i_{\Lambda^{op}}(Ext_\Lambda(M, \Lambda), \Lambda^{op}) = 0$ for $i \neq n$.

We have a complex: $0 \to P^*_0 \to P^*_1 \to \cdots \to P^*_n \to 0$ with $n-th$ homology group $Ext^n_\Lambda(M, \Lambda)$, hence a presentation: $P^*_{n-1} \to P^*_n \to Ext^n_\Lambda(M, \Lambda) \to 0$.

We have the following commutative exact diagram:

$$\begin{array}{ccccccccc}
 & & 0 & & 0 & & & & \\
 & & \downarrow & & \downarrow & & & & \\
0 & \to & P & \to & P & \to & 0 & & \\
 & & \downarrow & & \downarrow & & \downarrow & & \\
0 & \to & H & \to & P_n^* & \to & Ext_\Lambda^n(M,\Lambda) & \to & 0 \\
 & & \downarrow & & \downarrow & & \downarrow & & \\
0 & \to & \Omega(Ext_\Lambda^n(M,\Lambda)) & \to & Q_n^* & \to & Ext_\Lambda^n(M,\Lambda) & \to & 0 \\
 & & \downarrow & & \downarrow & & \downarrow & & \\
 & & 0 & & 0 & & 0 & &
\end{array}$$

where P is a projective module.

Set $\hat{M} = Ext_\Lambda^n(M, \Lambda)$.

There exists a decomposition of H, as $H = P \oplus \Omega(\hat{M})$. The projective cover of H is $P \oplus Q_{n-1}^*$ and there exist an epimorphism: $P_{n-1}^* \to H \to 0$. It follows that P_{n-1}^* has a decomposition $P_{n-1}^* \cong P \oplus P' \oplus Q_{n-1}^*$ and there exists a commutative square:

$$\begin{array}{ccc}
P_{n-1}^* & \xrightarrow{f_{n-1}^*} & P_n^* \\
\downarrow \cong & & \downarrow \cong \\
P \oplus P' \oplus Q_{n-1}^* & \xrightarrow{\begin{pmatrix} 1 & 0 \\ 0 & t \end{pmatrix}} & P \oplus Q_n^*
\end{array}$$

Dualizing, we obtain a commutative square:

$$\begin{array}{ccc}
P_n & \xrightarrow{f_{n-1}} & P_{n-1} \\
\downarrow \cong & & \downarrow \cong \\
P_n^{**} & \xrightarrow{f_{n-1}^{**}} & P_{n-1}^{**} \\
\downarrow \cong & & \downarrow \cong \\
P^* \oplus Q_n^{**} & \xrightarrow{\begin{pmatrix} 1 & 0 \\ 0 & t \end{pmatrix}} & P^* \oplus P'^* \oplus Q_{n-1}^{**}
\end{array}$$

Since $\text{Im}\, f_{n-1} \subseteq rP$, it follows $P = 0$ and P_n^* is the projective cover of $\hat{M}$. We have an induced map of complexes:

$$\begin{array}{ccccccccccccc}
0 & \to & P_0^* & \to & P_1^* & \to & \cdots & P_{n-1}^* & \to & P_n^* & \to & \hat{M} & \to & 0 \\
 & & \downarrow h_0^* & & \downarrow h_1^* & & & \downarrow h_{n-1}^* & & \downarrow h_n^* & & \downarrow \cong & & \\
0 & \to & Q_0^* & \to & Q_1^* & \to & \cdots & Q_{n-1}^* & \to & Q_n^* & \to & \hat{M} & \to & 0
\end{array}$$

with h_n^* an isomorphism.

Dualizing, we get a commutative exact diagram:

$$0 \to Q_n^{**} \xrightarrow{g_n^*} Q_{n-1}^{**} \to \cdots \to Q_1^{**} \xrightarrow{g_1^*} Q_0^{**} \to Ext_{\Lambda^{op}}^n(\hat{M}, \Lambda^{op}) \to 0$$
$$\quad\;\; \downarrow h_n^{**} \qquad \downarrow h_{n-1}^{**} \qquad\qquad \downarrow h_1^{**} \quad\; \downarrow h_o^{**} \qquad\qquad\quad \downarrow h$$
$$0 \to P_n^{**} \xrightarrow{f_n^{**}} P_{n-1}^{**} \to \cdots \to P_1^{**} \xrightarrow{f_1^{**}} P_0^{**} \to Ext_{\Lambda^{op}}^n(\hat{M}, \Lambda^{op}) \to 0$$

The module M is torsion free and $Ext_{\Lambda^{op}}^n(\hat{M}, \Lambda^{op})$ of finite length, therefore: $h = 0$.

It follows, there exists a map $s : Q_0^{**} \to P_1^{**}$ with $f_1^{**} s = h_o^{**}$.

The equalities: $f_1^{**}(h_1^{**} - s.g_1) = f_1^{**} h_1^{**} - h_o^{**} g_1 = 0$ imply the existence of a map: $s_1 : Q_1^{**} \to P_2^{**}$ such that $f_2^{**} s_1 = h_1^{**} - s.g_1$.

By induction, there exist homotopies $s_i : Q_i^{**} \to P_{i+1}^{**}$ with $h_i^{**} = s_{i-1}.g_i + f_{i+1}^{**} s_i$, in particular, $h_{n-1}^{**} = s_{n-2}.g_{n-1} + f_n^{**} s_{n-1}$. It follows $h_{n-1}^{**} g_n = f_n^{**} s_{n-1} g_n = f_n^{**} h_n^{**}$.

The map f_n^{**} is a monomorphism, hence $s_{n-1} g_n = h_n^{**}$ and g_n splits. A contradiction. □

COROLLARY. Let Λ be a noetherian generalized Auslander regular Koszul algebra of global dimension n. Then any torsion free module M has $pd M < n$.

Proof. Let M be torsion free. Then $M = \varinjlim M_\alpha$ with M_α finitely generated and torsion free. By [9], $Ext_\Lambda^n(\varinjlim M_\alpha, \Lambda) \cong \varprojlim Ext_\Lambda^n(M_\alpha, \Lambda)$.

By the proposition, $Ext_\Lambda^n(M_\alpha, \Lambda) = 0$ for each α. Therefore $Ext_\Lambda^n(M, \Lambda) = 0$.

It follows, $p \dim M < n$. □

We can prove now theorem 6.1.

Proof. Let Λ be a noetherian generalized Auslander regular Koszul algebra and let $0 \to \Lambda \to E_0 \to E_1 \to E_2 \to \cdots \to E_n \to 0$ be the minimal injective coresolution of Λ.

We proved in [9], that $E_n \cong D(\Lambda)[n] \oplus E_n'$ with $soc\, E_n' = 0$.

Consider the exact sequence: $0 \to \Omega^{-n+1}(\Lambda) \to E_{n-1} \to E_n \to 0$ and assume $E_n' \neq 0$.

By the corollary, $Ext_\Lambda^n(E_n', \Lambda) = Ext_\Lambda^i(E_n', \Omega^{-n+1}(\Lambda)) = 0$.

This implies that the top row in the pull back:

$$0 \to \Omega^{-n+1}(\Lambda) \to W \to E_n' \to 0$$
$$\quad\;\; \downarrow id \qquad\;\; \downarrow \qquad \downarrow i$$
$$0 \to \Omega^{-n+1}(\Lambda) \to E_{n-1} \to E_n \to 0$$

splits.

Hence; E_n' is a summand of E_n, contradicting the minimality of the coresolution. We have proved $E_n' = 0$. □

REFERENCES

[1] K. Ajitabh, S.P. Smith, J.J. Zhang, *Auslander Gorenstein rings and their injective resolutions*, preprint, (1999).

[2] M. Auslander and M. Bridger, *Stable Module Theory*, Mem. of AMS 94, Providence 1969.

[3] A. Beilinson, V. Ginsburg, and W. Soergel, *Koszul duality patterns in representation theory*, J. Amer. Math. Soc. 9 (1996), 473–527.

[4] A.I. Bondal, E. Politshchuk, *Homological properties of associative algebras: the method of helices*, Russian. Acad. Sci. Izv. Math. 42, no. 2 (1994), 219–260.

[5] E. Cartan, S. Eilenberg, *Homological Algebra*, Princeton University Press, 1956.

[6] E. L Green, R. Martínez Villa, *Koszul and Yoneda algebras I*, Rep. Theory of Algebras, CMS Conference Proceedings, Vol. 18 (1996), 247–306.

[7] E. L Green, R. Martínez Villa, *Koszul and Yoneda algebras II*, in "Algebras and Modules II" CMS Conference Proceedings, Amer. Math. Soc. Providence, Vol. 24 (1998), 227–244.

[8] P. Jorgensen, J. Zhang, *Gourmet guide to Gorensteinness*, preprint (1999).

[9] R. Martínez-Villa, *Serre Duality for Generalized Auslander Regular Algebras*, Contemporary Math. Vol. 229 (1998), 237–263.

[10] R. Martínez-Villa, *Graded, Selfinjective and Koszul Algebras*, J. Algebra 215 (1999), 34–72.

[11] S. P. Smith, *Some finite dimensional algebras related to elliptic curves*, Rep. Theory of Algebras and Related Topics, CMS Conference Proceedings, Vol. 19 (1996), 315–348.

[12] J. Zhang, *Connected Graded Gorenstein Algebras with Enough Normal Elements*, J. Algebra, 189 (1997), 390–405.

Some remarks about the "double extension" algebra of a finite poset

TERESITA NORIEGA Dpto Ecuaciones Diferenciales, Fac Mat-Comp, Universidad de la Habana, San Lazaro y L Habana 4, Cuba, email: noriega@matcom.uh.cu

ABSTRACT In [4], Bautista and Martínez introduced what we will call the "double extension" algebra $\hat{\Lambda}$ of a finite poset S. We will denote by Λ the incidence algebra of S. The tilted algebras were introduced by Happel and Ringel in [5] and since then, have been proven to be a powerful tool in the study of different classes of algebras. The main result of this paper is a theorem that relates the property of $\hat{\Lambda}$ being tilted to the same condition in Λ. Some corollaries give more precise results in the cases: Λ is hereditary, Λ is tilted, Λ is tilted and of finite representation type.

1 INTRODUCTION

Throughout this paper, k will denote a fixed algebraically closed field. We will consider finitely generated right modules over a finite dimensional (associative with a unit) k-algebra A (or finitely generated left modules over A^{op}). If M is an A-module we denote by $End_A(M)$ the ring of endomorphisms of M, by $RadM$ its radical, by $SocM$ its socle and by $pd_A M$ ($id_A M$) its projective dimension (injective dimension). S will be considered a finite and connected poset. The incidence algebra Λ of S is the quotient of the path algebra corresponding to the Hasse diagram of S modulo the ideal generated by all commutativity relations.

The "double extension" algebra $\hat{\Lambda}$ of S is the quiver algebra $k\hat{Q}/I$ where $\hat{Q}$ is the Hasse diagram of $\hat{S} = S \cup \{m, f\}$ with $m \preceq s$ and $s \preceq f$ $\forall s \in S$ and I is the ideal generated by all differences of paths sharing the same initial and end points. $\hat{\Lambda}$ has a unique projective-injective $\hat{\Lambda}$-module $P_m = I_f$ and this we will denote by $\hat{P}$. We recall some basic definitions.

DEFINITION 1. *[1] Let A be a finite dimensional k-algebra, let T_A be a finitely generated A-module, we say that T_A is a tilting module if and only if it satisfies:*

1. $pdT_A \leq 1$

2. $Ext_A^1(T,T) = 0$

3. There exists a short exact sequence $0 \longrightarrow A_A \longrightarrow T'_A \longrightarrow T''_A \longrightarrow 0$ where T'_A and T''_A are direct sums of direct summands of T_A.

The third condition is equivalent to the following: "The number of indecomposable non-isomorphic summands of T_A is equal to the number of non-isomorphic simple modules in A."

There is also the dual notion of co-tilting module. We say that T_A is a cotilting module if and only if it satisfies:

1. $idT_A \leq 1$

2. $Ext_A^1(T,T) = 0$

3. There exists a short exact sequence $0 \longrightarrow T''_A \longrightarrow T'_A \longrightarrow D(A) \longrightarrow 0$ where T'_A and T''_A are direct sums of direct summands of T_A.

DEFINITION 2. [1] *An algebra A is called a tilted algebra of type $\vec{\Delta}$ (where $\vec{\Delta}$ is a finite connected quiver without oriented cycles) if there exists a tilting module T_B over the path algebra $B = k\vec{\Delta}$ such that $A = End(T_B)$ or equivalently if there exists a tilting A^{op}-module ${}_A U$ such that $End({}_A U) = k\vec{\Delta}$.*

REMARK 1. *[1] If P_A is an indecomposable projective-injective A-module then it is a direct summand of any tilting module T_A.*

DEFINITION 3. *[1][5] A class S of non-isomorphic indecomposables A-modules is called a complete slice in mod A if it satisfies:*

1. *$U = \bigoplus_{M \in S} M$ is a sincere module (That is, $Hom_A(P,U) \neq 0$ for any non-zero projective A-module P).*

2. *If $M_0 \longrightarrow M_1 \longrightarrow \ldots \longrightarrow M_n$ is a sequence of non-zero non isomorphisms in mod A, with $M_0, M_n \in S$, then $M_i \in S$ for all $0 \prec i \prec m$.*

3. *If $0 \longrightarrow L \longrightarrow M \longrightarrow N \longrightarrow 0$ is an almost split sequence, then at most one of L and N lies in S. Furthermore, if an indecomposable summand of M lies in S, then either L or N lies in S.*

If S is a complete slice in A then $U = \bigoplus_{M \in S} M$ is a tilting A-module and it is called the slice module of S. A complete slice and its existence characterize tilted algebras according to the following theorem:

THEOREM 1. *[6] If B is hereditary and T_B is a tilting module with $A = EndT_B$, then the class of all indecomposable A-modules of the form $Hom_B(T,I)$ with I_B indecomposable injective is a complete slice in mod A. Conversely, if S is a complete slice in mod A, then $U_A = \bigoplus_{M \in S} M$, is a tilting module with $B = EndU_A$ hereditary and thus S is isomorphic to a complete slice of the previous form.*

In Λ it will be important to consider the indecomposable sincere module H with k at each vertex and all maps the identity.

2 MAIN THEOREM

Our main result is the following:

THEOREM 2. *$\hat{\Lambda}$ is tilted if and only if Λ is tilted and there is a tilting Λ-module T such that H is a direct summand of T and $End_\Lambda(T)$ is hereditary.*

Proof. 1.-Suppose that $\hat{\Lambda}$ is tilted.

Since $\hat{\Lambda}$ is tilted its Auslander-Reiten quiver contains a complete slice $\hat{S}$ as a connected full sub-quiver.
Let $\hat{T}$ be the slice module for $\hat{S}$. As $\hat{P}$ is a projective-injective $\hat{\Lambda}$-module it is a direct summand of every tilting $\hat{\Lambda}$-module and so lies on $\hat{S}$.
The only almost split sequence of $\hat{\Lambda}$-modules which contains $\hat{P}$ is

$$0 \to \operatorname{rad}\hat{P} \to \hat{P} \oplus \operatorname{rad}\hat{P}/\operatorname{soc}\hat{P} \to \hat{P}/\operatorname{soc}\hat{P} \to 0.$$

Thus $\hat{S}$ contains either $\operatorname{rad}\hat{P}$ or $\hat{P}/\operatorname{soc}\hat{P}$.
Let I be an injective $\hat{\Lambda}$-module. Then $\operatorname{top}I = \operatorname{top}\hat{P}$ is not a composition factor of $\operatorname{rad}\hat{P}$ and so there is no non-zero map from I to $\operatorname{rad}\hat{P}$. Dually, there is no nonzero map from $\hat{P}/\operatorname{soc}\hat{P}$ to a projective $\hat{\Lambda}$-module. It follows that, without loss of generality, we may assume that

$$\hat{T} = \hat{P} \oplus \operatorname{rad}\hat{P} \oplus \operatorname{rad}\hat{P}/\operatorname{soc}\hat{P} \oplus T' \qquad (1)$$

for some $\hat{\Lambda}$-module T'.
Write $T = \operatorname{rad}\hat{P}/\operatorname{soc}\hat{P} \oplus T'$. Then $\hat{T} = \hat{P} \oplus \operatorname{rad}\hat{P} \oplus T$ is a tilting module if and only if $\hat{T}' = \hat{P} \oplus \hat{P}/\operatorname{soc}\hat{P} \oplus T$ is a tilting module. We denote by $\hat{S}'$ the slice corresponding to $\hat{T}'$.
We shall show that T is a tilting Λ-module with $End_\Lambda(T)$ hereditary.

T is a Λ-module.

We have to prove that T can't have $Top\hat{P}$ or $Soc\hat{P}$ as composition factors, it is enough to prove that if X is a summand of T, then $Hom_{\hat{\Lambda}}\left(\hat{P}, X\right) = 0$ and $Hom_{\hat{\Lambda}}\left(X, \hat{P}\right) = 0$. Suppose that there is a summand X of T such that $Hom_{\hat{\Lambda}}\left(\hat{P}, X\right) \neq 0$. Let o be a non-zero map from $\hat{P}$ to X, then o factorizes through $\hat{P}/Soc\hat{P}$ and hence $\hat{P}/Soc\hat{P}$ will belong to the slice $\hat{S}$, but this is impossible as $Rad\hat{P} = \tau\left(\hat{P}/Soc\hat{P}\right)$ belongs to $\hat{S}$.
Suppose that $Hom_{\hat{\Lambda}}\left(X, \hat{P}\right) \neq 0$ with X a direct summand of T.
From (1) we had that $\hat{T}$ is tilting if and only if $\hat{T}'$ is tilting. Consider now $\hat{T}'$ and let η be a non-zero map from X to $\hat{P}$. Then η factors through $Rad\hat{P} = \tau\left(\hat{P}/Soc\hat{P}\right)$ which is impossible as $\hat{P}/Soc\hat{P}$ belongs to $\hat{S}'$.

REMARK 2. *Notice that $Rad\hat{P}/Soc\hat{P}$ is isomorphic to H as a Λ-module.*

$pd_\Lambda T \leq 1$.

As $pd_{\hat\Lambda} \hat T = Sup\{pd_{\hat\Lambda} \hat P, pd_{\hat\Lambda} Rad\hat P, pd_{\hat\Lambda} T\}$ and $pd_{\hat\Lambda} \hat T \leq 1$, we have $pd_{\hat\Lambda} T \leq 1$.
So we have a projective resolution:

$$0 \longrightarrow_{\hat\Lambda} P_1 \longrightarrow_{\hat\Lambda} P_0 \longrightarrow_{\hat\Lambda} T \longrightarrow 0 \qquad (2)$$

of T as $\hat\Lambda$-module.
Using Proposition 1.1 of [2], we get

$$_\Lambda T \simeq Hom_{\hat\Lambda}(\overline P, T)$$

where $\overline P = \bigoplus_{x \in \Lambda_0} P_x$ and thus $pd_\Lambda T \leq 1$.

$\underline{Ext_\Lambda(T,T) = 0}$.

It suffices to prove that $Ext_\Lambda^1(T_i, T_j) = 0$ $\forall i,j$, where the T_i's are the direct summands of T.
Since $pd_\Lambda T_i \leq 1$, we have $Ext_\Lambda^1(T_i, T_j) \simeq D\, Hom_\Lambda(T_j, \tau_\Lambda T_i)$.
It will be enough to prove that there is no path from T_j to $\tau_\Lambda T_i$. Suppose the contrary, that is, there is a path from T_j to $\tau_\Lambda T_i$. Let's consider

$$0 \longrightarrow \tau_\Lambda T_i \xrightarrow{u} F \xrightarrow{v} T_i \longrightarrow 0 \qquad (3)$$

the almost split sequence in mod Λ.

$$0 \longrightarrow \tau_{\hat\Lambda} T_i \xrightarrow{f} E \xrightarrow{g} T_i \longrightarrow 0 \qquad (4)$$

the almost split sequence in mod $\hat\Lambda$.
The sequence (3) considered as a sequence in mod $\hat\Lambda$ is exact non split, so we will have the following conmutative diagram:

$$\begin{array}{ccccccccc} 0 & \longrightarrow & \tau_\Lambda T_i & \xrightarrow{u} & F & \xrightarrow{v} & T_i & \longrightarrow & 0 \\ & & w \downarrow & & h \downarrow & & \| & & \\ 0 & \longrightarrow & \tau_{\hat\Lambda} T_i & \xrightarrow{f} & E & \xrightarrow{g} & T_i & \longrightarrow & 0 \end{array}$$

Then we can find a path

$$T_j \longrightarrow \cdots \longrightarrow \tau_\Lambda T_i \longrightarrow \tau_{\hat\Lambda} T_i \longrightarrow * \longrightarrow T_i$$

Since T_i and T_j belong to the slice $\hat S$, it follows that $\tau_{\hat\Lambda} T_i$ should belong also, which contradicts the fact that slices are sectional.

The number i of indecomposable non-iso summands of T is equal to the number s of simple non-iso Λ-modules

We know [5] that the number of indecomposable non-iso summands of $\hat T$ is equal to the number of simple non-iso $\hat{\Lambda}$-modules. Let us call both of them n. Since $\hat{T} = \hat{P} \oplus Rad\hat{P} \oplus T$, we have $i = n - 2$ and by construction of Λ, we have $s = n - 2$ too, so $i = s$.

$End_\Lambda(T)$ is hereditary

Let $\hat{T} = \hat{P} \oplus Rad\hat{P} \oplus T = T'' \oplus T$. Then:

$$\hat{H} = End_{\hat{\Lambda}}\left(\hat{T}\right) = \begin{pmatrix} End_{\hat{\Lambda}}(T'') & 0 \\ Hom_{\hat{\Lambda}}(T'', T) & End_{\hat{\Lambda}}(T) \end{pmatrix}$$

The bound quiver of $End_\Lambda(T) = End_{\hat{\Lambda}}(T)$ is a full convex bound subquiver of $\hat{H}$. Hence, $End_\Lambda(T)$ is hereditary.

2.-Suppose that Λ is tilted and that there is a tilting Λ-module T such that H is a direct summand of T and $End_\Lambda T$ is hereditary.

We will first establish two lemmas.

LEMMA 1. *Let Λ be an algebra and T a tilting Λ-module such that $End_\Lambda(T)$ is hereditary. Let T_0 be a summand of T. If $\Sigma = \Lambda[T_0]$ is the one-point extension [7] of Λ by T_0 and P is the indecomposable projective Σ-module which is not a Λ-module, then:*

1. *$T \oplus P$ is a tilting Σ-module.*

2. *$End_\Sigma(T \oplus P)$ is hereditary.*

Note. Results similar to 1. are known, see [6].

Proof. $pd_\Sigma(T \oplus P) \leq 1$.

$pd_\Sigma(T \oplus P) = Sup\{pd_\Sigma T, pd_\Sigma P\} = pd_\Sigma T = pd_\Lambda T \leq 1$

$Ext^1_\Sigma(T \oplus P, T \oplus P) = 0$.

$$Ext^1_\Sigma(T \oplus P, T \oplus P) = Ext^1_\Sigma(T, T \oplus P) \oplus Ext^1_\Sigma(P, T \oplus P).$$

As P is Σ-projective, we have $Ext^1_\Sigma(P, T \oplus P) = 0$ so we have to prove that

$$Ext^1_\Sigma\left(T, T \oplus P\right) = 0,$$

which is the same as $Ext^1_\Sigma(T, T) = 0$ and $Ext^1_\Sigma(T, P) = 0$.
Let us consider the sequence:

$$0 \longrightarrow T \longrightarrow E \longrightarrow T \longrightarrow 0 \tag{5}$$

an exact sequence of Σ-modules. Λ is a full convex subcategory of Σ, so mod Λ is closed under extensions and (5) will be an exact sequence of Λ-modules.
As $Ext_\Lambda^1(T,T) = 0$, we have $Ext_\Sigma^1(T,T) = 0$.
We will consider now the exact sequence in mod Σ:

$$0 \longrightarrow T_0 \longrightarrow P \longrightarrow S \longrightarrow 0$$

(*notice that $T_0 \simeq RadP$*). Applying to it $Hom_\Sigma(T,-)$, we get:

$$\cdots \longrightarrow Ext_\Sigma^1(T,T_0) \longrightarrow Ext_\Sigma^1(T,P) \longrightarrow Ext_\Sigma^1(T,S) \longrightarrow 0$$

$Ext_\Sigma^1(T,S) = 0$, because S is Σ-injective, $Ext_\Sigma^1(T,T_0) = 0$ because T_0 is a direct sumand of T and $Ext_\Sigma^1(T,T) = 0$, hence $Ext_\Sigma^1(T,P) = 0$.

The number $\tilde{\imath}$ of indecomposable non-iso summands of $T \bigoplus P$ is equal to the number $\tilde{s}$ of non-iso simples in Σ.

As the number n of indecomposable non-iso summands of T is equal to the number of non-iso simples in Λ, then by definition of one-point extension we get: $\tilde{\imath} = n + 1 = \tilde{s}$.

$End_\Sigma(T \bigoplus P)$ is hereditary.

We have:

$$End_\Sigma(T \bigoplus P) = \begin{pmatrix} End_\Sigma(T) & Hom_\Sigma(T,P) \\ Hom_\Sigma(P,T) & End_\Sigma(P) \simeq k \end{pmatrix}$$

$Hom_\Sigma(P,T) = 0$, because if there is a non-zero morphism $f: P \longrightarrow T$, then T will have $topP$ as composition factor and that is impossible.
So $End_\Sigma(T \bigoplus P)$ is the one point extension of $End_\Sigma(T)$ by $Hom_\Sigma(T,P)$. As $End_\Sigma(T) = End_\Lambda(T)$ and is hereditary, we need only to show that $Hom_\Sigma(T,P)$ is $End_\Sigma(T)$-projective.
We claim that $Hom_\Sigma(T,P) = Hom_\Sigma(T,T_0)$
Let $f: T \longrightarrow P$ be a Σ-linear map. Then f is not epi, because if f is epi, as P is Σ-projective, f splits and P will be a direct summand of T, which is impossible. P is indecomposable projective with a unique maximal submodule $RadP$, so $Im\, f \subseteq Rad\, P \simeq T_0$, and then:

$$Hom_\Sigma(T,P) = Hom_\Sigma(T,T_0) = Hom_\Lambda(T,T_0)$$

But $Hom_\Lambda(T,T_0)$ is $End_\Lambda(T')$-projective, since $T = T_0 \bigoplus T'$ and

$$End_\Lambda(T) = Hom_\Lambda(T,T_0) \bigoplus Hom_\Lambda\left(T,T'\right).$$

Therefore $Hom_\Sigma(T,P)$ is $End_\Sigma(T)$-projective and hence $End_\Sigma(T \bigoplus P)$ is hereditary. □

We will now give the following lemma, dual of Lemma 1, without proof.

LEMMA 2. *If Γ is tilted with U a cotilting module and $\text{End}_\Gamma(U)$ hereditary, let U_0 be a direct summand of U, then the one-point coextension [7] $\Gamma' = [U_0]\Gamma$ of Γ by U_0 is tilted with cotilting module $U \oplus I'$, where I' is the indecompasable innjective Γ'-module which is not a Γ-module and $\text{End}_\Gamma(U \oplus I')$ is hereditary.*

We will now prove that $\hat{\Lambda}$ is tilted.
First we consider $\Lambda' = \Lambda[H]$ the one-point extension of Λ by H. Let us call P' the new projective. By Lemma 1, Λ' is tilted with tilting Λ'-module $T \oplus P'$ and $\text{End}_{\Lambda'}(T \oplus P')$ hereditary.
Now by Lemma 2 we can consider the one-point coextension $[P']\Lambda'$ of Λ' by P'. The indecomposable injective $[P']\Lambda'$-module which is not a Λ'-module, will be a cotilting module over $[P']\Lambda' \simeq \hat{\Lambda}$, by abuse of language, we will call it $\hat{P}$.
But:

$$pd_{\hat{\Lambda}}\left(T \oplus P' \oplus \hat{P}\right) = pd_{\hat{\Lambda}}(T \oplus P') = pd_\Lambda(T \oplus P') \leq 1$$

So $T \oplus P' \oplus \hat{P}$ is a tilting $\hat{\Lambda}$-module with $\text{End}_{\hat{\Lambda}}\left(T \oplus P' \oplus \hat{P}\right)$ hereditary and hence $\hat{\Lambda}$ is tilted. □

The following corollaries can be obtained from the theorem.

COROLLARY 1. *If Λ is hereditary then $\hat{\Lambda}$ is tilted $\iff H$ is not regular.*

Proof. By [5], H not regular $\iff H$ is postprojective (in [5] they are called preprojective) or preinjective.
As H is an indecomposable and sincere Λ-module, there exists a complete slice in $\mod \Lambda$ on which H lies. Then by the theorem, $\hat{\Lambda}$ is tilted. □

COROLLARY 2. *If Λ is tilted, then $\hat{\Lambda}$ is tilted $\iff H$ is directing.*

Proof. If H is directing, since it is an indecomposable and sincere Λ-module, it belongs to a complete slice and applying the theorem $\hat{\Lambda}$ is tilted.
If $\hat{\Lambda}$ is tilted, then H belongs (by the theorem) to a complete slice, then also to a connecting component and then by [7], H is directing. □

COROLLARY 3. *If Λ is tilted and of finite representation type, then $\hat{\Lambda}$ is tilted.*

Proof. This follows from Corollary 2, since if Λ is of finite representation type, all indecomposable Λ-modules are directing. □

ACKNOWLEDGEMENTS

The author would like to thank Dr. Sheila Brenner and Dr Ibrahim Assem for many usefull and enlightning discusions and sugestions.

REFERENCES

[1] I. Assem. Tilting Theory, an introduction. Banach Center Bulletin. Vol 26, part 1, 1990, pp 127-179.

[2] I. Assem and P. Brown. Strongly simply connected Auslander algebras. Université de Sherbrooke. preprint No 160, June 1995.

[3] M. Auslander, I. Reiten and S. Smalo. Representation theory of Artin algebras. Cambridge Studies in Advanced Mathematics, Vol 36. Cambridge University Press. 1994.

[4] R. Bautista and R. Martínez. Representation of posets and 1-Gorenstein Artin algebras. Proceedings of the 1978 Antwerp Conference. Marcel Dekker Inc., 1979, pp 385-433.

[5] D. Happel and C.M. Ringel. Tilted Algebras. Transactions of the American Mathematical Society. Vol 274, No 2, 1982. pp 399-443.

[6] C.M. Ringel. Tame algebras and integral cuadratic forms. Lect. Notes in Math. 1099, Springer Verlag. 1984.

[7] C.M. Ringel. Representation theory of finite dimensional algebras. London Math. Soc. Lect. Notes 116, 1985. pp 7-81.

Coil algebras which are derived-tame

JOSE ANTONIO DE LA PEÑA Instituto de Matemáticas, UNAM, Ciudad Universitaria, México 04510, D. F., México, E-mail: jap@penelope.matem.unam.mx

BERTHA TOMÉ Depto. de Matemáticas, Facultad de Ciencias, UNAM, Ciudad Universitaria, México 04510, D. F., México, E-mail: bta@hp.fciencias.unam.mx

ABSTRACT Let k be an algebraically closed field and A be a finite dimensional k-algebra. A classification of the coil algebras A having non-negative Euler form χ_A is presented in order to prove our main result: the repetitive algebra $\hat{A}$ of a coil algebra A is tame, if and only if the Euler form χ_A of A is non-negative.

Let A be a finite-dimensional algebra over an algebraically closed field k. Following [14], we say that A is *derived-tame* if $gl\dim A < \infty$ and the repetitive category $\hat{A}$ of A is tame. It has been shown that the class of derived-tame algebras is closed under derived equivalence and includes the hereditary tame algebras, the domestic tubular and tubular algebras and the pg-critical algebras (see [2,10,11,14]).

If $gl.\dim A < \infty$, the *homological bilinear form* is given, for the classes $\dim X = x$ and $\dim Y = y$ of the modules X, Y in the Grothendieck group $K_0(A)$ of A, by $\langle x, y \rangle_A = \sum_{i=0}^{\infty} (-1)^i \dim_k \operatorname{Ext}_A^i(X, Y)$. The corresponding quadratic form, denoted by χ_A, is called the *Euler form* of A. We shall denote by $(-,-)$ the symmetrization of the form $\langle -, - \rangle$.

For several classes of algebras (see [6,7,14]), it has been shown that A is derived-tame if and only if χ_A is a non-negative quadratic form. In this work we establish a result of this kind for the class of algebras obtained as coil enlargements of critical algebras not of type $\tilde{A}_n$. Coil enlargements of critical algebras were introduced in [3,4] for the study of polynomial growth algebras and henceforth are considered as important families of algebras. In proving our theorem we shall obtain a complete description of the considered algebras.

The results of this work were presented by the second author during the Conference in Representation Theory of Algebras held in Sao Paulo in July 1999. Both authors acknowledge support from CONACyT, México.

1 PRELIMINARIES

1.1. Let A be a basic, finite-dimensional algebra over an algebraically closed field k. It is well known that A may be written as $A = kQ/I$, where Q is a finite quiver and I is an admissible ideal of the path algebra kQ. We assume moreover that Q is connected and has no oriented cycle. We consider A as a k-category whose objects are the vertices Q_0 of Q and in which the morphism space $A(x,y)$ from x to y is $e_y A e_x$, where e_x denotes the primitive idempotent associated to the vertex x. See [9].

The *repetitive category* $\hat{A}$ is the k-category with objects $Q_0 \times \mathbb{Z}$ (which are denoted by $s[i]$ for $s \in Q_0$ and $i \in \mathbb{Z}$) and in which the only possibly non-zero morphism spaces are $\hat{A}(r[i], s[i]) = A(r,s) \times \{i\}$ and $\hat{A}(r[i], s[i+1]) = DA(s,r) \times \{i\}$, where $D = \mathrm{Hom}_k(-, k)$ denotes the usual duality. See [10].

1.2. Let $A = kQ/I$ be as above. We recall that the one-point extension $A[M]$ of the algebra A by a module M is the k-category with objects $Q_0 \cup \{s\}$ and morphism spaces $A[M](s, x) = M(x)$ for $x \neq s$, $A[M](s,s) = k$, and such that A is a full convex subcategory. See [16].

Let j be a sink in Q. The *reflection* $S_j^+ A$ of A at j is the quotient of the one-point extension $A[I_j]$ by the two-sided ideal generated by e_j, where I_j denotes the injective envelope of the simple S_j at j. The reflection $S_i^- A$ of A at a source i of Q is defined dually.

It was shown in [18] that $S_j^+ A$ and A are *tilting-cotilting equivalent*, that is, there is a sequence of algebras $A = A_0, A_1, \ldots, A_m = S_j^+ A$ and a sequence of modules $_{A_i}T^i$ with $0 \leq i < m$ such that $A_{i+1} = \mathrm{End}_{A_i} T^i$ and T^i is either a tilting or a cotilting module. In particular, this implies that $S_j^+ A$ and A are derived equivalent, that is, they have triangle equivalent derived categories.

Let B be any finite-dimensional k-algebra. Then by [12], the repetitive categories $\hat{A}$ and $\hat{B}$ are isomorphic if and only if A and B are *reflection-equivalent*, that is, there is a sequence of algebras $A = A_0, A_1, \ldots, A_m = B$ where $A_{i+1} = S_{\sigma(i)}^- A_i$ or $S_{\tau(i)}^+ A_i$ for some source $\sigma(i)$ or some sink $\tau(i)$ of A_i ($0 \leq i < m$).

1.3. An algebra A is *tame* if for every $d \in \mathbb{N}$ there is a family $M_1, \ldots, M_{\kappa(d)}$ of $A - k[t]$-bimodules which are finitely generated and free as right $k[t]$-modules and such that almost every indecomposable A-module X of dimension d is of the form $M_i \otimes_{k[t]} S_\lambda$ for some $1 \leq i \leq \kappa(d)$ and $\lambda \in k$, where $S_\lambda = k[t]/(t-\lambda)$ is a simple $k[t]$-module.

An infinite category, such as the repetitive category $\hat{A}$ of an algebra A, is said to be *tame* if every full finite subcategory is tame. Clearly, if $\hat{A}$ is derived-tame, then A is tame.

1.4. We recall that a *2-tubular extension* is a one-point extension $D[M]$ of a critical algebra D of type $\tilde{\mathbb{D}}_m$ by an indecomposable regular module M of regular length 2 and period $m - 2$. For basic definitions see [16].

Therefore, we call an iterated one-point extension $D[M_1,\ldots,M_r]$ of a critical algebra D of type $\tilde{\mathbb{D}}_m$ by indecomposable regular modules of regular length 2 and period $m - 2$ such that $M_i \not\cong \tau_D M_{i+1}$ for $1 \leq i \leq r - 1$, and $M_r \not\cong \tau_D M_1$, an *iterated 2-tubular extension*.

1.5. For an indecomposable module M in an inserted tube of a domestic tubular algebra, the notion of *level* was defined in [17], where it was used indistinctively for domestic tubular or cotubular algebras. Here we will distinguish the two cases.

Let D be a domestic branch extension of a critical algebra C, and let $\mathcal{T}$ be an inserted tube in Γ_D. The ray modules in $\mathcal{T}$ will be called of level 1, and a module $M \in \mathcal{T}$ will be called of level $n > 1$ if there is a sectional path in $\mathcal{T}$ of length $n - 1$

$$M = M_n \to M_{n-1} \to \cdots \to M_2 \to M_1 = E$$

with E a ray module. The notion of colevel is defined dually.

Let D and $\mathcal{T}$ be as above, and let $M \in \mathcal{T}$. By [16], if B is any critical algebra of tubular type the extension type of D over C, then there exist a tilting B-module $T = T_1 \oplus T_2$, with T_1 preprojective and T_2 regular, such that $D = \text{End}_B(T)$, and an indecomposable regular B-module X such that $M = \Sigma X$, where $\Sigma = \text{Hom}_B(T, -)$. It was shown in [17] that the level of M is n if and only if the regular length $\ell(X)$ of X is n.

The following result is also taken from [17].

LEMMA. *Let D and M be as above. If $\chi_{D[M]}$ is non-negative then level $(M) \leq 2$. Moreover, if level $(M) = 2$, then C is of type $\tilde{\mathbb{D}}_m$, and the extension type of D over C is $(n - 2, 2, 2)$, where $n > m \geq 4$.*

1.6. Let B be an algebra, $\mathcal{C}$ be a standard component of Γ_B and X be an indecomposable module in $\mathcal{C}$. In [3], three *admissible operations* (ad 1), (ad 2) and (ad 3) were defined depending on the shape of the support of $\text{Hom}_B(X, -)|_{\mathcal{C}}$ in order to obtain a new algebra B'.

(ad 1) If the support of $\text{Hom}_B(X, -)|_{\mathcal{C}}$ is of the form

$$X = X_0 \to X_1 \to X_2 \to \cdots$$

we set $B' = (B \times D)[X \oplus Y_1]$, where D is the full $t \times t$ lower triangular matrix algebra and Y_1 is the indecomposable projective-injective D-module.

(ad 2) If the support of $\text{Hom}_B(X, -)|_{\mathcal{C}}$ is of the form

$$Y_t \leftarrow \cdots \leftarrow Y_1 \leftarrow X = X_0 \to X_1 \to X_2 \to \cdots$$

with $t \geq 1$, so that X is injective, we set $B' = B[X]$.

(ad 3) If the support of $\text{Hom}_B(X, -)|_{\mathcal{C}}$ is of the form

$$\begin{array}{ccccccc}
Y_1 & \to & Y_2 & \to & \cdots & \to & Y_t \\
\uparrow & & \uparrow & & & & \uparrow \\
X = X_0 & \to & X_1 & \to & \cdots & \to & X_{t-1} & \to & X_t & \to & \cdots
\end{array}$$

with $t \geq 2$, so that X_{t-1} is injective, we set $B' = B[X]$.

In each case, the module X and the integer t are called, respectively, the *pivot* and the *parameter of the admissible operation*. Moreover, the component C' of $\Gamma_{B'}$ containing X is standard under certain conditions satisfied in this work.

The dual operations are denoted by (ad 1*), (ad 2*) and (ad 3*).

Following [5], an algebra A is a *coil enlargement* of the critical algebra C if there is a sequence of algebras $C = A_0, A_1, \ldots, A_m = A$ such that for $0 \leq i < m$, A_{i+1} is obtained from A_i by an admissible operation with pivot in a stable tube of Γ_C or in a component (coil) of Γ_{A_i} obtained from a stable tube of Γ_C by means of the admissible operations done so far. When A is tame, we call A a *coil algebra*.

If A is a coil enlargement of a critical algebra C, then there is a maximal branch coextension A^- of C inside A which is full and convex in A, and such that A is obtained from A^- by a sequence of admissible operations of types (ad 1), (ad 2) and (ad 3). Dually, there is a maximal branch extension A^+ of C inside A which is full and convex in A, and such that A is obtained from A^+ by a sequence of admissible operations of types (ad 1*), (ad 2*) and (ad 3*).

1.7. For a coil enlargement A of a critical algebra C, we define the type $t(A)$ of A as follows:

Let $\mathcal{T} = (\mathcal{T}_\lambda)_{\lambda \in \mathbb{P}_1(k)}$ be the separating tubular family of mod C. For each $\lambda \in \mathbb{P}_1(k)$, let n_λ be the rank of $\mathcal{T}_\lambda$, r_λ (respectively, c_λ) be the number of rays (respectively, corays) inserted in $\mathcal{T}_\lambda$ by the sequence of admissible operations that leads from C to A, and $t_\lambda = n_\lambda + r_\lambda + c_\lambda$. Finally, let $t(A) = (t_\lambda)_{\lambda \in \mathbb{P}_1(k)}$, where we write down only those $t_\lambda \neq 1$.

1.8. A coil enlargement A of a critical algebra C is called a *branched-critical algebra* if A is obtained from C by a sequence of admissible operations of types (ad 1) and (ad 1*) such that the pivot of each operation is both a ray and a coray module (see [8]).

We observe that Ringel's branch extensions or coextensions of critical algebras are branched-critical algebras A for which A^- or A^+ is trivial. Moreover, if A is branched-critical and A^- and A^+ are non-trivial, then A can be written as $A = A^-[M_i, K_i]_{i=1}^t$, where the M_i are both ray and coray A^--modules and the K_i are branches.

2 COIL ALGEBRAS WITH NON-NEGATIVE EULER FORM

2.1. We devote this section to the characterization of the coil enlargements A of a critical algebra C, not of type $\tilde{A}_n$, having non-negative Euler form. The following proposition together with [5] and [15] shows that these algebras are tame, that is, coil algebras.

PROPOSITION. *Let A be a coil enlargement of a critical algebra C, not of type $\tilde{A}_n$, such that both A^- and A^+ are non-trivial. If χ_A is non-negative, then A^- and A^+ are domestic.*

Proof. Since χ_A is non-negative, so are χ_{A^-} and χ_{A^+}. Thus, by [15], A^- and A^+

are either domestic or tubular. Assume that A^- is a tubular algebra which is a branch extension of C_0 and a branch coextension of $C_\infty = C$. Then

$$\text{ind } A^- = \mathcal{P}_0 \vee \mathcal{T}_0 \vee \bigvee_{\gamma \in \mathbb{Q}^+} \mathcal{T}_\gamma \vee \mathcal{T}_\infty \vee \mathcal{I}_\infty.$$

Let z_0 be the minimal positive generator of rad χ_{C_0}, and let E be a simple regular C_0-module of period 1. Without loss of generality, we may assume that A is obtained from A^- by a single admissible operation. Let $X \in \text{ind } A^-$ be the pivot of the admissible operation and let P_w be the indecomposable projective A-module such that X is a direct summand of rad P_w. Since $E \in \mathcal{T}_0$, $X \in \mathcal{T}_\infty$, $\mathcal{T}_0$ is separating, and $p\dim_A E = p\dim_{A^-} E = 1$,

$$\begin{aligned}(\dim P_w, z_0)_A &= \langle \dim E, \dim P_w \rangle \\ &= \dim_k \text{Hom}_A(E, P_w) - \dim_k \text{Ext}^1_A(E, P_w) \\ &= \dim_k \text{Hom}_{A^-}(E, X) > 0.\end{aligned}$$

Then $\chi_A(\dim P_w - 2z_0) = 1 - 2(\dim P_w, z_0) < 0$, a contradiction. Therefore A^- is domestic.

Dually, one shows that A^+ is domestic. □

2.2. Since the coil algebra A can be obtained from A^- by a sequence of admissible operations of types (ad 1), (ad 2) and (ad 3), and χ_A is non-negative, (2.1) and (1.5) show that the pivots of such operations which belong to mod A^- must be of colevel at most 2.

Assume first that A is branched-critical, that is, $A = A^-[M_i, K_i]_{i=1}^t$, where the M_i are both ray and coray A^--modules and the K_i are branches.

Let $A^- = \text{End}_B(T)$, where B is a critical algebra of tubular type the coextension type of A^- over C, and $T = T_1 \oplus T_2$ is a cotilting B-module with T_1 regular and T_2 preinjective. For $1 \le i \le t$, let $M_i = \Sigma X_i$, where X_i is a simple regular B-module and $\Sigma = D\text{Hom}_B(-, T)$. Finally, let $B' = B[X_i, K_i]_{i=1}^t$.

LEMMA. *With the notation introduced above, there is a cotilting B'-module T' such that $A = \text{End}_{B'}(T')$.*

Proof. By induction on t. When $t = 1$, $A = A^-[M_1, K_1]$ can be obtained from A^- by a sequence of one-point extensions and coextensions. The proof of this case then follows from [17]. □

2.3. For branched-critical algebras, we achieve our goal with the following result.

PROPOSITION. *If A is a branched-critical algebra, then χ_A is non-negative if and only if $t(A)$ is Dynkin or Euclidean.*

Proof. We may assume that both A^- and A^+ are non-trivial. Under any of the two hypothesis, A^- is domestic, and hence A is cotilting equivalent to a branch extension $B' = B[X_i, K_i]_{i=1}^t$, where B is a critical algebra whose tubular type is the same as the coextension type of A^- over C. Therefore $t(A)$ is the extension type of B' over B. The assertion then follows from [15]. □

2.4. Assume now that A is not branched-critical, that is, the sequence of admissible operations that leads from A^- to A contains an operation of one of the following kinds: an operation (ad 1) whose pivot is not a coray module, an operation (ad 2), an operation (ad 3). We consider the first operation of this kind that appears in the sequence, and we denote its pivot by N. By [5], we may assume that all the operations of type (ad 1) whose pivot is a coray module and which can be carried out before the operation with pivot N precede it in the sequence.

Let A' be the algebra obtained from A^- by the operations mentioned above, that is, $A' = A^-[M_i, K_i]_{i=1}^r[N]$, where $A^-[M_i, K_i]_{i=1}^r$ is branched-critical. Note that the coil containing N in $\Gamma_{A^-[M_i,K_i]_{i=1}^r}$ has no exceptional mesh (see [3]). Hence we can keep the notions of level and colevel as defined in (1.5).

As in (2.2), $A^-[M_i, K_i]_{i=1}^r = \operatorname{End}_{B'}(T')$, where $B' = B[X_i, K_i]_{i=1}^r$ is a branch extension of the critical algebra B whose tubular type is the coextension type of A^- over C, and T' is a cotilting B'-module.

If N is an A^--module, then $N = \Sigma Z$, where Z is an indecomposable regular B-module with regular length $\ell(Z) = \operatorname{colevel}(N) > 1$. If N is not an A^--module, then $N = \overline{R} = (R, \operatorname{Hom}_{A^-}(M_i, R), 1)$ for some indecomposable A^--module R and some $1 \leq i \leq r$. As above, $R = \Sigma Z$, where Z is an indecomposable regular B-module with $\ell(Z) = \operatorname{colevel}(R) = \operatorname{colevel}(N) > 1$. Then $N = \overline{R} = \Sigma \overline{Z}$, where $\overline{Z} = (Z, \operatorname{Hom}_B(X_i, Z), 1)$ is an indecomposable B'-module with $\operatorname{colevel}(\overline{Z}) = \ell(Z) > 1$. Hence, in both cases, we obtain that $N = \Sigma Y$ for some indecomposable B'-module Y with $\operatorname{level}(Y) \geq \operatorname{colevel}(Y) = \operatorname{colevel}(N) > 1$.

Let $B'' = B'[Y]$, then there is a cotilting B''-module T'' such that $A' = \operatorname{End}_{B''}(T'')$. Since $\chi_{A'}$ is non-negative, so is $\chi_{B''}$. The next result, whose proof is similar to that in (2.1), shows that B' is domestic.

LEMMA. *Let B' be a branch extension of a critical algebra B and Y be an indecomposable B'-module in the separating tubular family of mod B' obtained from the tubular family of mod B by ray insertions. If $\chi_{B'[Y]}$ is non-negative, then B' is domestic.*

2.5. The next two lemmas show that the colevel of N, as defined in (2.4), is 2.

LEMMA. *Let A be obtained from a branched-critical algebra by an admissible operation of one of the following kinds:*

i) *(ad 1) with parameter $t = 0$ and pivot a module of colevel greater than 1,*

ii) *(ad 2),*

iii) *(ad 3).*

Let N be the pivot of such operation. If χ_A is non-negative, then the colevel of N is 2.

Proof. As in (2.4), A is cotilting equivalent to $B'' = B'[Y]$, where B' is a branch extension of a critical algebra B, and Y is an indecomposable B'-module lying in an inserted tube of $\Gamma_{B'}$ and satisfying $\operatorname{level}(Y) \geq \operatorname{colevel}(Y) = \operatorname{colevel}(N) > 1$.

Since χ_A is non-negative, B' is domestic, and therefore tilting equivalent to a critical algebra D whose tubular type is the extension type of B' over B. Moreover,

Coil Algebras that Are Derived-Tame

$Y = \Sigma U$, where U is an indecomposable regular D-module with $\ell(U) = $ level (Y). Then $B'' = B'[Y]$ is tilting equivalent to $D[U]$, and so $\chi_{D[U]}$ is non-negative. By [13], $\ell(U) = 2$. Therefore colevel $(N) = 2$. □

2.6. LEMMA. *Let A be obtained from a branched-critical algebra by an operation (ad 1) whose pivot N is not a coray module. If χ_A is non-negative, then the parameter t of the operation is 0.*

Proof. Let $A^-[M_i, K_i]_{i=1}^r$ denote the branched-critical algebra from which A is obtained. If $t \geq 1$, then A is obtained from $A' = A^-[M_i, K_i]_{i=1}^r[N]$ by a sequence of t one-point coextensions. Without loss of generality, we may assume that $t = 1$.

Let 0 denote the extension vertex of A'. Then $A = [I_0']A'$, where I_0' is the simple injective A'-module corresponding to the vertex 0. As in (2.5), A' is cotilting-tilting equivalent to $D_1 = D[U]$, which is a 2-tubular extension. Then A is cotilting-tilting equivalent to $D_2 = [I_0]D_1$, where I_0 is the simple injective D_1-module corresponding to the vertex 0. Since χ_A is non-negative, so is χ_{D_2}.

Since D_1 is a 2-tubular extension, rad χ_{D_2} has 2 generators: the minimal positive generator z of rad χ_D and $\dim W + e_0$, where W is an indecomposable preinjective D-module such that $\dim_k \mathrm{Hom}_D(U, W) = 2$ (see [17]). Let w be the coextension vertex of $D_2 = [I_0]D_1$. Then $\chi_{D_2}(e_2 + 2(\dim W + e_0)) < 0$ for

$$(e_w, \dim W + e_0) = \langle \dim W + e_0, e_w \rangle = \langle e_0, e_w \rangle = -1.$$

Therefore $t = 0$. □

2.7. Note that the admissible operations in the sequence that leads from $A' = A^-[M_i, K_i]_{i=1}^r[N]$ to A fall into two classes:

a) those operations whose pivot arises from the one-point extension by N, and hence cannot be performed before this operation,

b) those operations which can be performed before the one-point extension by N.

A case by case inspection shows that there is no operation of the first kind.

LEMMA. *In the sequence of admissible operations that leads from $A' = A^-[M_i, K_i]_{i=1}^r[N]$ to A there is no operation whose pivot arises from the one-point extension by N.*

Proof. We analyze only the case in which N is an (ad 2)-pivot. The other cases are treated similarly.

Since the colevel of N is 2, the parameter t of the operation (ad 2) is 1, and there is an arrow from N to a simple module S lying on the mouth of the coil $\mathcal{C}'$ that contains N in $\Gamma_{A'}$. Any operation whose pivot arises from the one-point extension by N is either of type (ad 1) with pivot S, or has pivot a module of the form $\overline{R} = (R, \mathrm{Hom}_{A^-[M_i, K_i]}(N, R), 1)$, where R is an indecomposable $A^-[M_i, K_i]_{i=1}^r$-module lying on the ray starting at N.

First, we may assume that $A = A'[S]$. Then, keeping the notation introduced in (2.5) and (2.6), A is cotilting-tilting equivalent to $D_2 = D_1[V]$, where V is the inverse translate in mod D of the regular socle of U. Since χ_A is non-negative, so is χ_{D_2}. Let w be the extension vertex of D_2 and $\dim W + e_0$ be one of the generators of rad χ_{D_1}. Since $\text{pdim}_{D_1} V = \text{pdim}_D V = 1$, then

$$(e_w, \dim W + e_0)_{D_2} = -\langle \dim V, \dim W \rangle + \langle e_w, e_0 \rangle =$$
$$= -\dim_k \text{Hom}_D(V, W) + \dim_k \text{Ext}^1_D(V, W) = -1,$$

and therefore $\chi_{D_2}(e_w + 2(\dim W + e_0)) < 0$.

Next, we may assume that $A = A'[R]$. Then A is cotilting-tilting equivalent to $D_2 = D_1[\overline{V}]$, where $\overline{V} = (V, \text{Hom}_D(U, V), 1)$ and V is an indecomposable D-module lying on the ray starting at U. Thus χ_{D_2} is non-negative. Let w and $\dim W + e_0$ be as above. Since $\text{pdim}_{D_1} \overline{V} \leq 2$ and $\tau_{D_1} \overline{V}$ is a regular D-module,

$$(e_w, \dim W + e_0)_{D_2} =$$
$$= -\langle \dim \overline{V}, \dim W \rangle + \langle e_w, e_0 \rangle =$$
$$= -\dim_k \text{Hom}_{D_1}(\overline{V}, W) + \dim_k \text{Ext}^1_{D_1}(\overline{V}, W) - \dim_k \text{Ext}^2_{D_1}(\overline{V}, W) - 1 =$$
$$= -\dim_k \text{Ext}^2_{D_1}(\overline{V}, W) - 1 < 0$$

and therefore $\chi_{D_2}(e_w + 2(\dim W + e_0)) < 0$. □

2.8. Thus far we have shown that the sequence of admissible operations that leads from the branched-critical algebra $A^-[M_i, K_i]_{i=1}^r$ to A consists of operations of type (ad 1) with parameter 0 and operations of types (ad 2) and (ad 3), all of which have as pivot an $A^-[M_i, K_i]_{i=1}^r$-module of colevel 2 (and hence the operations commute between them).

We are now able to prove the following result which completes our classification.

PROPOSITION. *Let A be a coil enlargement of a critical algebra C not of type $\tilde{A}_n$. If A is not branched-critical, then χ_A is non-negative if and only if*

i) *C is of type $\tilde{D}_m$, $m \geq 4$,*

ii) *$t(A) = (n-2, 2, 2)$ with $n > m$, and*

iii) *A is cotilting-tilting equivalent to an iterated 2-tubular extension.*

Proof. Let $A' = A^-[M_i, K_i]_{i=1}^r[N]$ be as in the discussions above. If χ_A is non-negative, we know that A' is cotilting equivalent to $B'' = B[X_i, K_i]_{i=1}^r[Y]$, where B is a critical algebra with tubular type the coextension type of A^- over C. In turn, B'' is tilting equivalent to the 2-tubular extension $D_1 = D[U]$, where D is a critical algebra with tubular type the extension type of $B' = B[X_i, K_i]_{i=1}^r$ over B. Hence, the extension type of B' over B (which is the type of the branched-critical algebra $A^-[M_i, K_i]_{i=1}^r$) is $(p-2, 2, 2)$ for some $p \geq 4$. Then i) follows from the fact that C is not of type $\tilde{A}_m$, ii) follows from the considerations above the proposition, and iii) from the same considerations and repeated application of [17].

Conversely, we may assume that A is cotilting-tilting equivalent to one of the following iterated 2-tubular extensions whose Euler forms are easily seen to be non-negative.

$$\chi_D(x) = \frac{1}{4}(2x_1 - x_3)^2 + \frac{1}{4}(2x_2 - x_3)^2 + \frac{1}{2}(x_3 - x_4)^2 + \cdots + \frac{1}{2}(x_{a-1} - x_a)^2$$
$$+ \frac{1}{2}(x_a - x_{a+1} - x_{w_1})^2 + \frac{1}{2}(x_{a+1} - x_{a+2} - x_{w_1})^2 + \frac{1}{2}(x_{a+2} - x_{a+3})^2$$
$$+ \cdots + \frac{1}{2}(x_{b-1} - x_b)^2 + \frac{1}{2}(x_b - x_{b+1} - x_{w_2})^2$$
$$+ \frac{1}{2}(x_{b+1} - x_{b+2} - x_{w_2})^2 + \frac{1}{2}(x_{b+2} - x_{b+3})^2 + \cdots$$
$$+ \frac{1}{2}(x_{n-3} - x_{n-2})^2 + \frac{1}{2}(x_{n-2} - x_{n-1} - x_{w_k})^2$$
$$+ \frac{1}{2}(x_{n-1} - x_n - x_{n+1} - x_{w_k})^2 + \frac{1}{2}(x_n - x_{n+1})^2$$

D':   [diagram with vertices $w_1, 1, 2, 3, \ldots, a, a+1, a+2, \ldots, b, b+1, b+2, \ldots, n-2, n-1, n, n+1$ and relations $\varepsilon\beta\alpha = \varepsilon\delta\gamma$]

$$\chi_{D'}(x) = \frac{1}{4}(2x_1 - x_3 - x_{w_1})^2 + \frac{1}{4}(2x_2 - x_3 - x_{w_1})^2 + \frac{1}{2}(x_3 - x_4 - x_{w_1})^2$$
$$+ \frac{1}{2}(x_4 - x_5)^2 + \cdots + \frac{1}{2}(x_{a-1} - x_a)^2 + \frac{1}{2}(x_a - x_{a+1} - x_{w_2})^2$$
$$+ \frac{1}{2}(x_{a+1} - x_{a+2} - x_{w_2})^2 + \frac{1}{2}(x_{a+2} - x_{a+3})^2 + \cdots$$
$$+ \frac{1}{2}(x_{n-3} - x_{n-2})^2 + \frac{1}{2}(x_{n-2} - x_{n-1} - x_{w_k})^2$$
$$+ \frac{1}{2}(x_{n-1} - x_n - x_{n+1} - x_{w_k})^2 + \frac{1}{2}(x_n - x_{n+1})^2$$

□

2.9. Combining (2.3) and (2.8) we obtain:

THEOREM. *Let A be a coil enlargement of a critical algebra C not of type $\tilde{A}_n$. Then χ_A is non-negative if and only if A is branched-critical and $t(A)$ is Dynkin or Euclidean, or A is cotilting-tilting equivalent to an iterated 2-tubular extension and $t(A) = (n-2, 2, 2)$ for some $n \geq 4$.*

3 ALGEBRAS DERIVED EQUIVALENT TO COIL ALGEBRAS

3.1. This section contains the main result of our work. We start by proving that the coil algebras that appear in the classification given in (2.9) are derived-tame.
PROPOSITION. *For a branched-critical algebra A, the following are equivalent:*

i) χ_A is non-negative.

ii) $t(A)$ is Dynkin or Euclidean.

iii) A is derived-tame.

Proof. The equivalence between i) and ii) was established in (2.3). There we showed that if $t(A)$ is Dynkin or Euclidean, then A is cotilting equivalent to a branch extension $B' = B[X_i, K_i]_{i=1}^t$ of a critical algebra B, and the extension type of B' over B is $t(A)$. Therefore B' is either a domestic tubular or a tubular algebra and hence A, which is derived equivalent to B', is derived-tame. Thus we have shown that i) implies iii).

Finally, assume that $t(A)$ is neither Dynkin nor Euclidean. We shall prove that A is not derived-tame. We may assume that both A^- and A^+ are non-trivial and, as in (1.8), we may write $A = A^-[M_i, K_i]_{i=1}^t$, where the M_i are both ray and coray A^--modules and the K_i are branches. Moreover, we may assume that A^- is tame, for otherwise A is not derived-tame. Therefore A^- is either a domestic tubular or a tubular algebra.

If A^- is a tubular algebra, then

$$\text{ind } A^- = \mathcal{P}_0 \vee \mathcal{T}_0 \vee \bigvee_{\gamma \in \mathbb{Q}^+} \mathcal{T}_\gamma \vee \mathcal{T}_\infty \vee \mathcal{I}_\infty$$

and the extension modules M_i belong to the coinserted tubular family $\mathcal{T}_\infty$. The algebra $[M_i]A^-$ is a full subcategory of the repetitive category $\hat{A}$. By [13] (see also [1]), $[M_i]A^-$ is of wild type and hence A is not derived-tame.

If A^- is domestic, then as in (2.2), A is tilting equivalent to a branch extension $B' = B[X_i, K_i]_{i=1}^t$ of a critical algebra B, and the type $t(B') = t(A)$. By [15], B' is of wild type and consequently neither B nor A are derived-tame. □

3.2. LEMMA. *If A is an iterated 2-tubular extension then A is derived-tame.*

Proof. As in (2.8), we may assume that A has the following shape (the other case being similar)

The algebra $S_{w_1}^- S_{a-1}^- \ldots S_3^- S_2^- S_1^- A$ obtained by reflections from A is tilting equivalent to the algebra B_1 below

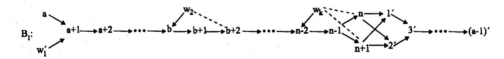

with commutative squares. The algebra $S_{w_2}^- S_{b-1}^- \ldots S_{a+1}^- S_a^- S_{w_1'}^- B_1$ is tilting equivalent to the algebra B_2 below with commutative squares.

Repeating the procedure, we get that A is derived equivalent to an algebra D of the shape

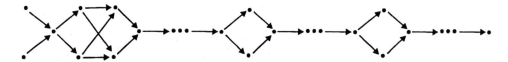

with all squares commutative. Clearly, every convex subcategory of $\hat{B}$ has again the above shape and hence $\hat{B}$ is tame as observed in [14]. □

3.3. THEOREM. *Let A be a coil enlargement of a critical algebra C not of type $\tilde{\mathbb{A}}_n$. Then A is derived-tame if and only if χ_A is non-negative.*

Proof. Assume first that χ_A is non-negative. Because of the classification established in section 2, either A is branched-critical with $t(A)$ Dynkin or Euclidean or A is tilting-cotilting equivalent to an iterated 2-tubular extension. By (3.1) and (3.2), A is derived-tame.

Assume now that χ_A is non-negative. If A is branched-critical, then A is not derived-tame by (3.1). Otherwise, we may assume that $A = A'[N]$, where A' is branched-critical and derived-tame, and N is either an (ad 2) or (ad 3)-pivot or an (ad 1)-pivot which is not a coray module. Since $\chi_{A'}$ is non-negative, A' is cotilting equivalent to a tame branch extension B' of a critical algebra whose tubular type is the coextension type of A^- over C. Moreover, A is cotilting equivalent to $B = B'[Y]$, where Y is an indecomposable module belonging to the separating tubular family of mod B' that contains projectives.

If B' is tubular, then B is wild by [1,13], and so neither B nor A is derived-tame. If B' is domestic, then A is tilting-cotilting equivalent to a one-point extension $D[U]$ of a critical algebra D by an indecomposable regular module U, and $\chi_{D[U]}$ is not non-negative. By [13], $D[U]$ is wild and A is not derived-tame. □

REFERENCES

[1] I. Assem and A. Skowroński. *Quadratic forms and iterated tilted algebras.* J. Algebra 128 (1990) 55–85.

[2] I. Assem and A. Skowroński. *On tame repetitive algebras.* Fund. Math. 142 (1993) 59–84.

[3] I. Assem and A. Skowroński. *Indecomposable modules over multicoil algebras.* Math. Scand. 71 (1992) 31–61.

[4] I. Assem and A. Skowroński. *Multicoil algebras.* Proc. ICRA VI Canadian Math. Soc. Conference Proc. 14 (1993) 29–67.

[5] I. Assem, A. Skowroński and B. Tomé. *Coil enlargements of algebras.* Tsukuba J. Math. 19 No. 2 (1995) 453–479.

[6] M. Barot. *Representation-finite derived tubular algebras*. Arch. Math. 74 (2000) 89–94.

[7] M. Barot and J. A. de la Peña. *Derived tubular strongly simply connected algebras*. To appear in the Proceedings of the Euroconference on Computer Algebra in Representations of groups and algebras.

[8] F. Coelho, J. A. de la Peña and B. Tomé. *Algebras whose Tits form weakly controls the module category*. J. of Algebra 191 (1997) 89–108.

[9] P. Gabriel and A. V. Roiter. *Representations of finite-dimensional algebras*. Algebra VIII, Encyclopaedia of Math. Sc. Vol 73. Springer (1992).

[10] D. Happel. *Triangulated categories in the Representation Theory of finite dimensional algebras*. London Math. Soc. LN 119 (1988).

[11] D. Happel and C. M. Ringel. *The derived category of a tubular algebra*. In Representation Theory I. Springer LNM 1177 (1984) 156–180.

[12] D. Hughes and J. Waschbüsch. *Trivial extensions of tilted algebras*. Proc. London Math. Soc. 46 (3) (1983) 347–364.

[13] J. A. de la Peña. *On the representation type of one-point extensions of tame concealed algebras*. Manuscr. Math. 61 (1988) 183–194.

[14] J. A. de la Peña. *Algebras whose derived category is tame*. Contemporary Mathematics Vol. 229 (1998) 117–127.

[15] J. A. de la Peña and B. Tomé. *Iterated tubular algebras*. J. Pure Appl. Alg. 64 (1990) 303–314.

[16] C. M. Ringel. *Tame algebras*. Representation Theory I. Proc. ICRA II (Ottawa, 1979), Springer LNM 831 (1980) 137–287.

[17] B. Tomé. *One-point extensions of algebras with complete preprojective components having non-negative Tits forms*. Communications in Algebra, 22 (5) (1994) 1531–1549.

[18] T. Wakamatsu. *Stable equivalence between universal covers of trivial extensions*. Tsukuba J. Math. 9 (1985) 299–316.

One-point extensions of quasitilted algebras by modules on stable tubes

JOSE ANTONIO DE LA PEÑA Instituto de Matemáticas, UNAM., Ciudad Universitaria, México 04510, D. F, México, E-mail: jap@penelope.matem.unam.mx

SONIA E. TREPODE Dpto de Matemáticas. FCEyN, Universidad de Mar del Plata, 7600 Mar del Plata. Pcia. Bs. As., Argentina, E-mail: strepode@mdp.edu.ar

ABSTRACT In this note we study some necessary and sufficient conditions for the one-point extension of a quasitilted algebra by an indecomposable module to be again quasitilted. In particular we show that if A is a quasitilted algebra with a convex standard stable tube $\mathcal{T}$, then the one point extension by a module of the mouth of $\mathcal{T}$ is again quasitilted. We characterize the one-point extension of quasitilted algebras by simple modules which are quasitilted. In particular we show that in case that a tilted algebra A has a strong sink, then the one-point extension of a tilted algebra by the simple projective associated with this strong sink is always quasitilted. Finally we study the double extension of a tame hereditary algebra by indecomposable modules.

Quasitilted algebras were introduced in [4] as a natural generalization of two classes of algebras extensively studied in the last years: the tilted algebras and the canonical algebras. We are going to consider only finite dimensional algebras over an algebraically closed field k. Let $\mathcal{H}$ be a locally finite *hereditary* k-category, that is, for every couple of objects X and Y in $\mathcal{H}$ the spaces $\text{Hom}_{\mathcal{H}}(X,Y)$ and $\text{Ext}^1_{\mathcal{H}}(X,Y)$ are finite dimensional vector spaces and Ext^2 vanishes on $\mathcal{H}$. An object T in $\mathcal{H}$ is a *tilting object* if $\text{Ext}^1_{\mathcal{H}}(T,T) = 0$ and if $\text{Hom}_{\mathcal{H}}(T,X) = 0 = \text{Ext}^1_{\mathcal{H}}(T,X)$ implies that $X = 0$. For a tilting object T the k-algebra $A = End_{\mathcal{H}}(T)^{op}$ is said to be a *quasitilted algebra*. In case $\mathcal{H} = modH$ for a finite dimensional hereditary k-algebra H, then A is said to be a *tilted algebra*. In case $\mathcal{H}$ is derived equivalent to $Coh\mathbb{X}$ the category of coherent sheaves on a weighted projective line $\mathbb{X}$, then A is said to be *quasitilted of canonical type*. It has been conjectured that every quasitilted algebra is either tilted or quasitilted of canonical type. The conjecture has received a positive answer in several special cases, see [5], [10], [12], [14].

We recall that by definition the *one-point extension* of an algebra A by a A-module M is the algebra $A[M] = \begin{pmatrix} A & M \\ 0 & k \end{pmatrix}$ with the usual matrix operations.

It is known that any quasitilted k-algebra is a one-point extension of a quasitilted algebra. Many authors have studied the one-point extensions $A[M]$, where A is an indecomposable quasitilted algebra and M is a decomposable A-module ([1], [2]) and also when A is a decomposable quasitilted algebra ([7], [8]). Along this work A will be an indecomposable k-algebra. In [10] we showed that the conjecture can be stated in the following way:

CONJECTURE: *For every quasitilted algebra A there exists an indecomposable A-module M such that $A[M]$ is quasitilted.*

In this note we study, in certain cases, conditions for the one-point extension of a quasitilted algebra by an indecomposable module to be again quasitilted, using these results and [10] we obtain our main result:

THEOREM: *Let A be a quasitilted algebra with a convex standard stable tube in $\mathcal{L}_A \cap \mathcal{R}_A$. Then either A is a tilted algebra or a quasitilted algebra of canonical type.*

The note is organized in the following way: In section 1 we discuss some basic facts about quasitilted algebras. In section 2 we show our main theorem. In section 3, as an application we caracterize the one-point extension of quasitilted algebras by simple modules which are again quasitilted. In particular we show that in case that a tilted algebra A has a strong sink, then the one-point extension of a tilted algebra by the simple projective associated with this strong sink is always quasitilted. Finally we study some cases concerning double extensions. In particular, we discuss the double extension of a tame hereditary algebra by a simple regular module and by a postprojective module, in those cases such extensions are tilted or quasitilted of canonical type.

We thank the referee for several suggestions which clarified the work. The authors acknowledge finantial support from CONACyT, México.

1 PRELIMINARIES

In the following, A will be a finite dimensional indecomposable k-algebra.

It has been proven in [4] that A is a quasitilted algebra if and only if A satisfies the following homological properties: $gldim A \leq 2$ and for every indecomposable A-module either $id_A X \leq 1$ or $pd_A X \leq 1$. We recall from [4] the definition of two important classes of modules in $mod A$.

We say that Y is a *predecessor* of X if there exists a path $Y \longrightarrow X_1 \longrightarrow \cdots \longrightarrow X_n \longrightarrow X$ of non-zero morphisms which are non-isomorphism between indecomposable modules. Dually we define a successor of a module. Then we have

$$\mathcal{L}_A = \{ X \in ind A : \text{for all } Y \text{ predecessor of } X \text{ we have that } pd_A Y \leq 1 \}.$$

One-Point Extensions of Quasitilted Algebras

$\mathcal{R}_A = \{ X \in indA$: for all Y successor of X we have that $id_A Y \leq 1\}$

Let A be a quasitilted algebra. According to [4], the following holds:

i) $\mathcal{R}_A \cup \mathcal{L}_A = indA$.

ii) $\mathcal{R}_A$ contains all indecomposable injective A-modules, and $\mathcal{L}_A$ contains all indecomposable projective modules.

iii) $\mathrm{Hom}_A(\mathcal{R}_A - \mathcal{L}_A, \mathcal{R}_A \cap \mathcal{L}_A) = 0 = \mathrm{Hom}_A(\mathcal{R}_A \cap \mathcal{L}_A, \mathcal{L}_A - \mathcal{R}_A)$
$= \mathrm{Hom}_A(\mathcal{R}_A, \mathcal{L}_A - \mathcal{R}_A)$.

iv) If $\mathcal{L}_A$ contains an injective A-module, then A is a tilted algebra.

We recall that a map $f : M^r \longrightarrow X$ is said to be *decomposable* is there exist a number $0 < s \leq r$ and maps $f_1 : M^s \longrightarrow X_1$ and $f_2 : M^{r-s} \longrightarrow X_2$ such that either $s < r$ or $X_2 \neq 0$ and isomorphisms Φ_1 and Φ_2 making the following diagram commutative:

$$\begin{array}{ccc} M^r & \xrightarrow{\Phi_1} & M^s \oplus M^{r-s} \\ \downarrow f & & \downarrow \left(\begin{smallmatrix} f_1 & 0 \\ 0 & f_2 \end{smallmatrix}\right) \\ X & \xrightarrow{\Phi_2} & X_1 \oplus X_2 \end{array}$$

REMARK 1. *a)* $f : M^r \longrightarrow X$ *is an indecomposable map if and only if the $A[M]$-module $Z = (k^r, X, f)$ is an indecomposable module.*

b) Let $Z = (k^r, X, f)$ be an indecomposable $A[M]$-module with $f \neq 0$. If $Z \in \mathcal{R}_{A[M]}$, then $X \in add\mathcal{R}_A$.

c) Let $Z = (k^r, X, f)$ be an indecomposable $A[M]$-module. If $Z \in \mathcal{L}_{A[M]}$, then $X \in add\mathcal{L}_A$.

Proof. a): is clear. We give the proof of b) and c).

b): Suppose $X \notin add\mathcal{R}_A$. Then there exists X_1 an indecomposable direct summand of X such that $X_1 \notin \mathcal{R}_A$. Then there exists Y successor of X_1 such that $id_A Y = 2$. We are going to construct a successor Z' of Z with $id_{A[M]} Z' = 2$ which is a contradiction with $Z \in add\mathcal{R}_A$. Since Y is a successor of X_1 there exists a path of non-zero, non-isomorphism maps between indecomposable modules of the form:

$$X_1 \xrightarrow{\Phi_1} X_2 \xrightarrow{\Phi_2} \cdots \xrightarrow{\Phi_n} X_{n+1} = Y$$

If $\Phi_n \cdots \Phi_1 f \neq 0$ then any indecomposable summand $Z' = (k^s, Y, f_1)$ of $(k^r, Y, \Phi_n \cdots \Phi_1 f)$ is a successor of Z with $id_{A[M]} Z' = 2$. Otherwise the module $Z' = (0, Y, 0)$ is a successor of Z with $id_{A[M]} Z' = 2$.

c): Suppose $X \notin add\mathcal{L}_A$ then there exists X_1 an indecomposable direct summand of X such that $X_1 \notin \mathcal{L}_A$. So there exists L a predecessor of X_1 such that $pd_A L = 2$. Hence the $A[M]$-module L is also a predecessor of Z, and $Z \notin \mathcal{L}_{A[M]}$. □

LEMMA 1. *Let M be an indecomposable A-module such that $A[M]$ is quasitilted. If there exists a non-zero indecomposable morphism $f : M^r \longrightarrow X$ such that f is a monomorphism and $X \notin add\mathcal{R}_A$ or $Ext^1_A(M, X) \neq 0$, then $pd_A coker f \leq 1$.*

Proof. Observe that the $A[M]$-module $Z = (k^r, X, f)$ is indecomposable. Using the same commutative diagrams in [4], III.2.1 and the fact that f is a monomorphism, we get that $pd_{A[M]}Z = 2$. Since $X \notin add\mathcal{R}_A$ or $Ext^1_A(M,X) \neq 0$, then $Z \notin \mathcal{R}_{A[M]}$. So we obtain that $A[M]$ is not quasitilted, a contradiction against our assumption. $\square$

2 MAIN RESULTS

We recall the following lemma from [4]. Here $\alpha(Y)$ denotes the number of indecomposable direct summands in the middle term of the almost split sequence ending at Y.

LEMMA 2. *Let A be an artin algebra and*

$$M = N_1 \xrightarrow{u_1} N_2 \cdots N_{n-1} \xrightarrow{u_{n-1}} N_n = N$$

be a sequence of irreducible monomorphisms between indecomposable non-injective modules and assume $\alpha(TrDM) = 1$ and $\alpha(TrDN_i) = 2$ for $2 \leq i < n$. Let $f : M \longrightarrow X$ be a map where X is indecomposable and not isomorphic to N_i for $1 \leq i < n$. Then there is some $h : N \longrightarrow X$ such that $ht = f$ where $t = u_{n-1} \cdots u_1$.

We recall that a component $\mathcal{C}$ of the Auslander-Reiten quiver of A is said to be convex if for every pair of non-zero maps between indecomposable modules $f : X \longrightarrow Y$ and $g : Y \longrightarrow Z$ with $X, Z \in \mathcal{C}$ we have that Y is also in $\mathcal{C}$. In a similar way than in [4] we get the following lemma:

LEMMA 3. *Let A be a quasitilted algebra with a convex standard stable tube $\mathcal{T}$ in $\mathcal{L}_A \cap \mathcal{R}_A$. Let M be an indecomposable module in the mouth of $\mathcal{T}$. Suppose that there is a non-zero indecomposable map $f : M^r \longrightarrow X$ such that $Ext^1_A(M,X) \neq 0$. Then this morphism is of the form $f : M \longrightarrow X$ with X an indecomposable module in $\mathcal{T}$. Moreover, $pd_A coker f \leq 1$.*

Proof. Let $f : M^r \longrightarrow X$ be a non-zero indecomposable map such that $Ext^1_A(M,X) \neq 0$. Then there exists X_1 an indecomposable direct summand of X such that $Ext^1_A(M,X_1) \neq 0$. Hence there exist non-zero maps $g : X_1 \longrightarrow \tau M$ and $f_1 : M \longrightarrow X_1$. Since $\mathcal{T}$ is convex it follows that $X_1 \in \mathcal{T}$.

Write $X = X_1 \oplus L$ where we can choose X_1 in $\mathcal{T}$ such that the length of X_1 is maximal among the indecomposable direct summands Y of X such that $Ext^1_A(M,Y) \neq 0$.

Write $f = \begin{pmatrix} a & b \\ c & d \end{pmatrix} : M \oplus M^{r-1} \longrightarrow X_1 \oplus L$, where we can assume that $a : M \longrightarrow X_1$ is not zero. Then a must be a composition

$$M = M_1 \xrightarrow{u_1} M_2 \cdots \xrightarrow{u_{n-1}} M_n = X_1$$

of irreducible monomorphisms between indecomposables modules in $\mathcal{T}$, since $\mathcal{T}$ is standard. We know that $\alpha(TrDM) = 1$ and $\alpha(TrDM_i) = 2$, for $2 \leq i < n$. Since no summand of L is isomorphic to any M_i for $i < n$, we get from Lemma 1 that $c : M \longrightarrow L$ factors through $a : M \longrightarrow X_1$, i.e there exists $e : X_1 \longrightarrow L$ such that

$c = ea$. Further $b : M^{r-1} \longrightarrow X_1$ can be written as ah where $h : M^{r-1} \longrightarrow M$, since $\dim_k \mathrm{Hom}_A(M, X_1) = 1$. Then we get the following commutative diagram:

$$\begin{array}{ccc} M \oplus M^{r-1} & \stackrel{\begin{pmatrix} 1 & h \\ 0 & 1 \end{pmatrix}}{\longrightarrow} & M \oplus M^{r-1} \\ \downarrow f & & \downarrow \begin{pmatrix} a & 0 \\ 0 & d-eb \end{pmatrix} \\ M_1 \oplus L & \stackrel{\begin{pmatrix} 1 & 0 \\ -e & 1 \end{pmatrix}}{\longrightarrow} & X_1 \oplus L \end{array}$$

Since f is an indecomposable map, then $r = 1$ and $L = 0$. The rest of the proof is easy. □

PROPOSITION 1. *Let A be a quasitilted algebra with a convex standard stable tube $\mathcal{T}$ in $\mathcal{L}_A \cap \mathcal{R}_A$. Let M be an A-module in the mouth of $\mathcal{T}$. Then $A[M]$ is quasitilted.*

Proof. It is clear that $\mathrm{gldim} A[M] \leq 2$. Any indecomposable $A[M]$-modules of the form $(0, X, 0)$ does not have both projective and injective dimension equal 2, by [4], III.2.5. So, let $Z = (k^r, X, f)$ be an indecomposable $A[M]$-module with $f \neq 0$. Since $M \in \mathcal{R}_A$ and $f \neq 0$ is an indecomposable map, we have that $X \in \mathrm{add}\mathcal{R}_A$, i.e, $id_A X \leq 1$. If $\mathrm{Ext}^1_A(M, X) = 0$, then by [4] III.2.2 it follows that $id_{A[M]} Z \leq 1$. However, if $\mathrm{Ext}^1_A(M, X) \neq 0$, by Lemma 3 we obtain that $r = 1$, X is an indecomposable module in $\mathcal{T}$, and $pd_A \mathrm{coker} f \leq 1$. Note that f is a monomorphism. Then by [4], III.2.1, it follows that $pd_{A[M]} Z \leq 1$ and so $A[M]$ is quasitilted. □

We recall a theorem in [10] (see also [7]):

THEOREM 1. *Let M be an A-module such that $A[M]$ is quasitilted. Then either A is a tilted algebra or a quasitilted algebra of canonical type.*

Using this result we get:

THEOREM 2. *Let A be a quasitilted with a convex standard stable tube $\mathcal{T}$ in $\mathcal{L}_A \cap \mathcal{R}_A$. Then either A is a tilted algebra or a quasitilted algebra of canonical type.*

Proof. Take M in the mouth of the tube $\mathcal{T}$. By Proposition 1 we have that $A[M]$ is quasitilted. Theorem 1 implies that A is a tilted algebra or a quasitilted algebra of canonical type. □

REMARK 2. *In the situation above note that if $\mathcal{T}$ is sincere the result follows from the main theorem in [13].*

3 APPLICATIONS

We are going to show now some further aplications of the results in section 1.

PROPOSITION 2. *Let A be a quasitilted algebra, and S be a simple module in $\mathcal{L}_A$. Then $A[S]$ is quasitilted if and only if for any non-zero indecomposable map $f : S^r \longrightarrow X$ with $X \notin \mathrm{add}\mathcal{R}_A$ or $\mathrm{Ext}^1_A(S, X) \neq 0$, we have $pd_A \mathrm{coker} f \leq 1$.*

Proof. Suppose that $A[S]$ quasitilted. Let $f: S^r \longrightarrow X$ be a non-zero indecomposable map with $X \notin add\mathcal{R}_A$ or $Ext^1_A(S,X) \neq 0$. Since f is an indecomposable map and S^r is semisimple then f is a monomorphism. Under the hypothesis it follows from Lemma 1 that $pd_A coker f \leq 1$.

For the converse observe that $gldim A[S] \leq 2$. Any indecomposable $A[S]$-module of the form $(0,X,0)$ does not have both projective and injective dimension equal 2, by [4], III.2.5. So, let $Z = (k^r, X, f)$ be an indecomposable $A[S]$-module with $f \neq 0$. Note that f is a monomorphism. In case $X \notin add\mathcal{R}_A$ or $Ext^1_A(S,X) \neq 0$, then by our assumptions and [4], III.2.1., it follows that $pd_{A[S]} Z \leq 1$. Otherwise $id_A Z \leq 1$ by [4], III. 2.2. Hence $A[S]$ is quasitilted. □

REMARK 3. *a) Let A be a tilted algebra of Dynkin or Euclidean type and S be a simple A-module such that $A[S]$ is quasitilted. Then $A[S]$ is tilted or quasitilted of canonical type.*

b) Let A be a quasitilted not tilted algebra and S be a simple A-module such that $A[S]$ is quasitilted. Then $A[S]$ is quasitilted of canonical type.

Proof. a): It follows from our results in [11], prop 2.1, prop 2.2 (see also Proposition 6 below) and prop 2.4, that if A is tilted of Dynkin or Euclidean type then $A[S]$ is tilted or quasitilted of canonical type.

b): If A is quasitilted but not tilted, then it follows from [7] Th.13 that $A[S]$ is quasitilted of canonical type. □

Let a be a sink of Q_A, we say that a is a *strong sink* if does not exist a path of non-isomorphisms between indecomposable modules

$$I \longrightarrow X_1 \longrightarrow X_2 \longrightarrow \cdots \longrightarrow X_n \longrightarrow I_a$$

with I an injective module and I_a the injective envelope of the simple module S_a associated with the vertex a.

Note that our definition of strong sink is not the standard definition, since the injective I could coincide with I_a.

LEMMA 4. *Let A be a quasitilted algebra such that there exists a strong sink. Then A is a tilted algebra.*

Proof. We are going to show that if a is a strong sink then $I_a \in \mathcal{L}_A$, which implies that A is tilted.

Suppose that $I_a \notin \mathcal{L}_A$, then there exists X a predecessor of I_a such that $pd_A X = 2$. Hence there exists an indecomposable injective module I such that $Hom_A(I, \tau X) \neq 0$ and we have a path:

$$I \longrightarrow \tau X \longrightarrow E \longrightarrow X \longrightarrow \cdots \longrightarrow I_a,$$

which contradicts the fact that a is a strong sink. □

REMARK 4. *In case A is a tilted algebra there is not always a strong sink in Q_A.*

Proof. Let $A = [N]C$ with C tame concealed and N simple regular such that A is tilted. We call I_w the injective associated with the new vertex w, then I_w belongs to a coray tube and there exists a path from I_w to I_w, then w is not a strong sink. □

PROPOSITION 3. *Let A be a tilted algebra with a strong sink a in Q_A. Then $A[S_a]$ is quasitilted.*

Proof. It is clear that $gldim A[S_a] \leq 2$. Moreover the indecomposable $A[S_a]$-modules of the form $Z = (0, X, 0)$ have projective or injective dimension at most 1. Let $Z = (k^r, X, f)$ be an indecomposable $A[S_a]$-module, with $f \neq 0$. Since $f \neq 0$ is an indecomposable map and S^r is semisimple, f is a monomorphism and $ker f$ is projective.

We show first that if $f : S_a^r \longrightarrow X$ is an indecomposable map then $pd_A X \leq 1$. Suppose $pd_A X = 2$ then there exists X_1 an indecomposable direct summand of X such that $pd_A X_1 = 2$. Then $Hom_A(I, \tau X_1) \neq 0$ for some indecomposable injective module I. Since f is an indecomposable map, it follows that $Hom_A(S_a, X_1) \neq 0$ and $Hom_A(X_1, I_a) \neq 0$. Then we have the following path:

$$I \longrightarrow \tau X_1 \longrightarrow E \longrightarrow X_1 \longrightarrow I_a,$$

contradicting the fact that a is a strong sink. Hence $pd_A X \leq 1$. Now we are showing that $pd_A coker f \leq 1$. Since f is monomorphism then we obtain the following inequality:

$$pd_A coker f \leq max\{pd_A X, pd S_a^r + 1\} = 1.$$

$\square$

Now we are going to point out some results about double extensions of quasitilted algebras by an indecomposable module.

PROPOSITION 4. *Let A be a quasitilted algebra and M an indecomposable directing module in $\mathcal{L}_A \cap \mathcal{R}_A$. Then $A[M][M]$ is tilted.*

Proof. By [9] it is enough to show that $A[M]$ is a quasitilted algebra and M is a directing module in $\mathcal{L}_{A[M]} \cap \mathcal{R}_{A[M]}$. Observe that, in fact, we are showing that $A[M]$ is tilted (using again [9]).

By [9], A is tilted. Since M is a directing module in $\mathcal{L}_A \cap \mathcal{R}_A$, by [4], III.2.6, it follows that $A[M]$ is quasitilted. Now we see that M is a directing module in $\mathcal{L}_{A[M]} \cap \mathcal{R}_{A[M]}$. Let P be the new indecomposable projective $A[M]$-module. Since $A[M]$ is quasitilted we have that $P \in \mathcal{L}_{A[M]}$. Since M is a predecessor of P, we get that $M \in \mathcal{L}_{A[M]}$. Suppose now that M is not a directing $A[M]$-module. So, M belongs to a cycle in mod $A[M]$. Then there exists a proper $A[M]$-module N in that cycle (otherwise M would not be a directing module in mod A). Hence, there is a non-zero morphism from P to N, and therefore P is not a directing $A[M]$-module. Then [6] implies that M is not a directing A-module.

Finally we show that $M \in \mathcal{R}_{A[M]}$. Since M is an indecomposable A-module, it is easy to verify that all successor X of M in mod A satisfy $Ext_A^1(M, X) = 0$. Hence, since $M \in \mathcal{R}_A$, it follows that each indecomposable $A[M]$-module $Z = (k^r, X, f)$ with $f \neq 0$, or $Z = (0, Y, 0)$ with Y be successor of M in mod A, is such that $id_{A[M]} Z \leq 1$. In conclusion $M \in \mathcal{R}_{A[M]}$, and this finishes the proof. $\square$

As an application of Proposition 1 we prove the following result.

PROPOSITION 5. *Let A be a tame hereditary algebra, N be a simple regular A-module and M be a postprojective A-module. Let $C = A[N]$. Assume the following conditions hold:*

a) *There exist non-zero indecomposable maps* $g : M^r \longrightarrow X$ *where* X *is an indecomposable C-module and* $id_C X = 2$, *and for any such map we have that* $\ker g$ *is projective.*

b) *For any non-zero indecomposable map* $f : M^r \longrightarrow X$ *with* $id_C X = 2$ *we have* $pd_C \operatorname{coker} f \leq 1$.

Then $A[N][M]$ *is quasitilted.*

Proof. First observe that $C = A[N]$ is quasitilted, by Proposition 1. Moreover, it is easy to verify that M also is a directing C-module in $\mathcal{L}_C$. So is clear that gl dim $C[M] \leq 2$, and that each indecomposable $C[M]$-module of the form $(0, X, 0)$ has projective or injective dimension at most 1. Let $Z = (k^r, X, f)$ be an indecomposable $C[M]$-module with $f \neq 0$. Since $f \neq 0$ is an indecomposable map and M is an indecomposable directing C-module, then $Ext^1_A(M, X) = 0$. If $id_C X \leq 1$, then by [4]. III.2.2, we have that $id_{C[M]} Z \leq 1$. But, if $id_C X = 2$, then there exists an indecomposable direct summand X' of X such that $id_C X' = 2$. Hence, taking the canonical projection $\pi : X \longrightarrow X'$, we have that $g = \pi f : M^r \longrightarrow X'$ is a non-zero indecomposable C-morphism, and from a) it follows that $\ker g$ is a projective C-module. Then $\ker f$ as a submodule of $\ker g$ is also projective. From b) it follows that $pd_C \operatorname{coker} f \leq 1$, and therefore in that case, by [4], III.2.1, $pd_{C[M]} Z \leq 1$. So $A[N][M]$ is quasitilted. □

We recall the following result from [11]:

PROPOSITION 6. *Let A be a representation infinite tilted algebra of euclidean type and let M an indecomposable A-module such that $A[M]$ is quasitilted. The following holds:*

i) *If M is postprojective or preinjective, then $A[M]$ is tilted.*

ii) *If M belongs to a semiregular tube, then $A[M]$ is quasitilted of canonical type.*

REMARK 5. *Let A, M and N be as in Proposition 5.*

(a) If there is no non-zero indecomposable map $g : M^r \longrightarrow X$ *with X indecomposable and $id_C X = 2$ then $A[N][M]$ is quasitilted.*

(b) Under the assumption of Proposition 5 and Remark 5.1 , $A[N][M]$ is tilted or quasitilted of canonical type.

Proof. a): The proof is contained in the one of Proposition 5, since any non-zero indecomposable C-morphism $f : M^r \longrightarrow X$ we have that $id_C X \leq 1$.

b): If $A[N]$ is quasitilted but not tilted, then by [7] Th.13 we get that $A[N][M]$ is quasitilted of canonical type. If $A[N]$ is tilted, then it is tilted of euclidean type. Since M is a postprojective $A[N]$-module it follows from Proposition 6 above, that $A[N][M]$ is tilted. □

NOTE ADDED IN PROOF. After the submission of this paper we learned that Happel has proved the Conjecture recalled in the Introduction.

REFERENCES

[1] F. Coelho, I. Martins and J. A. de la Peña. *Quasitilted one-point extensions I.* Proc. of AMS, to appear.

[2] F. Coelho, I. Martins and J. A. de la Peña. *Quasitilted one-point extensions II.* Journal of Algebra 227 (2000) 582-594.

[3] D. Happel. *Quasitilted Algebras.* Canadian Math. Soc. Conf. Proc. 23 (1998), 55-82.

[4] D. Happel, I. Reiten, and S. Smaloe. *Tilting in abelian categories and quasitilted algebras.* Memoirs AMS, 575 (1996).

[5] D. Happel and I. Reiten. *Hereditary categories with tilting objects..* Math.Z. 232 no. 3, (1999), 559-588.

[6] D. Happel and C. Ringel. *Directing projectives modules.* Arch. Math., Vol. 60 (1993), 237-246.

[7] D. Happel and H. Slungard. *On quasitilted algebras which are one-point extensions of hereditary algebras.* Coll. Math., vol 81, no. 1, (1999) 141-152.

[8] D. Happel and H. Slungard. *One-point extensions of hereditary algebras.* Canadian Math. Soc. Proc. 24 (1998), 285-291.

[9] J. A. de la Peña and I. Reiten. *Trisection of modules categories.* In preparation.

[10] J. A. de la Peña and S. Trepode. *Algebras with quasitilted one-point extensions.* Preprint (1999).

[11] J. A. de la Peña and S. Trepode. *Quasitilted one-point extensions of tilted algebras.* Preprint (1999).

[12] H. Lenzing. *Hereditary noetherian categories with tilting complex.* Proc. AMS 125(1997), 1893-1901.

[13] I. Reiten and A. Skowroński. *Sincere stable tubes.* To appear in J. Algebra.

[14] A. Skowroński. *Tame quasitilted algebras.* J. Algebra 203 (1998), 470-490.

Combinatorial Partial Tilting Complexes for the Brauer Star Algebras

MARY SCHAPS Department of Mathematics and Computer Science, Bar-Ilan University, 52900, Ramat-Gan, Israel.

EVELYNE ZAKAY-ILLOUZ[1] Department of Mathematics, Jordan Valley College, Jordan Valley, Israel

ABSTRACT In modular representation theory, the Brauer star algebras play a very special role, being the "local" blocks corresponding under the Brauer correspondence to the "global" blocks of a group ring of cyclic defect group. Every block of cyclic defect group can be obtained by tilting from a Brauer star algebra. Among the various tilting complexes for the Brauer star algeras, we distinguish a subclass, which we will call two-restricted tilting complexes, of a special combinatorial character. In this paper we determine necessary conditions for a two-restricted complex to be a partial tilting complex for the Brauer star algebra.

1 INTRODUCTION

This paper is intended to be the first of three based on the doctoral dissertation of the second author. Since we cannot refer most readers to the original dissertation, which is in Hebrew, we intend to give a fairly detailed presentation in the published version.

The general subject is tilting complexes and deformations for blocks of cyclic defect group. The work was motivated by two problems: First, to find a Donald-Flanigan deformation of blocks of cyclic defect which would be more natural than that given in [S2]. Second, to try to understand the combinatorics of the Brauer tree in terms of the tilting theory. Rickard had already shown that a connection existed in [R2], where he constructed a tilting complex from a block of the "global" Brauer tree to the Brauer star. We chose to work in the opposite direction; to find

---

[1] Work done for a doctoral dissertation at Bar-Ilan University, and partially supported by the Bar-Ilan Research Authority

tilting complexes for the Brauer star algebra from which one can "read off" the global Brauer tree directly.

Since not every tree is the Brauer tree of a block of a group ring, our results are not restricted to blocks of a group ring, and in fact, we will work with the larger class of algebras called Brauer tree algebras, over fields of arbitrary characteristic. However, the main application is to modular representation theory of finite groups. This paper corresponds to Chapters III and IV of the thesis, classifying two-restricted partial tilting complexes. The second paper, corresponding to Chapter V, will demonstrate a one-to-one correspondence between Brauer trees with an additional structure, called a pointing, and equivalence classes of two-restricted tilting complexes modulo cyclic rotation of the Brauer star. The third paper, Chapter VI of the thesis, will give a "natural" deformation based on tilting theory.

Let e, m be natural numbers, with $e > 1$. Let K be any field containing a primitive e-th root of unity ξ. Let $\hat{n} = em + 1$. Then the cyclic group C_e of order e acts on the truncated polynomial ring $A = K[x]/x^{\hat{n}}$ by letting the generator d of C_e send x to ξx.

We now construct the skew group ring

$$b = A[C_e],$$

where the algebra generation x of A and the group generator d of C_e commute according to the relation

$$xd = \xi dx$$

or

$$d^{-1}xd = \xi x.$$

The most important case of this construction occurs when K is a field of characteristic p, e is a divisor of $p - 1$, and $m = (p^c - 1)/e$. In this case

$$b \xrightarrow{\sim} K[C_{p^c} \rtimes C_e].$$

The generator d of C_e induces an action on $C_{p^c} = \langle a \rangle$ given by $d^{-1}ad = a^r$, where r is an integer which is a primitive root of unity modulo e. The isomorphism between this case and the previous, more general case, is established by setting

$$x = \frac{1}{e} \sum_{j=0}^{e-1} r^{-j} a^{r^j}.$$

Then $d^{-1}xd = rx$, as desired.

This algebra b has been intensively studied in the context of modular representation theory. For a discussion of the quiver of b in a more general context, see [MS] or [SSS]. Alternatively, one can see b as a special case of the Brauer tree algebra, which we will discuss below.

Returning to the situation where K has an arbitrary characteristic, we set $A = K[x]/x^{\hat{n}}$ as before, and define idempotents of $b = A[C_e]$ as

$$f_i = \sum_{j=0}^{e-1} \xi^{-ij} d^j, \quad i = 1, ..., e.$$

Partial Tilting Complexes

This is a complete set of primitive orthogonal idempotents of b, and thus the indecomposable projectives are

$$P_i = bf_i, \quad i = 1, ..., e.$$

We will denote the identity map on P_i by id_i.

LEMMA 1.1. *The set $\{x^s f_i \mid s = 0, \ldots, \hat{n} - 1; i = 1, \ldots, e\}$ is a basis of eigenvectors, with eigenvalues ξ^s, for the action of d. It satisfies*

$$x^s f_i = f_k x^s, \quad \text{where} \quad k \equiv i - s \quad (\mathrm{mod}\ e).$$

Proof: Since $\{1, x, \ldots, x^{n-1}\}$ is a basis for A, and $\{1, \ldots, d^{e-1}\}$ are the elements of C_e, a basis for the skew group algebra $A[C_e]$ is given by $\{x^s d^i \mid s = 0 \ldots, \hat{n} - 1; i = 1, \ldots, e\}$. We can make a change of basis for each s to $\{x^s f_i \mid s = 0, \ldots, \hat{n} - 1, i = 1; \ldots, e\}$.

Now,

$$\begin{aligned} x^s f_i &= x^s \left(\frac{1}{e} \sum_{j=0}^{e-1} \xi^{-ij} d^j \right) \\ &= \left(\frac{1}{e} \sum_{j=0}^{e-1} \xi^{-ij} \xi^{sj} d^j \right) x^s \\ &= f_k x^s, \quad \text{where} \quad k \equiv i - s \quad (\mathrm{mod}\ e). \end{aligned}$$

DEFINITION. The basis $\{x^s f_i\}$ will be called the *normalized basis*. The natural grading of $K[x]/(x^n)$ induces a grading of $A[C_e]$ in which the space of elements of degree s is spanned by $x^s f_1, \ldots, x^s f_e$. The unique b-homomorphism from P_k to P_i sending f_k to $x^s f_i$ is said to be *normal homogeneous of degree s*.

COROLLARY. *If $k = i$, then there are $m + 1$ normal homogeneous maps $q : P_k \to P_i$, for $s = 0, e, 2e, \ldots, me$. If $k \neq i$, and $\{i - k\}_e$ is the residue mod e, then there are m normal homogeneous maps, for $s = \{i - k\}_e + \ell e, \ell = 0, \ldots, m - 1$.*

2 BRAUER TREES AND TILTING MODULES

DEFINITION. Let T be any tree, together with a cyclic orientation on the edges adjacent to each vertex. Assume that T has e edges, and that there is one designated vertex, called the *exceptional* vertex which is assigned a multiplicity $m \geq 1$. T is called a *Brauer tree of type (e, m)*.

To any field k and any Brauer tree T, we can associate a k-algebra A_T called the Brauer tree algebra. We describe the algebra A_T in terms of composition series.

For each edge f_{r_1} and each vertex v of the edge which is not a non-exceptional terminal vertex, we define the *standard circuit* at v denoted by $S(v, f_{r_1})$ to be the sequence of edges in the cyclic ordering around v either one time or, if v is exceptional, m times.

We describe the composition series of the projective modules P_i, $i = 1,\ldots,e$. The top and socle of P_i are the same simple module S_i. $\text{Rad}(P_i) / \text{Soc}(P_i)$ is the direct sum of either one or two uniserial components, whose composition factors are the simples corresponding to the edges encountered in a standard circuit of the vertices at either end of the edge. If one of those vertices is a non-exceptional terminal vertex, then the corresponding factor is empty. A detailed description in the case of blocks of a group ring is given in [A].

In [R2], Rickard proved that any two blocks of a group ring with the same Brauer tree are derived equivalent. By the various reductions in [R1], this means that there is a complex $Q^\cdot$ of projective A-modules such that $B = \text{End}_A(Q)$, where $Q^\cdot$ is a tilting complex according to the following definition. The notation $Q^\cdot[n]$ denotes shifting the complex n places to the left.

DEFINITION. Let R be a Noetherian ring. A bounded complex of finitely-generated projective R-modules is called a *tilting complex* if

(1) $\text{Hom}_{D^b(R)}(Q^\cdot, Q^\cdot[n]) = 0$ whenever $n \neq 0$.

(2) For any indecomposable projective P, define the stalk complex to be the complex
$$P^\cdot : 0 \to P \to 0.$$
Then every such $P^\cdot$ is in the triangulated category generated by the direct summands of direct sums of $Q^\cdot$.

$Q^\cdot$ satisfying (1) and (2) is called a *tilting complex* for R. If it satisfies (1), it is called a *partial tilting complex*.

Rickard proved his result [R2] on derived equivalence of blocks of a group ring by showing that every block of a group ring which is a Brauer tree algebra can be tilted to an algebra Morita equivalent to the Brauer tree algebra of a star, i.e., to an algebra like the b defined above. Membrillo proved a similar result with regard to the generalized Brauer tree algebras, in which each vertex is allowed arbitrary multiplicity [M].

Our interest in the subject of Brauer tree algebras came from deformation theory. We wanted to transfer the very easily described deformation of b given in [S2] to an arbitrary Brauer star algebra B, using the Rickard result [R3] on deformations of tilting complexes. Where Rickard and Membrillo were passing from B to b, we need to go from b to B. Furthermore, we wanted to get deformations which would be "homogeneous" with regard to the deformation parameters in a sense which will be made explicit in the sequel [SZ2]. Thus we restricted ourselves to tilting complexes of a particularly combinatorial form.

DEFINITION. A complex of projectives over $b = A[C_e]$ will be called *two-restricted* if it is a direct sum of shifts of complexes involving no more than two indecomposable projectives of the form

$$S_r : \quad 0 \to P_r \to 0$$
$$T_{rt} : \quad 0 \to P_r \xrightarrow{h} P_t \to 0 \quad r \neq t.$$

where the map h is homogeneous of minimal degree. The complexes $S_r[n]$, $T_{rt}[n]$ will be called *elementary*.

NOTE. One can show that any indecomposable partial tilting complex containing at most two terms is elementary, and Theorem 2 below will establish that the elementary complexes are all indecomposable.

We recall some generalities about tilting complexes from [R1]. If A has e simple modules, and $Q^{\cdot}$ is a tilting complex over A, then $Q^{\cdot}$ is homotopic to a complex which is the direct sum of indecomposable complexes from exactly e distinct isomorphism classes. A partial tilting complex has no more than e distinct elementary complexes. If we take a tilting complex $Q^{\cdot}$ over A, and replace it by a complex $Q^{\cdot\prime}$ with the same distinct indecomposables but a different number of copies of each, we get an algebra $B' = \text{End}_A(Q^{\cdot\prime})$ which is Morita equivalent to $B = \text{End}_A(Q^{\cdot})$. We will generally work with tilting complexes in which there is a single copy of each indecomposable, so that the tilted algebra B will be basic.

An indecomposable complex can be recognized because its endomorphism ring is local. It will be a corollary of the results in this paper that the complexes S_r and T_{rt} are indecomposable. Thus if we can find a sum of e complexes S_r and T_{rt} which form a partial tilting complex, it will be a tilting complex. There are many other indecomposable complexes from which tilting complexes could be constructed; for example $P_r \xrightarrow{h} P_t \xrightarrow{s} P_t \xrightarrow{s} ... \xrightarrow{s} P_t$, where the map s takes the top of P_t to the socle of P_t. We concentrate on the complexes S_r and T_{rt} because they are particularly "combinatorial" in the following sense: they provide a minimal encoding of the information contained in an edge of the Brauer graph.

NOTATION. The set of all two-restricted partial tilting complexes will be denoted by PTC_2, and the subset of two-restricted tilting complexes by TC_2.

In this first paper, we will classify the elements of PTC_2, and determine the homomorphisms between their indecomposable components. In the second paper, we will show that there is a one-to-one correspondence between elements of TC_2 modulo rotation of b and Brauer trees with an additional structure we will call a pointing. This will show that TC_2 is adequate to produce all Brauer trees, but also will show that there are generally many different elements of TC_2 which will produce a given Brauer tree. Finally, in the third paper, we will derive the existence of many different representations of A_T as a graded algebra, all equally "natural", and all giving the same deformation.

3   STATEMENT OF RESULTS

In order to state the results in full, we need to introduce one more concept, that of a short sequence of indices.

DEFINITION. Let b be the Brauer star algebra $K[x]/(x^{em+1})[C_e]$, as above. We let $\{x^s f_i\}_{s=0}^{em}$ be a basis for P_i, and define the following maps

$$\varepsilon_i : P_i \to P_i \qquad \varepsilon_i(f_i) = x^e f_i$$

$$\tilde{h}_{ij} : P_i \to P_j \qquad \tilde{h}_{ij}(f_i) = x^k f_i, \qquad k \equiv j - i \pmod{e} \qquad 0 \le k < e.$$

For $i \ne j$, we will denote $\tilde{h}_{ij}$ by h_{ij}, and for $i = j$ by id_i. The homomorphism $\varepsilon_i^m : P_i \to P_i$ will be called the *socle* map, for the obvious reason that it maps the top of P_i into its socle $\langle x^{em} f_i \rangle$.

DEFINITION. Consider a sequence $\{r_i\}_{i=1}^{\ell}$ of elements of $\{1, ..., e\}$. Set

$$h = \tilde{h}_{r_{\ell-1} r_\ell} \circ ... \circ \tilde{h}_{r_1 r_2} = \varepsilon_{r_\ell}^\alpha \tilde{h}_{r_1 r_\ell}.$$

Then the sequence is *short* if $\alpha = 0$ and *long* if $\alpha > 0$. We generally represent the sequence in the form $r_1 \to r_2 \to ... \to r_\ell$.

EXAMPLES.

$r \to s \to r$	is long
$r \to r \to s$	is short
$1 \to 2 \to 3$	is short
$1 \to 3 \to 2$	is long.

We gather together the following facts about short and long sequences.

(1) *Cyclic permutation property:* If the sequence $r_1 \to r_2 \to ... \to r_\ell$ is short (or long), then the cyclic permutation

$$r_i \to r_{i+1} \to ... \to r_\ell \to r_1 \to ... \to r_{i-1}$$

is short (or long), respectively.

(2) *Restriction property:* If $r_1 \to ... \to r_\ell$ is short, so is every subsequence.

(3) *Mixing property:* If all the r_i are distinct, then up to cyclic permutation there is a unique permutation σ of indices such that $r_{\sigma(1)} \to r_{\sigma(2)} \to ... \to r_{\sigma(\ell)}$ is short.

(4) *Refinement property:* If $r_1 \to ... \to r_\ell$ is short, and $\alpha^1, ..., \alpha^{\ell-1}$ are sequences such that for $k = 1, ..., \ell - 1$

$$r_k \to \alpha^k \to r_{k+1}$$

is short, then

$$r_1 \to \alpha^1 \to r_2 \to \alpha^2 \to ... \to r_\ell$$

is short.

Proof. The proofs of (1), (2) and (4) are simply repeated applications of the associability law for the maps $\tilde{h}_{r_i r_{i+1}}$. The proof of (3) is by induction, using the fact that if $r \to s \to t$ is short, then $r \to t \to s$ is long. $\square$

Partial Tilting Complexes

The first theorem will describe which pairs of indecomposable elements of PTC_2 can occur together in a partial tilting complex.

THEOREM 1. *If $Q^{\cdot}$ is in PTC_2, then:*

(a) *All elementary components of $Q^{\cdot}$ of the form $S_r[n]$ must have their non-zero term in the same degree, which we will assume to be zero.*

(b) *If two elementary components of $Q^{\cdot}$ have a common term P_r, it must occur with the same degree in each.*

(c) *If $T_{rs}[n]$ and S_k occur together in $Q^{\cdot}$ then $r \to k \to s$ is long.*

(d) *If $T_{rs}[n]$ and $T_{tu}[n']$ occur in $Q^{\cdot}$, with r, s, t, u all distinct, then one of the following sequences must be short.*

   (1) $r \to s \to t \to u$

   (2) $r \to t \to u \to s$

   (3) $r \to s \to u \to t$.

REMARK. We divide the circle into e segments with end points numbered counterclockwise from 1 to e, and let $\overline{rs}$ and $\overline{tu}$ be directed counterclockwise segments. Then (1) represents the case where the ones $\overline{rs}$ and $\overline{tu}$ are disjoint with compatible orientations, (2) the case where I_u is included in $\overline{rs}$, and (3) the case $\overline{rs}$ included in $\overline{tu}$.

The three excluded cases represent partial intersections

$$r \to t \to s \to u$$
$$r \to u \to s \to t$$
$$r \to u \to t \to s.$$

This approach in terms of segments of a circle originated with [KZ1]. The actual problem treated by König and Zimmermann is different, and the two term sequences represent projective resolution in a hereditary order. Also, in their case all the tilting complexes are two-restricted, which is not the case for Brauer tree algebras. However, the combinatorics which were developed independently, turned out virtually equivalent. The case of Green orders treated by Zimmermann in [KZ2] presumably provides a link, since it uses methods similar to [KZ1] and reduces, modulo the prime, to the case of Brauer tree algebras.

The proof of Theorem 1 will depend on Propositions 1 - 5 in the next section. In these propositions, we will determine all homogeneous chain maps between indecomposable complexes which are not homotopic to zero. This information about the non-trivial chain maps will also be important in the sequel for constructing the endomorphism ring of a tilting complex $Q^{\cdot}$ in TC_2, so we will summarize in Theorem 2. However, since the statement of Theorem 2 will also include information about the degree of these chain maps, we first pause to show that this is a well-defined concept for chain maps in TC_2.

DEFINITION. A chain map $\ell^{\cdot}$ between $C^{\cdot}$ and $D^{\cdot}$ is called *normal homogeneous* if each vertical map is normal homogeneous.

LEMMA 3.1. *Any chain map $\ell^{\cdot}$ between irreducible elements of PTC_2 is a linear combination of normal homogeneous chain maps.*

Proof: If there is only one non-zero vertical arrow, this is obvious. Thus we need only consider the case

$$\begin{array}{ccccccc} 0 & \longrightarrow & P_r & \xrightarrow{h_{rs}} & P_s & \longrightarrow & 0 \\ & & \downarrow \ell_1 & & \downarrow \ell_2 & & \\ 0 & \longrightarrow & P_t & \xrightarrow{h_{tu}} & P_u & \longrightarrow & 0. \end{array}$$

Set

$$\ell_1 = \left(\sum_{i=0}^{m-1} a_i \, \varepsilon_t^i \, \tilde{h}_{rt}\right) + \delta_{rt} \, a_m \, \varepsilon_t^m \, \tilde{h}_{rt}$$

$$\ell_2 = \left(\sum_{i=0}^{m-1} b_i \, \varepsilon_u^i \, \tilde{h}_{su}\right) + \delta_{su} \, b_m \, \varepsilon_u^m \, \tilde{h}_{su}.$$

Since $\ell^{\cdot}$ is a chain map, the diagram is commutative, and thus

(*) $$\ell_2 h_{rs} = h_{tu} \ell_1.$$

Define α and β by

$$h_{tu} \, \tilde{h}_{rt} = \varepsilon_u^\alpha \, \tilde{h}_{ru}$$

$$\tilde{h}_{su} \, h_{rs} = \varepsilon_u^\beta \, \tilde{h}_{ru}.$$

We compute the two compositions in (*):

$$\begin{aligned} h_{tu} \ell_1 &= h_{tu} \left(\sum_{i=0}^{m-1} a_i \, \varepsilon_t^i \, \tilde{h}_{rt} + \delta_{rt} \, a_m \, \varepsilon_t^m \, \tilde{h}_{rt}\right) \\ &= \sum_{i=0}^{m-1} a_i \, \varepsilon_u^{i+\alpha} \, \tilde{h}_{ru} + \delta_{rt} \, \delta_{ru} \, a_m \, \varepsilon_u^{m+\alpha} \\ &= \sum_{i=0}^{m-1} a_i \, \varepsilon_u^{i+\alpha} \, \tilde{h}_{ru} \quad \text{since } \delta_{rt} \, \delta_{ru} = 0 \quad \text{because } t \neq u. \end{aligned}$$

Similarly, $\ell_2 \, h_{rs} = \sum_{j=0}^{m-1} b_j \, \varepsilon_u^{j+\beta} \, \tilde{h}_{ru}$. Set $a_{-1} = b_{-1} = 0$. Then for $0 \leq k \leq m - 1 + \delta_{ru}$, we get

$$a_{k-\alpha} = b_{k-\beta}.$$

Dividing into cases according to α_1 and β_1, we get the following decompositions of

Partial Tilting Complexes

$\ell = (\ell_1, \ell_2)$:

$\alpha = 0, \beta = 0$:
$$(\ell_1, \ell_2) = \sum_{i=0}^{m-1} a_i \left(\varepsilon_t^i \tilde{h}_{rt}, \varepsilon_u^i \tilde{h}_{su}\right) + \delta_{rt} a_m \left(\varepsilon_r^m, 0\right) + \delta_{su} b_m (0, \varepsilon_s^m)$$

$\alpha = 0, \beta = 1$:
$$(\ell_1, \ell_2) = \sum_{i=0}^{m-1} a_i \left(\varepsilon_t^i \tilde{h}_{rt}, \varepsilon_u^{i-1} \tilde{h}_{su}\right) + \delta_{rt} a_m \left(\varepsilon_r^m, 0\right) + b_{m-1} \left(0, \varepsilon_u^{m-1} h_{su}\right)$$

$\alpha = 1, \beta = 0$:
$$(\ell_1, \ell_2) = \sum_{j=0}^{m-1} b_j \left(\varepsilon_t^{j-1} \tilde{h}_{rt}, \varepsilon_u^j \tilde{h}_{su}\right) + a_{m-1} \left(\varepsilon_t^{m-1} h_{rt}, 0\right) + \delta_{su} b_m \left(0, \varepsilon_s^m\right))$$

$\alpha = 1, \beta = 1$: In this case, if $r = u$, then $a_{m-1} = b_{m-1}$. Thus,
If $r \neq u$,
$$(\ell_1, \ell_2) = \sum_{i=1}^{m-2} b_i \left(\varepsilon_t^i h_{rt}, \varepsilon_u^i h_{su}\right) + a_{m-1} \left(\varepsilon_t^{m-1} h_{rt}, 0\right) + b_{m-1} \left(0, \varepsilon_u^{m-1} h_{su}\right)$$
If $r = u$,
$$(\ell_1, \ell_2) = \sum_{i=1}^{m-1} a_i \left(\varepsilon_t^i h_{rt}, \varepsilon_u^i h_{su}\right).$$

□

This lemma means that if there is any non-zero homomorphism between indecomposable complexes, then there is a normal homogeneous non-zero homomorphism, so it suffices to study the normal homogeneous chain maps, of which there are a finite number.

Since we have reduced ourselves to the study of normal homogeneous chain maps, it will also be useful to know that if such a map is homotopic to zero, then homotopy can also be chosen with the same property.

LEMMA 3.2. *If a normal homogeneous chain map between two indecomposable elements of PTC_2 is homotopic to zero, then we may choose the homotopy to be normal homogeneous.*

Proof: Let $C^{\cdot}$ and $D^{\cdot}$ be indecomposable elements of PTC_2. If there is a homogeneous non-zero chain map $\ell^{\cdot}$ from C to D which is homotopic to zero with homotopy map $T^{\cdot}$, we have

$$\begin{array}{ccccccccc} 0 & \to & C_1 & \to & C_2 & \to & C_3 & \to & 0 \\ & & \downarrow \ell_1 & T_1 \swarrow & \downarrow \ell_2 & T_2 \swarrow & \downarrow \ell_3 & & \\ 0 & \to & D_1 & \to & D_2 & \to & D_3 & \to & 0 \end{array}$$

where $T^{\cdot}$ is a non-zero homotopy between $\ell^{\cdot}$ and the zero map. We may assume that we have removed from $T^{\cdot}$ any homogeneous components whose composition with the relevant horizontal maps is zero.

Since each of $C^{\cdot}$ and $D^{\cdot}$ is of width no greater than two, we have four cases with non-zero homotopy : a square, a triangle with the base at the top or at the bottom and a parallelogram. In the first three cases it is not hard to show that the mild assumption made on $T^\cdot$, that it not contain any irrelevant homogeneous factors, insures that it is normal homogeneous. Thus we will give the proof only for the more difficult case of a parallelogram. We have

$$\begin{array}{ccccccc}
0 & \to & P_r & \xrightarrow{h_{rs}} & P_s \\
\downarrow 0 & T_1 \swarrow & \downarrow \ell_2 & T_2 \swarrow & \downarrow 0 \\
P_t & \xrightarrow[h_{tu}]{} & P_u & \to & 0
\end{array}$$

where $h_{tu} T_1 + T_2 h_{rs} = \ell_2$. We have assumed that $\ell^\cdot$ is normal homogeneous, so

$$\ell_2 = \varepsilon_u^{i_0} \tilde{h}_{ru} \ , \ 0 \le i_0 \le m - 1 + \delta_{ru}.$$

We have
$$T_1 = \sum_{i=0}^{m-1} a_i \, \varepsilon_t^i \, \tilde{h}_{rt} + \delta_{rt} \, a_m \, \varepsilon_t^m$$
$$T_2 = \sum_{j=0}^{m-1} b_j \, \varepsilon_u^j \, \tilde{h}_{su} + \delta_{su} \, a_m \, \varepsilon_u^m.$$

Choose α, β such that
$$\begin{aligned} h_{tu}\tilde{h}_{rt} &= \varepsilon_u^\alpha \tilde{h}_{ru} \\ \tilde{h}_{su} h_{rs} &= \varepsilon_u^\beta \tilde{h}_{ru}. \end{aligned}$$

Note that $\delta_{rt} = 1 \Rightarrow \alpha = 0$ and $\delta_{su} = 1 \Rightarrow \beta = 0$

$$\ell_2 = \varepsilon_u^{i_0} \tilde{h}_{ru} = \left(\sum_{i=0}^{m-1} a_i \, \varepsilon_u^{i+\alpha} \, \tilde{h}_{ru} \right) + \left(\sum_{j=0}^{m-1} b_j \, \varepsilon_u^{j+\beta} \, \tilde{h}_{ru} \right).$$

Thus, for every $k \ne i_0$, $0 \le k \le m - 1 + \delta_{ru}$, we have

$$a_{k-\alpha} + b_{k-\beta} = 0.$$

Thus we may substitute for $T^\cdot$ a new homotopy $T'^\cdot$ in which $a_{k-\alpha} = b_{k-\beta} = 0$ for $k \ne i_0$. For $k = i_0$, we get
$$a_{i_0-\alpha} + b_{i_0-\beta} = 1.$$

$i_0 - \alpha$ and $i_0 - \beta$ cannot both be -1. Thus if we set one of $a_{i_0-\alpha}$ or $b_{i_0-\beta}$ equal to one and the other equal to zero, we get the desired normal homogeneous homotopy $\tilde{T}^\cdot$.

DEFINITION. If $\ell^\cdot : C^\cdot \to D^\cdot$ is a normal homogeneous chain map with one non-zero map, then the degree of $\ell^\cdot$ is the degree of its non-zero component. If $\ell^\cdot$ has two non-zero maps, one of which is a map between two identical projectives, the degree of $\ell^\cdot$ is the degree of the map between the identical projectives.

REMARK. It will be a consequence of the main theorems of this paper that every normal homogeneous chain map between two indecomposable components of an element of PTC_2 fulfills one of these two conditions and thus has a well-defined degree.

THEOREM 2. *Between two elementary complexes in PTC_2, there are chain maps not homotopic to zero in the following cases:*

(i) *From S_i to S_j there are $m + \delta_{ij}$ normal homogeneous maps. If we let $\{j - i\}_e$ represent the residue of $j - i$ modulo e, then the maps have degree $\{j - i\}_e + ke$, $0 \leq k \leq m + \delta_{ij}$.*

(ii) *From T_{ij} to itself there are two normal homogeneous maps, the identity and the socle map.*

(iii) *Between two elementary components with one common index, there is a normal homogeneous map of degree zero in one direction and of degree em in the opposite direction.*

4 HOMOMORPHISMS BETWEEN INDECOMPOSABLE COMPLEXES

Our eventual aim, summarized in Theorem 1 given in the previous section, is to describe all the partial tilting complexes $Q^{\cdot} \in PTC_2$, as direct sums of shifts of the complexes S_i and T_{jk} of projective b modules. The condition for $Q^{\cdot}$ to be in PTC_2 is that
$$\mathrm{Hom}_{D^b(b)}(Q^{\cdot}, Q^{\cdot}[n]) = 0 \quad \text{for} \quad n \neq 0.$$
By the standard properties of homomorphisms of direct sums, this is equivalent to showing that for any pair R and R' of elementary indirect summands of $Q^{\cdot}$, we have
$$\mathrm{Hom}_{D^b(b)}(R, R^{\cdot}[n]) = 0 \quad \text{for} \quad n \neq 0.$$
Since our elementary complexes are all of a length less than or equal to two, there are at most three possible relative shifts for which a non-zero projective of R is positioned over a non-zero projective of R. Where a non-zero chain map exists we must decide whether or not it is homotopic to zero.

Since, for those partial tilting complexes which are actually tilting complexes, we will be interested in calculating the endomorphism ring, we want to determine all maps in $\mathrm{Hom}_b(R, R')[n])$ which are not homotopic to zero. This information, which will be summarized in Theorem 2, will be determined from the same case-by-case study of homomorphisms between elementary components. We consider the following five cases, which will be treated in the subsequent five propositions.

Case 1: S with S.

Case 2: S with T, and no common indices.

Case 3: S with T, and a common index.

Case 4: T with T, and no common indices.

Case 5: T with T, and common indices.

Once we have determined all non-trivial maps, we can combine the results into Theorem 1 by using the principle that if there are two or more relative positions which produce non-trivial maps, then the two elementary complexes cannot appear together at all in partial tilting complex. If there is exactly one, then that must be the relative position of the two elementary complexes in any element of PTC_2. Finally, if there are no non-trivial maps between the two elementary complexes, then they can appear together without any restrictions on their relative position.

Since in classifying all the maps we also show that they have a well-defined degree and determine what it is, we then get Theorem 2 more or less for free. We will also establish that the elementary complexes are in fact indecomposable, since their endomorphism rings will be local.

PROPOSITION 1. *Two complexes, $S_r[n]$ and $S_t[n']$, can appear together in a partial tilting complex in PTC_2 if and only if $n = n'$. The normal homogeneous maps from $S_r[n]$ to $S_t[n]$ are the following*

$$\varepsilon_t^k \tilde{h}_{rt} \quad , \quad 0 \leq k \leq m - 1 + \delta_{rt}.$$

If $\{t - r\}_e$ is the residue modulo e, the degrees are

$$\{t - r\}_e + ke \quad , \quad 0 \leq k \leq m - 1 + \delta_{rt}.$$

Proof: There are no non-zero chain maps from $S_r[n]$ to $S_t[n']$ unless $n = n'$. If $n \neq n'$, then there is a map after a shift of $n' - n$, so $S_r[n]$ and $S_r[n']$ cannot occur together. When $n = n'$, the normal homogeneous maps and their degree were calculated in Lemma 2.1. □

Henceforward we will assume that $Q^{\cdot}$ has been shifted so that if there is any indecomposable S_r, the non-zero projective module appears in degree zero.

PROPOSITION 2. *If S_r and $T_{st}[n]$ have no common index, then they can appear together in a partial tilting complex in PTC_2 if and only if $r \to s \to t$ is short.*

Proof. ($\Leftarrow$) Assume that $r \to s \to t$ is short. We will show that every non-zero chain map in either direction is homotopic to zero.
Case 1. $\ell^{\cdot} : S_r \to T_{st}$. If $\ell_1 = \varepsilon_s^i h_{rs}$ is non-zero, so is $h_{st} \circ \ell_1 = \varepsilon_t^i h_{rt}$, and thus the following diagram does not commute.

$$\begin{array}{ccccccc}
S_r : & 0 & \to & P_r & \to & 0 & \\
 & & & \downarrow \ell_1 & & \downarrow & \\
T_{st} : & 0 & \to & P_s & \xrightarrow{h_{st}} & P_t & \to & 0.
\end{array}$$

Case 2. $\ell^{\cdot} : S_r \to T_{st}[1]$. If $\ell_1 = \varepsilon_t^i h_{rt}$ is non-zero, then $T_1 = \varepsilon_s^i h_{rs}$ is a homotopy

$$\begin{array}{ccccc}
 & 0 & \to & P_r & \to & 0 \\
 & \downarrow & T_1 \swarrow & \downarrow \ell_1 & & \\
P_s & \xrightarrow{h_{st}} & & P_t & \to & 0.
\end{array}$$

Partial Tilting Complexes

Case 3. $\ell^{\cdot} : T_{st} \to S_r$. As in Case 2, there is a non-trivial homotopy.

Case 4. $\ell^{\cdot} : T_{st}[1] \to S_r$. As in Case 1, there are no well-defined chain maps.

($\Rightarrow$) Now suppose that $r \to s \to t$ is long, which implies, by the cyclic permutation property, that $s \to t \to r$ is long. We obtain that these are two different non-trivial maps from S_r to T_{st}, with different shifts, so S_r and T_{st} cannot appear together in any relative position.

Case (1). $\ell'^{\cdot} : S_r \to T_{st}$.

$$\begin{array}{ccccc} 0 & \to & P_r & \to & 0 \\ & & \downarrow \varepsilon_s^{m-1} h_{rs} & & \\ 0 & \to & P_s & \to & P_t \end{array}$$

This map is well defined because $h_{st} \circ (\varepsilon_s^{m-1} h_{rs}) = 0$, and no homotopy is possible.

Case (2). $\ell'^{\cdot} : S_r \to T_{st}[1]$.

$$\begin{array}{ccccccc} & & 0 & \to & P_r & \to & 0 \\ & & \downarrow & & \downarrow h_{rt} & & \\ 0 & \to & P_s & \to & P_t & \to & 0 \end{array}$$

This map is clearly well defined, and no homotopy is possible because $h_{st} \circ (\varepsilon_s^i h_{rs}) = \varepsilon_t^{i+1} h_{rt}$, and this cannot equal h_{rt}. □

PROPOSITION 3. *If S_r and T_{st} or $T_{st}[1]$ have a common index, then they can appear together in a partial tilting complex in PTC_2 if and only if the common index appears in the same degree in each. The normal homogeneous homomorphisms are then the following:*

(i) *If $r = s$* : $\ell^{\cdot} : S_r \to T_{rt}$ has $\ell_1 = \varepsilon_r^m$ of degree em.
  $\bar{\ell}^{\cdot} : T_{rt} \to S_r$ has $\bar{\ell}_1 = \mathrm{id}_r$ of degree 0.

(ii) *If $r = t$* : $S_r \to T_{sr}[1]$ has $\ell_2 = \mathrm{id}_r$ of degree 0.
  $\bar{\ell}^{\cdot} : T_{sr}[1] \to S_r$ has $\bar{\ell}_2 = \varepsilon_r^m$ of degree em.

Proof. We will first show that the four maps given in the proposition are well-defined and not homotopic to zero. This will establish that the common projectives must be in the same degree in any partial tilting module, since otherwise we would get a map to a shift.

(i) If $\ell^{\cdot} : S_r \to T_{rt}$ is normal homogeneous, then ℓ_1 must be ε_r^i. In order for the diagram to commute, we must have $h_{rt}\, \varepsilon_r^i = 0$, which will only be true if $i = m$. By the shape of the diagram, there is no possible homotopy, so $\ell_1 = \varepsilon_r^m$ is a non-trivial homomorphism of degree em.

$\bar{\ell}^{\cdot} : T_{rt} \to S_r$ is well-defined for any $\bar{\ell}_1 = \varepsilon_r^i$, but if $i > 0$, then we have a homotopy $T_1 = \varepsilon_r^{i-1} h_{tr}$. When $i = 0$, there is no normal homogeneous T such that

$$T \circ h_{rt} = \mathrm{id}_r\,.$$

(ii) is dual to (i).

It remains to show that if the corresponding indices are not positioned one over the other, then there is no non-trivial chain map.

(i') $\underline{r=s}$: Consider $\ell\dot{} : S_r \to T_{rt}[1]$, with $\ell_1 : P_r \to P_t$ equal to $\varepsilon^j h_{rt}$. This has a homotopy $T = \varepsilon_r^j$ for any j.

Now take the opposite direction $\bar{\ell}\dot{} : T_{rt}[1] \to S_r$. If $\bar{\ell}\dot{} \neq 0$, then $\ell_1 = \varepsilon_r^j h_{tr}$, with $j < m$. Then, $\bar{\ell}$ cannot be well-defined because $\varepsilon_r^j h_{tr} \circ h_{rt} = \varepsilon_r^{j+1} \neq 0$.

$$\begin{array}{ccc} P_r & \xrightarrow{h_{rt}} & P_t \\ \downarrow & \searrow & \downarrow \ell_1 = \varepsilon_r^j h_{tr} \\ 0 & \to & P_r. \end{array}$$

(ii') $\underline{r=t}$: Dual to (i').

$\square$

The following is the most difficult of the five propositions:

PROPOSITION 4. *If $T_{rs}[n]$ and $T_{tu}[n']$ appear together in a partial tilting complex in PTC_2, with all four indices distinct, then at least one of the following three sequences of indices must be short:*

(1) $r \to s \to t \to u$

(2) $r \to t \to u \to s$

(3) $r \to s \to u \to t$

This implies, by the mixing property in Lemma 3.1, that none of the following can be short.

(4) $r \to t \to s \to u$

(5) $r \to u \to s \to t$

(6) $r \to u \to t \to s$

Proof. We will first show that in Cases (1), (2) and (3), there are no non-trivial homomorphisms, and thus these elementary complexes can appear together in some $Q\dot{} \in PTC_s$ with any desired shift. Since interchanging r, s with t, u produces the same three cases, it suffices to show that there are no non-trivial homomorphisms from T_{rs} to $T_{tu}[n]$, for $n = -1, 0, 1$.

Case (1). $r \to s \to t \to u$ is short.

n = −1 There is no well-defined map, as in case (i') of Proposition 3. The first square does not commute.

$$\begin{array}{ccccccccc} 0 & \to & P_r & \xrightarrow{h_{rs}} & P_s & \to & 0 & & \\ & & \downarrow & \searrow{\varepsilon_t^i h_{rt}} & \downarrow{\varepsilon_t^i h_{st}} & & & & \\ & & 0 & \to & P_t & \xrightarrow[h_{tu}]{} & P_u & \to & 0. \end{array}$$

Partial Tilting Complexes

n = 0

$$0 \longrightarrow P_r \xrightarrow{h_{rs}} P_s \longrightarrow 0$$
$$\varepsilon_t^i h_{rt} \downarrow \quad T \swarrow \quad \downarrow \varepsilon_u^i h_{su}$$
$$0 \longrightarrow P_t \xrightarrow[h_{tu}]{} P_u \longrightarrow 0$$

This is well defined if $i = j$, but then $T = \varepsilon_t^i h_{st}$ is a homotopy.

n = 1

$$0 \longrightarrow P_r \xrightarrow{h_{rs}} P_s \longrightarrow 0$$
$$T_1 \swarrow \quad \varepsilon_u^i h_{ru} \downarrow \quad T_2 \swarrow$$
$$0 \longrightarrow P_t \xrightarrow[h_{tu}]{} P_u \longrightarrow 0$$

$T_1 = \varepsilon_t^i h_{rt}$, $T_2 = \varepsilon_u^i h_{su}$ is a homotopy.

Case (2). $r \to t \to u \to s$ is short.

n = −1 This is not well defined, as in Case 1.

n = 0

$$0 \longrightarrow P_r \xrightarrow{h_{rs}} P_s \longrightarrow 0$$
$$\varepsilon_t^i h_{rt} \downarrow \quad T \swarrow \quad \downarrow \varepsilon_u^j h_{su}$$
$$0 \longrightarrow P_t \xrightarrow[h_{tu}]{} P_u \longrightarrow 0$$

This is well defined only if $i = j + 1$, but in that case $T = \varepsilon$.

n = 1 There is a homotopy, as in $n = 1$ for Case 1.

Case (3). $r \to s \to n \to t$ is short. If we rewrite this as $t \to r \to s \to u$, it is clearly dual to Case 2, where r is interchanged with t, and s with n.

We now show that the Cases (4), (5), and (6) cannot occur in $Q' \in PTC_2$, by exhibiting, in each case, two different shifts which produce non-trivial homomorphisms. Once again the cases are symmetric under the interchange of T_{rs} with T_{tu}. We will in general leave to the reader the verification that the maps are not homomorphic to zero.

Case (4). $r \to t \to s \to u$ is short.

$$0 \longrightarrow P_r \xrightarrow{h_{rs}} P_s$$
$$\downarrow \varepsilon_t^{m-1} h_{st}$$
$$0 \longrightarrow P_t \xrightarrow[h_{tu}]{} P_u \longrightarrow 0$$

$$
\begin{array}{ccccccc}
0 & \longrightarrow & P_r & \xrightarrow{h_{rs}} & P_s & \longrightarrow & 0 \\
& & h_{rt}\downarrow & & \downarrow h_{su} & & \\
0 & \longrightarrow & P_t & \xrightarrow{h_{tu}} & P_u & \longrightarrow & 0
\end{array}
$$

The second map is well defined because $h_{su} \circ h_{rs} = h_{ru} = h_{tu} \circ h_{rt}$. There is no homotopy because $h_{st} \circ h_{rs} = \varepsilon_r h_{rt}$.

Case (5). $t \to r \to u \to s$ is short.

$$
\begin{array}{ccccccc}
& & 0 & \longrightarrow & P_r & \longrightarrow & P_s \\
& & \downarrow & & \downarrow h_{ru} & & \downarrow \\
0 & \longrightarrow & P_t & \xrightarrow{h_{tu}} & P_u & \longrightarrow & 0
\end{array}
$$

$$
\begin{array}{ccccccc}
0 & \longrightarrow & P_r & \xrightarrow{h_{rs}} & P_s & \longrightarrow & 0 \\
& & h_{rt}\downarrow & & \downarrow h_{su} & & \\
0 & \longrightarrow & P_t & \xrightarrow{h_{tu}} & P_u & \longrightarrow & 0
\end{array}
$$

The second homomorphism is well defined because $h_{su} \circ h_{rs} = \varepsilon_u h_{ru} = h_{tu} \circ h_{rt}$.

Case (6). $r \to u \to t \to s$ is short.

$$
\begin{array}{ccccc}
0 \longrightarrow & P_r & \xrightarrow{h_{rs}} & P_s \longrightarrow 0 \\
& \downarrow & & \downarrow \varepsilon_t^{m-1} h_{st} \\
& 0 \longrightarrow & & P_t \xrightarrow[h_{tu}]{} P_u \longrightarrow & 0
\end{array}
$$

$$
\begin{array}{ccccccc}
& & 0 & \longrightarrow & P_r & \xrightarrow{h_{rs}} & P_s & \longrightarrow & 0 \\
& & \downarrow & & \downarrow h_{ru} & & \downarrow & & \\
0 & \longrightarrow & P_t & \xrightarrow{h_{tu}} & P_u & \longrightarrow & 0
\end{array}
$$

□

The final proposition is parallel to Proposition 3, dealing with the case of common indices.

PROPOSITION 5. *T_{rs} can occur in a two-restricted partial tilting complex, but not together with any shift of itself or of T_{sr}. If T_{rs} and $T_{tu}[n]$ have unique common index, then they can occur together in a two-restricted partial tilting complex if and only if the common index occurs in the same degree and*

1. *If $n = -1$, so that $s = t$, then $r \to s \to u$ is short.*
   There is one normal homogeneous map in each direction

$$
\begin{array}{ccc}
P_r & \longrightarrow & P_s \\
& & \varepsilon_s^m \downarrow\uparrow id_s \\
& & P_s & \longrightarrow & P_u
\end{array}
$$

Partial Tilting Complexes

2. If $n = 0$, and $r = t$, then s and u are symmetrical, and we may assume $r \to s \to u$ is short. There is one normal homogeneous map in each direction

$$\begin{array}{ccc} P_r & \longrightarrow & P_s \\ id_s \downarrow & & \downarrow h_{su} \\ P_r & \longrightarrow & P_u \end{array}$$

$$\begin{array}{ccc} P_r & \longrightarrow & P_s \\ \varepsilon_r^m \uparrow & & \uparrow 0 \\ P_r & \longrightarrow & P_u \end{array}$$

If $n = 0$ and $s = u$, then r, t are symmetric so we may assume $r \to t \to u$ is short.

3. If $n = 1$, we get the case dual to $n = -1$, and must have $t \to r \to s$ short.

$$\begin{array}{ccc} & P_r & \longrightarrow \quad P_s \\ & id_r \downarrow \uparrow \varepsilon_r^m & \\ P_t \longrightarrow & P_r & \end{array}$$

There are two normal homogeneous maps from T_{rs} to itself, the identity and the socle maps.

Proof: Since we always have $id : T_{rs} \to T_{rs}$, in order to show that T_{rs} can occur in $Q' \in PTC_2$, it suffices to show that there are no non-trivial homomorphisms with a shift

$$\begin{array}{ccc} P_r \xrightarrow{h_{rs}} P_s & \qquad P_r \xrightarrow{h_{rs}} P_s \\ \downarrow \varepsilon_r^i h_{sr} & \qquad \varepsilon_r^i \swarrow \quad \downarrow \varepsilon_s^i h_{rs} \quad \swarrow \varepsilon_s^i \\ P_r \xrightarrow[h_{rs}]{} P_s & \qquad P_r \xrightarrow[h_{rs}]{} P_s \end{array}$$

The left hand map is never well defined for any i, and the right hand map always has the indicated homotopy.

T_{rs} cannot occur together with any shift of T_{sr} because the following two homomorphisms are non-trivial and thus whatever relative position is chosen, there is always a diagonal map

$$\begin{array}{ccc} P_r \longrightarrow P_s & \qquad P_r \longrightarrow P_s \\ \downarrow \varepsilon_s^m & \qquad \downarrow id_r \\ P_s \longrightarrow P_r & \qquad P_s \longrightarrow P_r \end{array}$$

We are now prepared to consider a pair of elementary complexes, T_{rs} and $T_{tu}[n]$, for $n = -1, 0, 1$, with one common index. The case $n = 1$ is dual to the case $n = -1$, and for $n = 0$, the case $r = t$ is dual to $s = u$, so we will give the details only for $n = -1$ and $n = 0$ with $s = u$. We will first show that for the cases given in the proposition there are no non-trivial homomorphisms for other shifts. In addition, in the case $n = -1$, we must show that if $r \to s \to u$ is long, there are two possible shifts which give non-trivial maps.

n = −1   We first suppose that $r \to s \to u$ is short, and show that the maps indicated in the proposition are non-trivial

$$
\begin{array}{ccccc}
0 & \to & P_r & \to & P_s & \to & 0 \\
 & & & & \varepsilon_s^m \downarrow\uparrow \mathrm{id}_s & & \\
 & & & & P_s & \to & P_u & \to & 0
\end{array}
$$

The map from top to bottom is well defined because $\varepsilon_s^m h_{rs} = 0 = \varepsilon_u^m h_{su}$, and no homotopy is possible. The map from top to bottom is automatically well defined, and there is no homotopy because id_s does not factor through h_{rs} or h_{su}.

Any homomorphism

$$
\begin{array}{ccccccccc}
0 & \to & P_r & \xrightarrow{h_{rs}} & P_s & \to & 0 \\
 & & \varepsilon_s^i h_{rs} \downarrow & \varepsilon_s^i \swarrow & \downarrow \varepsilon_u^i h_{su} & & \\
0 & \to & P_s & \xrightarrow{h_{su}} & P_u & \to & 0
\end{array}
$$

has the indicated homotopy and so does not give a non-trivial map

$$
\begin{array}{ccccc}
 & & P_r & \xrightarrow{h_{rs}} & P_s \\
 & \varepsilon^i h_{rs} \swarrow & \downarrow & \swarrow \varepsilon_s^j h_{su} & \\
0 & \to \quad P_s \quad \xrightarrow{h_{su}} & P_u & &
\end{array}
$$

This is a homotopy with $i = j$ if $r \to s \to u$ is short. If $r \to s \to u$ is long, it is a homotopy when $i = j - 1$ for $j > 0$, but the case $j - 0$ gives a non-trivial homomorphism. Thus, if $r \to s \to u$ is long, T_{rs} and $T_{su}[-1]$ cannot appear together because there is a non-trivial homomorphism with a different shift, and if $r \to s \to u$ is short, they can.

n = 0   We do the case $s = u$, and assume rts is short.

$$
\begin{array}{ccc}
P_r & \xrightarrow{h_{rs}} & P_s \\
h_{rt} \downarrow & \swarrow & \downarrow \mathrm{id}_s \\
P_t & \xrightarrow{h_{ts}} & P_s
\end{array}
\qquad
\begin{array}{ccc}
P_t & \xrightarrow{h_{ts}} & P_s \\
0 \downarrow & T \swarrow & \downarrow \varepsilon_s^m \\
P_r & \xrightarrow{h_{rs}} & P_s
\end{array}
$$

In the left hand map there is no homotopy because id_s does not factor through h_{ts}. In the right hand case, if $T = \varepsilon_r^{m-1} h_{sr}$, then $T \circ h_{ts} \neq 0$.

It remains to show that there are no well-defined shifted maps

$$
\begin{array}{ccc}
P_r & \to & P_s \\
\downarrow & \searrow & \\
P_t & \to & P_s
\end{array}
\qquad
\begin{array}{ccc}
 & & P_r & \xrightarrow{h_{rs}} & P_s \\
 & 0 \swarrow & \varepsilon^i h_{rs} \downarrow & \swarrow \varepsilon_s^i & \\
P_t & \to & P_s & &
\end{array}
$$

The left hand map is not well defined because the indicated diagonal is non-zero, and the right hand map has the indicated homotopy since we did not need the condition $r \to t \to s$ is short. The same is true for maps from T_{ts} to T_{rs}. □

5 PROOF OF THEOREMS

Now that we have Propositions 1 – 5, the proofs of Theorems 1 and 2 are simply a matter of organization.

Proof of Theorem 1
(a) By Proposition 1, all $S_i[n]$ have their non-zero term in the same degree. Shifting Q' if necessary we may assume that this is degree zero.
(b) By Proposition 3, if S_r and $T_{tu}[m]$ have a common projective, it occurs in the same degree in both. By Proposition 5, this is also true for $T_{rs}[n']$ and $T_{tu}[n]$.
(c) By Proposition 4, if S_r and $T_{st}[n]$ occur in Q, then $r \to s \to t$ is short.
(d) This is Proposition 4. □

Proof of Theorem 2
(i) From Proposition 1.
(ii) From Proposition 5.
(iii) From Proposition 3 and Proposition 5. □

In conclusion, having shown that in a two-restricted partial tilting complex all occurrences of P_r are in the same degree, we want to summarize the information given by Propositions 1, 3, and 5 in a way that considers all S_r together, and all irreducibles with a common P_r together. As before, we let $\{t - s\}_e$ be the residue modulo e of $t - s$ between 0 and $e - 1$.

THEOREM 3. (Local order theorem)

1. If $\{S_{i_1}, \ldots, S_{i_r}\}$ are a set of S_i in $Q' \in PTC_2$, such that $i_1 \to i_2 \to \cdots \to i_r$ is short, then we have homomorphisms $h_{i_k i_{k+1}} : S_{i_k} \to S_{i_{k+1}}$ such that

$$h_{i_{k-1} i_k} \circ \cdots \circ h_{i_{k+1} i_{k+2}} \circ h_{i_k i_{k+1}} = \varepsilon_{i_k} \quad \text{for } k = 1, \ldots, r.$$

2. If $\{T_{s_i r}[1]\}_{i=1}^k \cup \{T_{r t_j}\}_{j=1}^\ell$ are a set of irreducibles in Q', we can arrange the indices so that

$$s_1 \to \cdots \to s_k \to r \to t_1 \to \cdots \to t_\ell$$

is short. All of the maps

$$T_{rt_1} \to T_{rt_2} \to \cdots \to T_{rt_\ell} \to T_{s_1 r}[1] \to \cdots \to T_{s_k r}[1]$$

will be identity on P_r, whereas $T_{s_k r}[1] \to T_{rt_1}$ will be ε^m on P_r.
If S_r is in Q', then S_r can be inserted between T_{rt_ℓ} and $T_{s_1 r}[1]$, with identity maps in P_r.

Proof. (1) This is direct from Proposition 1 and the definition of a short sequence.
(2) That each of the indicated maps is the identity on P_r is a consequence of Proposition 5, as is the fact that the map from T_{rt_ℓ} to $T_{s_1 r}[1]$ is ε_r^m. The possibility of inserting S_i comes from Proposition 1. □

6 APPLICATIONS

Our aim in studying these two-restricted partial tilting complexes was to give a classification to all TC_2 in combinatorial terms, which will be done in the sequel, [SZ1]. It was, furthermore, important to show that the homomorphisms between the irreducible complexes could be taken to be homogeneous, for this allows a study of a class of homogeneous deformations of the Brauer tree algebras [SZ2] which is much more natural than that given in [S2], in the sense that it is local rather than global, and also that it is derived from a tilting complex.

Although the "two-restricted" condition means that we have considered only a subclass of all possible tilting complexes, it is a sufficiently large subclass to allow us to reach every possible Brauer tree algebra. In fact, it is too large for it is possible to reach each Brauer tree algebra in many possible ways. In the next paper we consider all the different possible "foldings" of the tilting complex. It had already been shown by Rouquier [R] that each tree can be reached by a "completely folded" complex with only two non-zero terms. In [RS] we show that Rickard's combinatorial tree-to-star complex and the corresponding star-to-tree two-restricted complex give inverse equivalences, and reduce to the Rouquier two-term complex in the completely folded case. As Zimmermann pointed out in a very helpful conversation, if we go over to two-sided tilting complexes [Z] we can study the different possible tilting complexes which give the same Brauer tree algebra in terms of the group of two-sided self-equivalences of b. This group might be quite large; for the very simple case of the Brauer star algebra of type (2,1) Rouquier and Zimmermann [RoZ] calculated that, modulo shifts, it was the modular group, which is the free group on two generators of order 2 and order 3. The entire subject merits further investigation.

We also hope that it might be possible to use this approach for some generalization of the Brauer star algebra with abelian but not cyclic defect group. This might lead to some generalization of the Brauer tree. One such generalization has already been made by Benson [B], for dihedral defect groups.

REFERENCES

[A] Alperin J. *Local Representation Theory*, Cambridge Studies in Advanced Mathematics, vol. 11, 1986

[B] Benson D. *Representations and Cohomology* Cambridge Studies in Advanced Mathematics, vol. 30, 1991

[H] Happel D. *On the derived category of a finite dimensional algebra* Comment. Math. Helvetici, vol 62, 339-389, 1987

[KZ1] König S. and Zimmermann A. *Tilting hereditary orders* Comm. in Algebra vol 24, 1996, 1893-1913

[KZ2] König S. and Zimmermann A. *Derived equivsalence for group rings* LNM vol. 1685, 1998

[MS] Mejer C. and Schaps M. *Separable deformations of blocks with abelian defect group and of derived equivalent global blocks* Canadian Math. Soc. Conf. Proc. vol 18, 1996, 505-518

[M] Membrillo F.H. *Homological Properties of Finite Dimensional Algebras* Ph.D. Thesis, Oxford University, 6-13, 78, 1993

[R1] Rickard J. *Morita theory for derived categories* J. London Math. Soc., vol 39, 2, 436-456, 1989

[R2] Rickard J. *Derived categories and stable equivalence* J. of Pure and Applied Algebra, vol 61, 1989, 303-317

[R3] Rickard J.*Lifting theorems for tilting complexes* J. of Algebra, vol 142, 383-393, 1991

[RS] Rickard J. and Schaps M. *Folded titlting complexes for Brauer tree algebras* preprint

[Ro] Rouquier R. and Zimmermann A. *Picard groups for derived modular categories* preprint

[RoZ] Rouquier R. *From stable equivalence to Rickard equivalences for blocks with cyclic defect group* Proc. of Groups 93, Galway, St. Andrews, London Math. Soc. Lecture Notes Series, vol 212, 512-523

[S1] Schaps M. *Deformations of finite dimensional algebras and their idempotents* Trans. AMS, 1988, 843-856

[S2] Schaps M. *A modular version of Maschke's theorem for groups with cyclic p-Sylow subgroup* J. of Algebra, vol 163, 1994, 623-635

[SSS] Schaps M., Shapira D. and Shlomo D. *Quivers of blocks with normal defect group* preprint

[SZ1] M. Schaps and Zakay-Illouz E. *Pointed Brauer trees* preprint

[SZ2] M. Schaps and Zakay-Illouz E. *Homogeneous deformations of Brauer tree algebras* preprint

[ZI1] Zakay-Illouz E. *Basis-graphs and deformations for non-abelian groups of order $2^j \cdot 3^i \leq 24$, $i,j \geq 0$ with abelian p-Sylow subgroups* Master's thesis, Bar-Ilan Univesity, 1993

[ZI2] Zakay-Illouz E. *The Green Correspondence between Separable Deformations* Ph.D. dissertation, Bar-Ilan University, 1999

[Z] Zimmermann, A. *A two-sided tilting complex for Green orders and Brauer tree algebras* J. of Alg., vol 187, 1997, no. 2, 446-473

Almost split sequences in categories of Representations of Quivers II

SVERRE O. SMALØ Institutt for matematiske fag, NTNU, 7491 Trondheim, Norway, email: sverresm@math.ntnu.no

ABSTRACT Let k be a field, Q a connected quiver and $\mathrm{fd}(Q,k)$ be the category of finite dimensional representations of Q over k. In this note it is proved that for a quiver Q the subcategory $\mathrm{fd}_0(Q,k)$ of $\mathrm{fd}(Q,k)$ consisting of the representations having composition factors from the discrete simples, has almost split sequences if and only if Q is either path finite (without oriented cycles if Q is finite) or consists entirely of a single oriented cycle or is a subquiver of A_∞^∞ with linear orientation. [1]

INTRODUCTION

In this note a quiver Q is an oriented graph where only a finite number of arrows are adjacent to each vertex. A quiver is called path finite if there is no oriented path of infinite length.

Let k be a field and let $\mathrm{fd}(Q,k)$ denote the category of finite dimensional representations of Q over k. For each vertex p in Q, the representation given by a one-dimensional k-space at p, all other spaces being zero and all maps associated with the arrows being zero, is a simple representation. It will be denoted by S_p and called the discrete simple attached to the vertex p. Finally, let $\mathrm{fd}_0(k,Q)$ denote the full subcategory of $\mathrm{fd}(k,Q)$ consisting of representations having composition factors only among the discrete simples attached to vertices of Q.

The aim of this note is to prove the following result.

THEOREM 1 *Let Q be a connected quiver and k be a field.*

(a) *In case Q is finite, $\mathrm{fd}_0(Q,k)$ has almost split sequences if and only if one of the following two properties are satisfied.*

   (i) *Q contains no oriented cycles or*

(ii) Q is $\widetilde{A}_n$ for some n with cyclic orientation.

(b) In case Q is infinite, $\mathrm{fd}_0(Q,k)$ has almost split sequences if and only if one of the following two properties are satisfied.

(i) Q is path finite or

(ii) Q is either A_∞ with linear orientation or A_∞^∞ with linear orientation.

Let $\mathcal{C}$ be an abelian Krull–Schmidt-category, i.e. the indecomposable objects in $\mathcal{C}$ have local endomorphism rings and each object in $\mathcal{C}$ can be decomposed as a finite direct sum of indecomposables, and then in a unique way up to isomorphism by the Krull–Schmidt theorem.

The category $\mathcal{C}$ is said to have right almost split morphisms if for each indecomposable object X in $\mathcal{C}$, there is a morphism $f : Y \to X$ in $\mathcal{C}$ such that the cokernel, Coker($,f$), of the natural morphism $\mathrm{Hom}_\mathcal{C}(\,,f)$ as a functor on $\mathcal{C}$ takes the value 0 for all indecomposable objects Y not isomorphic to X and the value $\mathrm{Hom}_\mathcal{C}(X',X)/\mathrm{rad}(X',X)$ for the indecomposable objects X' isomorphic to X. Here $\mathrm{rad}(X',X)$ is the set of nonisomorphisms from X' to X which is a subgroup of $\mathrm{Hom}_\mathcal{C}(X',X)$ since the endomorphism ring of X is local. The morphisms are sent to their residues. A morphism $f : Y \to X$ satisfying the above property is called a right almost split morphism in $\mathcal{C}$.

Dually, the category $\mathcal{C}$ is said to have left almost split morphisms if for each indecomposable object X in $\mathcal{C}$, there is a morphism $f : X \to Y$ in $\mathcal{C}$ such that the cokernel, Coker($f,$), of the natural morphism $\mathrm{Hom}_\mathcal{C}(f,\,)$ satisfies the same condition as above, i.e. Coker($f,$) applied to an indecomposable object X' in $\mathcal{C}$ is $\mathrm{Hom}_\mathcal{C}(X,X')/\mathrm{rad}(X,X')$ where $\mathrm{rad}(X,X')$ is the subgroup of nonisomorphisms from X to X'. A morphism $f : X \to Y$ satisfying the above property is called a left almost split morphism in $\mathcal{C}$.

The category $\mathcal{C}$ is said to have almost split morphisms if it has both right and left almost split morphisms. It is said to have almost split sequences if in addition, for each indecomposable nonprojective module X in $\mathcal{C}$ there exists an exact sequence $0 \to Z \to Y \to X \to 0$ with Z indecomposable and where $Z \to Y$ is left almost split in $\mathcal{C}$ and where $Y \to X$ is right almost split in $\mathcal{C}$; and for each indecomposable noninjective module Z in $\mathcal{C}$ there exists an exact sequence $0 \to Z \to Y \to X \to 0$ with X indecomposable and where $Z \to Y$ is left almost split in $\mathcal{C}$ and where $Y \to X$ is right almost split in $\mathcal{C}$.

For background on the representation theory of artin algebras including finite dimensional algebras, and on the theory of representations of quivers, the reader is referred to the book [ARS].

A celebrated result of Auslander and Reiten is the following theorem (see [ARS]).

THEOREM 2 *If Λ is an artin algebra, then the category of finitely generated left Λ-modules has both left and right almost split morphisms as well as almost split sequences.*

THE PROOF OF THEOREM 1

Now to the proof of the result of this note.

Let us start by giving the arguments that if (i) or (ii) in (a) is satisfied, then $fd_0(Q,k)$ has almost split sequences. If (a) (i) is satisfied, then $fd_0(Q,k)$ is equal to $fd(Q,k)$, which again is equivalent to the category of finite dimensional modules over the path algebra kQ. Since the quiver is finite and has no oriented cycles, the path algebra kQ is finite dimensional and therefore the category of finite dimensional modules and the category of finitely generated modules coincide, and hence by the result of Auslander and Reiten quoted above, the category $fd_0(Q,k)$ has almost split sequences.

If (a) (ii) is satisfied, then $Q = \tilde{A}_n$ for some n with cyclic orientation, and $fd_0(Q,k)$ is a category with $n+1$ simple objects and where all indecomposable objects are uniserial. The indecomposable objects are then given by their socle and their length, so numbering the vertices of Q by $0, 1, ..., n$ one obtains a natural indexing of the indecomposable objects by $\{0, 1, ..., n\} \times \mathbb{N}$. With this indexing the indecomposable objects fit together into exact sequences

$$0 \to (j, m) \to (j, m+1) \amalg ((j-1), m-1) \to ((j-1), m) \to 0$$

where (s) denotes the residue modulo $n+1$, and the module $(t, 0)$ is zero. It is easy to see that these sequences are almost split (see [S] for more details on this). This shows that if either (i) or (ii) is satisfied, then $fd_0(Q,k)$ has right and left almost split morphisms as well as almost split sequences.

To prove the converse in (a) we prove the following somewhat stronger result.

PROPOSITION 3 *If Q is a finite connected quiver and contains a subquiver $\tilde{A}_i$ with cyclic orientation as a proper subquiver, then $fd_0(Q,k)$ has neither left nor right almost split morphisms.*

Proof: Assume that Q contains an oriented cycle and let $\tilde{A}_n$ be a fixed minimal subquiver of Q with cyclic orientation. Since Q is connected there is at least one additional arrow α starting or ending at a vertex q of $\tilde{A}_n$. Let $\tilde{Q}$ be the subquiver of Q consisting of the arrows in $\tilde{A}_n$ and α together with their initial and end vertices.

By duality, it is enough to consider the case when the arrow α ends in a vertex q of $\tilde{A}_n$. Let p be the start of α. Assume that $X \to S_p$ is a right almost split morphism in $fd_0(Q,k)$. Now, take the subcategory of $fd_0(Q,k)$ consisting of all objects where the maps corresponding to the arrows not in $\tilde{Q}$ are zero. This subcategory is closed with respect to sub-objects and quotient-objects in $fd_0(Q,k)$ and it is equivalent to the category $fd_0(\tilde{Q},k)$. The trace of this subcategory, τX, in X will induce a right almost split morphism $\tau X \to S_p$ in $fd_0(\tilde{Q},k)$. So in order to prove that there is no right almost split morphism $X \to S_p$ in $fd_0(Q,k)$, it is enough to prove that there is no right almost split morphism $Y \to S_p$ in $fd_0(\tilde{Q},k)$. Therefore we can without loss of generality, from now on assume that $Q = \tilde{Q}$.

Observe that $fd_0(Q,k)$ is the full subcategory of $fd(Q,k)$ consisting of the objects which, when viewed as modules over the path algebra, are annihilated by some power of the ideal generated by the arrows.

So for the right almost split morphism $X \to S_p$, there is some power, say $t > 0$, of the ideal I generated by the arrows in kQ that annihilates X. Then one knows by the Auslander-Reiten formulas that the minimal right almost split morphism $X \to S_p$ fits into an almost split sequence

$$0 \to D\mathrm{Tr}_{kQ/I^m} S_p \to X \to S_p \to 0$$

for all $m \geq t$. The projective resolution of S_p over kQ/I^m looks like

$$P_s \amalg P_q \to P_p \to S_p \to 0$$

where $P_s = 0$ if p is not in $\tilde{A}_n$ and, by the minimality of $\tilde{A}_n$ in Q, $s = q$ is the successor of p in $\tilde{A}_n$ if p is in $\tilde{A}_n$. Now the representation corresponding to the module $\text{Tr}_{kQ/I^m} S_p$ has a projective presentation

$$P_p^* \to P_q^* \amalg P_s^* \to \text{Tr}_{kQ/I^m} S_p \to 0$$

and its dimension is a function of m, which will tend to ∞ as m grows. This gives a contradiction to the existence of a right almost split morphism $X \to S_p$.

To prove that there do not exist left almost split morphisms in $\text{fd}_0(Q,k)$, we will use the simple representation S_q and show that there is no left almost split morphism starting in S_q. We can reduce to the situation where $Q = \tilde{Q}$ by using that if $S_q \to X$ is a left almost split morphism in $\text{fd}_0(Q,k)$, then $S_q \to X/(\text{rej}_{\text{fd}_0(\tilde{Q},k)} X)$ is a left almost split morphism in $\text{fd}_0(\tilde{Q},k)$, where $\text{rej}_{\mathcal{A}} X$, the reject of $\mathcal{A}$ in X, is the intersection of all kernels of all morphisms from the representation X to any representation in the subcategory $\mathcal{A}$.

Assume that $S_q \to X$ is a left almost split morphism in $\text{fd}_0(Q,k)$. Then there is a power, say $t > 0$ of the ideal I generated by the arrows in kQ that annihilates X, and we have by the Auslander–Reiten formulas an almost split sequence

$$0 \to S_q \to X \to \text{Tr}\, D_{kQ/I^m} S_q \to 0$$

for all $m \geq t$. Now the projective presentation of DS_q over $(kQ/I^m)^{\text{op}}$ is

$$P_p^* \amalg P_s^* \to P_q^* \to DS_q \to 0$$

where s is the immediate predecessor of q in $\tilde{A}_n$. Therefore the projective presentation of $\text{Tr}\, DS_q$ is

$$P_q \to P_p \amalg P_s \to \text{Tr}\, D_{kQ/I^m} S_q \to 0$$

which shows that the dimension of $\text{Tr}\, D_{kQ/I^m} S_q$ tends to ∞ as m grows. This gives the desired contradiction and completes the proof of the proposition as well as the proof of part (a) of the theorem.

The proof of part (b) of the theorem can be completed by using the following facts: (i) if Q is path finite, then $\text{fd}_0(Q,k)$ has enough projective and injective objects, and the construction of the dual of the transpose works locally to produce almost split sequences as in the situation with a finite quiver without oriented cycles. (ii) If Q is either A_∞ with linear orientation or A_∞^∞ with linear orientation, then all indecomposable modules are uniserial and fit into almost split sequences. Finally, if Q contains a subquiver of the form A_∞ and in addition there is a vertex p in this subquiver with two arrows starting or two arrows ending in p, then there is either an indecomposable representation (V, f) where no left almost split morphism starts or an indecomposable representation (V, f) where no right almost split morphism ends. Note that in this last situation one may end up with existence of left almost split morphisms starting in all indecomposable representations or right almost split morphisms ending in all indecomposable representation, but not both.

REFERENCES

[ARS] Auslander, M., Reiten, I. and Smalø, S. O. *Representation Theory of Artin Algebras*, Cambridge University Press, 1995.

[S] Smalø, S. O. *Almost split sequences in Categories of Representations of Quivers*, To appear in Proc. Amer. Math. soc..

Cotilting objects and dualities

ROBERT WISBAUER Mathematical Institute of the University, 40225 Düsseldorf, Germany

ABSTRACT Tilting modules generalize projective generators and may be characterized either by weakened generating and projectivity conditions or else by equivalences they define between certain subcategories. Dually cotilting modules generalize injective cogenerators and there are again principally two ways to describe them: first by weakened cogenerating and injectivity conditions, and second by dualities they induce between suitable subcategories. In this paper we begin with several characterizations related to the first point of view, and it turns out that for properties of the second type certain finiteness conditions are needed - similar to the situation for Morita dualities for rings.

INTRODUCTION

Dualizing tilting modules, cotilting modules Q in R-Mod are defined in [7] by the conditions

(1) inj dim $(_RQ) \leq 1$,

(2) $\text{Ext}_R^1(Q^\Lambda, Q) = 0$, for any set Λ,

(3) for all $N \in R$-Mod, $\text{Hom}_R(N, Q) = 0 = \text{Ext}_R^1(N, Q)$ implies $N = 0$.

In Section 1 various injectivity and cogenerating conditions are introduced for objects (in Grothendieck categories), which result from dualizing notions of interest in the study of (self-) tilting objects. Self-tilting modules M are those which are tilting in the category $\sigma[M]$, whose objects are submodules of M-generated modules, and they are precisely the $*$-modules (introduced by Menini-Orsatti, see [16]). For the characterization of (self-) cotilting modules we introduce the category $\pi[M]$, whose objects are factor modules of M-cogenerated modules. In Section 3 cotilting modules in $\pi[M]$ and R-Mod are introduced by injectivity and cogenerating conditions and it is shown in 3.5 that these modules coincide with those mentioned above.

Over artinian rings the two categories $\sigma[M]$ and $\pi[M]$ coincide and this is one of the reasons why in representation theory of finite-dimensional algebras tilting and cotilting modules are so closely connected by formal duality.

The interest in tilting modules arose from the fact that they provide equivalences between certain categories. So one expects certain dualities for the dual notion, the cotilting modules. However it is well known that there are no dualities between full module categories and one has to restrict to finitely closed subcategories. We will see in 4.8 and 4.10 that under some finiteness conditions cotilting modules yield such dualities.

Our techniques and results subsume and generalize previous work on the subject by Colby [4], Wang-Xu [13], Angeleri-Hügel, Colpi, Fuller, Tonolo, Trlifaj [8, 3, 5, 6], and others. For the special case of a faithfully balanced bimodule over an artinian algebra related results are obtained in Zhaoyong [18].

1 PRELIMINARIES

Throughout the paper $\mathcal{C}$ will denote a locally finite Grothendieck category, i.e., a cocomplete abelian category with exact direct limits and a generating set of finitely generated objects (e.g., [11, Chapter V]). Moreover R will be an associative ring with unit, R-Mod the category of unital left R-modules and R-mod the full subcategory of finitely generated R-modules.

1.1. Trace and reject. For any family $\mathcal{X}$ of objects in $\mathcal{C}$ and $N \in \mathcal{C}$, the *trace of $\mathcal{X}$ in N* is defined by

$$\mathrm{Tr}(\mathcal{X}, N) = \sum \{\mathrm{Im}(f) | f \in \mathrm{Hom}_\mathcal{C}(X, N), X \in \mathcal{X}\} \subset N,$$

and the *reject of $\mathcal{X}$ in N* is given by

$$\mathrm{Re}(N, \mathcal{X}) = \bigcap \{\mathrm{Ke}(f) | f \in \mathrm{Hom}_\mathcal{C}(N, X), X \in \mathcal{X}\} \subset N.$$

For $\mathcal{X} = \{X\}$ we simply write $\mathrm{Tr}(X, N)$ and $\mathrm{Re}(N, X)$, respectively.

1.2. P-generated and P-presented objects. Let $P \in \mathcal{C}$. An object $N \in \mathcal{C}$ is *(finitely) P-generated* if there is an epimorphism $P^{(\Lambda)} \to N$ (with Λ finite), and N is *P-presented* if there exists an exact sequence

$$P^{(\Lambda')} \to P^{(\Lambda)} \to N \to 0, \quad \Lambda', \Lambda \text{ some sets}.$$

We write $\mathrm{Gen}(P)$ and $\mathrm{Pres}(P)$ for the full subcategories of $\mathcal{C}$ consisting of P-generated, resp., P-presented modules.

1.3. The category $\sigma[M]$. For any object $M \in \mathcal{C}$, $\sigma[M]$ denotes the full subcategory of $\mathcal{C}$ whose objects are subobjects of M-generated objects. $\sigma[M]$ is again a locally finite Grothendieck category and the *trace functor*

$$\mathcal{T}^M : \mathcal{C} \to \sigma[M], \quad N \mapsto \mathcal{T}^M(N) := \mathrm{Tr}(\sigma[M], N),$$

is right adjoint to the inclusion functor $\sigma[M] \to \mathcal{C}$, i.e., $\sigma[M]$ is a coreflective subcategory of $\mathcal{C}$ (e.g., [14, 45.11]). Notice that for any injective $N \in \mathcal{C}$, $\mathcal{T}^M(N) = \mathrm{Tr}(M, N)$ is an injective object in $\mathcal{C}$.

Cotilting Objects and Dualities

For $P \in \mathcal{C}$, Add (P) (resp. add (P)) stands for the class of modules which are direct summands of (finite) direct sums of copies of P, and obviously

$$\text{add}(P) \subset \text{Add}(P) \subset \text{Pres}(P) \subset \text{Gen}(P) \subset \sigma[P] \subset \mathcal{C},$$

where all these inclusions may be proper.

1.4. Tilting objects. An object $P \in \mathcal{C}$ is called *tilting in* $\mathcal{C}$ if P is Gen(P)-projective, Gen$(P) = $ Pres(P), and P is a subgenerator in $\mathcal{C}$. P is said to be *self-tilting* if it is tilting in the category $\sigma[P]$. Self-tilting modules define an equivalence between Pres(P) and a suitable subcategory of End(P)-Mod (see [16]).

Now we consider notions which are dual to those presented above.

1.5. Q-cogenerated and Q-copresented objects. Let $N, Q \in \mathcal{C}$. Then N is *(finitely) Q-cogenerated* if there exists an embedding $N \to Q^\Lambda$ (with Λ finite), and N is *(finitely) Q-copresented* if there exists an exact sequence

$$0 \to N \to Q^\Lambda \to Q^{\Lambda'}, \quad \Lambda, \Lambda' \text{ some sets } (\Lambda \text{ finite}).$$

We write Cog(Q) and Cop(Q) for the full subcategories of $\mathcal{C}$ consisting of Q-cogenerated, resp., Q-copresented modules, and cop(Q) for the modules which are finitely copresented by Q (notice that for this - in the defining sequence - we do not require Λ' to be finite). Prod (Q) stands for the direct summands of arbitrary products of Q in $\mathcal{C}$.

Notice that the notions Cog(Q), Cop(Q) and Prod(Q) depend on the category $\mathcal{C}$. If it is necessary we will stress this by writing Cog$_\mathcal{C}(Q)$, Cop$_\mathcal{C}(Q)$, and Prod$_\mathcal{C}(Q)$ for clarity.

1.6. The category $\pi[M]$. For any object $M \in \mathcal{C}$, we denote by $\pi[M]$ the full subcategory of $\mathcal{C}$ whose objects are factor objects of M-cogenerated objects. By definition we have

$$\text{add}(M) \subset \text{Prod}(M) \subset \text{Cop}(M) \subset \text{Cog}(M) \subset \pi[M] \subset \mathcal{C}.$$

It is easy to see that $\sigma[M] \subset \pi[M]$ and $\pi[M]$ is also closed under direct sums, factor objects and subobjects and hence is a (locally finite) Grothendieck subcategory of $\mathcal{C}$. In fact:

For any generator G in $\mathcal{C}$ and $\Lambda := \text{Hom}_\mathcal{C}(G, M)$, we have

$$\pi[M] = \sigma[M^\Lambda] = \text{Gen}(G/\text{Re}(G, M)).$$

Proof. By the canonical monomorphism $G/\text{Re}(G, M) \to M^\Lambda$, we obviously have

$$\text{Gen}(G/\text{Re}(G, M)) \subset \sigma[M^\Lambda] \subset \pi[M].$$

For any $N \in \text{Cog}(M)$, there exists an epimorphism $G^{(\Omega)} \to N$ which clearly yields an epimorphism $G/\text{Re}(G, M)^{(\Omega)} \to N$. This implies $N \in \text{Gen}(G/\text{Re}(G, M))$ and Gen$(G/\text{Re}(G, M)) = \pi[M]$. $\square$

By the above equalities we know that $\pi[M]$ is a coreflective subcategory of $\mathcal{C}$ (see 1.3). For special categories $\mathcal{C}$ it is also a reflective subcategory:

Assume that in $\mathcal{C}$ products of epimorphisms are epimorphisms. Then

(1) $\pi[M]$ *is closed under products in* $\mathcal{C}$.

(2) *The functor* $\mathcal{C} \to \pi[M]$, $N \mapsto N/\mathrm{Re}(N,\pi[M])$,
is left adjoint to the inclusion functor $I : \pi[M] \to \mathcal{C}$.

(3) *For any projective object (generator)* $P \in \mathcal{C}$, $P/\mathrm{Re}(P,\pi[M])$ *is a projective object (generator) in* $\pi[M]$.

Proof. (1) Consider any family $\{N_\lambda\}_\Lambda$ of objects in $\pi[M]$. Then each N_λ is an image of some $U_\lambda \subset M^{\Omega_\lambda}$. By assumption, $\prod_\Lambda N_\lambda$ is a factor module of $\prod_\Lambda U_\lambda$ and hence belongs to $\pi[M]$.

(2) By (1) the functor is well defined and obviously
$$\mathrm{Hom}_\mathcal{C}(N, I(K)) \simeq \mathrm{Hom}_\mathcal{C}(N/\mathrm{Re}(N,\pi[M]), K), \text{ for any } N \in \mathcal{C}, K \in \pi[M].$$

(3) This is easily verified. □

1.7. AB4* categories. Abelian categories with products in which products of epimorphisms are epimorphisms are called *AB4* categories*. It is well known that for any associative ring R, R-Mod has this property (e.g., [14, 9.3]). In view of the relationship between the product in R-Mod and its coreflective subcategories (see 1.3) it is straightforward to prove:

Assume that for $M \in R$-Mod *the trace functor* $\mathcal{T}^M$ *is exact. Then in* $\sigma[M]$ *products of epimorphisms are epimorphisms.*

For characterizations of $\mathcal{T}^M$ being exact we refer to [15, 4.6].

Recall that $M \in \mathcal{C}$ is a *subgenerator* in $\mathcal{C}$, provided $\sigma[M] = \mathcal{C}$. We call M a *weak subgenerator* in $\mathcal{C}$, provided $\pi[M] = \mathcal{C}$.

1.8. Weak subgenerators. *For* $M \in \mathcal{C}$ *the following conditions are equivalent:*

(a) *M is a weak subgenerator in $\mathcal{C}$;*

(b) *for some Λ, M^Λ is a subgenerator in $\mathcal{C}$;*

(c) *every object in $\mathcal{C}$ is a subfactor of some object in $\mathrm{Cog}(M)$;*

(d) *every injective object in $\mathcal{C}$ is a factor of some object in $\mathrm{Cog}(M)$;*

(e) *$\mathrm{Cog}(M)$ contains a generator (a generating set of objects) of $\mathcal{C}$;*

(f) *$\mathrm{Cog}(M)$ contains a subgenerator of $\mathcal{C}$.*

Proof. This can be easily shown by standard arguments. □

Clearly any cogenerator of $\mathcal{C}$ is a weak subgenerator but not necessarily a subgenerator of $\mathcal{C}$. For example, $\mathbb{Q}/\mathbb{Z}$ is a cogenerator but is not a subgenerator in $\mathbb{Z}$-Mod ($\sigma[\mathbb{Q}/\mathbb{Z}]$ are just the torsion $\mathbb{Z}$-modules).

The following observations are obvious consequences.

1.9. Corollary.

(1) *If M is a weak subgenerator in $\mathcal{C}$, then $\mathrm{Cog}(M)$ contains all projectives of $\mathcal{C}$.*

Cotilting Objects and Dualities

(2) *Assume $\mathcal{C}$ has a (sub-) generating set of finitely cogenerated projective objects. Then Q is a weak subgenerator if and only if Q is a subgenerator in $\mathcal{C}$.*

For any R-module M, $\pi[M]$ has a particularly nice form:

1.10. $\pi[M]$ in R-Mod.

(1) *For any $M \in R\text{-Mod}$, $\pi[M] = R/\operatorname{An}(M)\text{-Mod}$.*

(2) *An R-module Q is a weak subgenerator in R-Mod if and only if Q is a faithful R-module.*

(3) *If ${}_RR$ is finitely cogenerated then every weak subgenerator is a subgenerator in R-Mod.*

(4) *R is left artinian if and only if, for any R-module M, $\sigma[M] = \pi[M]$.*

In studying dualities the following finitely closed subgategory turned out to be of importance (e.g., [14, 47.12]). Here it will also help to relate cotilting modules to dualities.

1.11. The category $\sigma_f[M]$. For any object $M \in \mathcal{C}$, let $\sigma_f[M]$ denote the full subcategory of $\mathcal{C}$ whose objects are subobjects of finitely M-generated objects. This is a finitely closed abelian subcategory of $\sigma[M]$ (see [14, 47.2]).

For $Q \in \mathcal{C}$, we denote by $\operatorname{Cog}_f(Q)$ (resp., $\operatorname{cog}(Q)$) the class of (finitely) Q-cogenerated objects in $\sigma_f[Q]$. Clearly $\operatorname{Cog}_f(Q) = \operatorname{Cog}(Q) \cap \sigma_f[Q]$, where $\operatorname{Cog}(Q)$ is defined in $\mathcal{C}$. $\operatorname{cog}(Q)$ is the same when formed in $\sigma_f[Q]$ or $\mathcal{C}$ and this also applies to $\operatorname{cop}(Q)$, the class of finitely Q-copresented modules (see 1.5).

1.12. Ext-functor in $\mathcal{C}$. By $\operatorname{Ext}^1_{\mathcal{C}}$ and $\operatorname{Ext}^2_{\mathcal{C}}$ we denote the first and second Ext-functor in $\mathcal{C}$. For $Q \in \mathcal{C}$ and any exact sequence $0 \to K \to L \to N \to 0$ in $\mathcal{C}$, we have the long exact sequence

$$\begin{aligned} 0 \to\ & \operatorname{Hom}_R(N, Q) \to \operatorname{Hom}_R(L, Q) \to \operatorname{Hom}_R(K, Q) \to \\ & \operatorname{Ext}^1_{\mathcal{C}}(N, Q) \to \operatorname{Ext}^1_{\mathcal{C}}(L, Q) \to \operatorname{Ext}^1_{\mathcal{C}}(K, Q) \to \operatorname{Ext}^2_{\mathcal{C}}(N, Q) \to \cdots . \end{aligned}$$

and we denote the kernel of $\operatorname{Ext}^1_{\mathcal{C}}(-, Q)$ by

$$\perp_{\mathcal{C}} Q := \{N \in \mathcal{C} \mid \operatorname{Ext}^1_{\mathcal{C}}(N, Q) = 0\}.$$

Applied to the Grothendieck category $\pi[M]$, for any $M \in \mathcal{C}$, we have the functors $\operatorname{Ext}^1_{\pi[M]}$ and $\operatorname{Ext}^2_{\pi[M]}$ and, for $Q \in \pi[M]$, we write

$$\perp_M Q := \{N \in \pi[M] \mid \operatorname{Ext}^1_{\pi[M]}(N, Q) = 0\}.$$

In particular, for $Q \in \mathcal{C}$ we use the notation

$$\perp_f Q := \{N \in \sigma_f[Q] \mid \operatorname{Ext}^1_{\sigma_f[Q]}(N, Q) = 0\}.$$

For $\mathcal{C} = R\text{-Mod}$ and $M = R$ we apply the usual notation Ext^1_R and Ext^2_R.

2 INJECTIVITY CONDITIONS

Dualizing projectivity conditions which came up in the study of tilting modules we recall and introduce injectivity properties which are of interest for our further investigations.

2.1. Definitions. Consider exact sequences $(*)$ $\;0 \to K \to L \to N \to 0$ in $\mathcal{C}$.
An object $Q \in \mathcal{C}$ is called   if $\mathrm{Hom}_{\mathcal{C}}(-, Q)$ is exact on $(*)$

$\mathcal{C}$-injective	for all sequences $(*)$ in $\mathcal{C}$;
self-Ext-injective	provided $N \in \mathrm{Cog}(Q)$;
$\mathrm{Cog}(Q)$-injective	provided $L, N \in \mathrm{Cog}(Q)$;
$\mathrm{cog}(Q)$-injective	provided $L \in \mathrm{cog}(Q)$, $N \in \mathrm{Cog}(Q)$;
w-Π-quasi-injective	provided $L = Q^\Lambda$, $N \in \mathrm{Cog}(Q)$;
w-Π_f-quasi-injective	provided $L = Q^k$, $k \in \mathbb{N}$, $N \in \mathrm{Cog}(Q)$;
self-pseudo-injective	provided $K = \mathrm{Re}(L, Q)$.

The notation w-Π-quasi-injective should indicate that this notion is dual to w-Σ-quasi-projective. Notice that for the definition of $\mathrm{cog}(Q)$-injective the object N in the sequence $(*)$ is not required to be in $\mathrm{cog}(Q)$ but only in $\mathrm{Cog}(Q)$. We have the obvious implications

$\mathcal{C}$-injective
$\Downarrow$
self-Ext-injective $\;\;\Rightarrow\;\;$ $\mathrm{Cog}(Q)$-injective $\;\;\Rightarrow\;\;$ w-Π-quasi-injective
$\Downarrow$ $\;\Downarrow\;\Downarrow$
self-pseudo-injective $\;\;\;\;\;\;\;$ $\mathrm{cog}(Q)$-injective $\;\;\Rightarrow\;\;$ w-Π_f-quasi-injective

Notice that in general these implications cannot be reversed. For example, any left hereditary Artin algebra A is self-Ext-injective but not necessarily injective in A-Mod; for a non-semisimple ring R, any cogenerator Q in R-Mod is trivially self-pseudo-injective but need not be self-Ext-injective; for such rings every semisimple R-module is $\mathrm{cog}(Q)$-injective (w-Π_f-quasi-injective) but not necessarily $\mathrm{Cog}(Q)$-injective (w-Π-quasi-injective). We will see in 2.3 that for a module Q, w-Π-quasi-injectivity is equivalent to $\mathrm{Cog}(Q)$-injectivity, provided $\mathrm{Cog}(Q) = \mathrm{Cop}(Q)$.

Next we give some characterizations and properties resulting from this notions.

2.2. Self-pseudo-injective modules in $\mathcal{C}$.

(1) *For $Q \in \mathcal{C}$ the following are equivalent:*

(a) Q *is self-pseudo-injective;*

(b) *for any* $N \in \mathcal{C}$, $\mathrm{Hom}_{\mathcal{C}}(\mathrm{Re}(N, Q), Q) = 0$;

(c) $\mathrm{Cog}(Q)$ *is closed under extensions in* $\mathcal{C}$;

(d) *any diagram with exact row in $\mathcal{C}$,*

$$\begin{array}{ccccccccc} 0 & \to & K & \to & L & \to & N & \to & 0 \\ & & & & \downarrow\beta & \vdots & & & \\ & & & & Q & \overset{\alpha}{\dashrightarrow} & Q & & \end{array}$$

where $N \in \mathrm{Cog}(Q)$, can be non-trivially commutatively extended by some $\alpha: Q \to Q$, $\beta: L \to Q$.

(2) *If Q is self-pseudo-injective then* $\text{Cog}(Q)$ *is closed under kernels.*

Proof. (1) These properties are known or easily verified ([12, 2.2], [15, 6.5]).

(2) This is proved in [9, Proposition 4.4], (d)⇒(e). □

Dual to the properties of w-Σ-quasi-projective-modules given in [16, 3.2] we have the following observations on w-Π-quasi-injective objects.

2.3. w-Π-quasi-injective objects.

(1) *For any $Q \in \mathcal{C}$ the following are equivalent:*

   (a) Q *is w-Π-quasi-injective;*

   (b) $\text{Hom}_\mathcal{C}(-, Q)$ *respects exact sequences* $0 \to K \to L \to N \to 0$, *where* $N \in \text{Cog}(Q)$ *and* $L \in \text{Cop}(Q)$.

(2) *If Q is w-Π-quasi-injective then* $\text{Cop}(Q)$ *is closed under kernels.*

(3) *If* $\text{Cog}(Q) = \text{Cop}(Q)$ *the following are equivalent:*

   (a) Q *is w-Π-quasi-injective;*

   (b) Q *is* $\text{Cog}(Q)$-*injective.*

Proof. (1) (b)⇒(a) is trivial.

(a)⇒(b) For any morphism $h : K \to Q$, we have a commutative diagram with exact row and columns,

$$\begin{array}{ccccccccc}
& & & & 0 & & 0 & & \\
& & & & \downarrow & & \downarrow & & \\
0 & \to & K & \xrightarrow{f} & L & \xrightarrow{g} & N & \to & 0 \\
& & h \downarrow & & p \downarrow & & q \downarrow & & \\
& & Q & \xleftarrow{\gamma} & Q^\Lambda & \xrightarrow{\alpha} & Q^\Omega & & \\
& & & & k \downarrow & & & & \\
& & & & Q^{\Lambda'} & & & &
\end{array}$$

where Λ, Λ', and Ω are suitable sets. By (a), there exists $\alpha : Q^\Lambda \to Q^\Omega$ with $p\alpha = gq$. It is easy to verify that (up to isomorphism) $Kfp = \text{Ke}\,k \cap \text{Ke}\,\alpha$ and hence $Q^\Lambda / Kfp \in \text{Cog}(Q)$. Again referring to (a) we obtain a morphism $\gamma : Q \to Q^\Lambda$ with $h = fp\gamma$ thus proving our assertion.

(2) This can be seen from the above proof. Notice that this is also proved in [9, Proposition 4.4], (c)⇒(e).

(3) follows immediately from (1). □

In the next proposition we dualize the properties of $\text{Gen}(Q)$-projective modules (e.g., [16, 3.3]).

2.4. $\text{Cog}(Q)$-injective objects. *If Q is a weak subgenerator in $\mathcal{C}$, the following are equivalent:*

(a) Q *is* $\text{Cog}(Q)$-*injective;*

(b) Q is self-Ext-injective;

(c) for each $K \in \mathrm{Cog}(Q)$, $\mathrm{Ext}^1_\mathcal{C}(K,Q) = 0$ (i.e., $\mathrm{Cog}(Q) \subset {}^{\perp c}Q$).

If $\mathcal{C}$ has a generator $G \in \mathrm{Cog}(Q)$ with $\mathrm{Ext}^2_\mathcal{C}(G,Q) = 0$ (e.g., a projective generator), then (a)-(c) are equivalent to:

(d) (i) $\mathrm{Ext}^1_\mathcal{C}(Q^\Lambda, Q) = 0$, for any set Λ;
    (ii) $\mathrm{Ext}^2_\mathcal{C}(N, Q) = 0$, for each $N \in \mathcal{C}$.

Proof. (a)$\Rightarrow$(b) Let $0 \to K \to L \to N \to 0$ be an exact sequence with $N \in \mathrm{Cog}(Q)$ and consider any morphism $f: K \to Q$. By assumption, every object in $\mathcal{C}$ is a factor of a subobject of some Q^Λ, and so there is an epimorphism $\alpha: X \to L$ with $X \in \mathrm{Cog}(Q)$. We use this to construct the commutative diagram with exact rows,

$$\begin{array}{ccccccccc} 0 & \to & K' & \to & X & \stackrel{\alpha g}{\to} & N & \to & 0 \\ & & \downarrow & & \alpha \downarrow & & \| & & \\ 0 & \to & K & \to & L & \stackrel{g}{\to} & N & \to & 0 \\ & & f \downarrow & & & & & & \\ & & Q & & & & & & \end{array},$$

where the upper exact sequence is in $\mathrm{Cog}(Q)$ and the left hand square is a pushout. By hypothesis we can extend the diagram commutatively by some morphism $X \to Q$, and the pushout property yields the desired morphism $L \to Q$, thus proving our assertion.

(b)$\Rightarrow$(a) and (b)$\Leftrightarrow$(c) are obvious.

(c)$\Rightarrow$(d) Clearly $\mathrm{Ext}^1_\mathcal{C}(Q^\Lambda, Q) = 0$, for any set Λ.

For any $N \in \mathcal{C}$, there exists an exact sequence $0 \to K \to P \to N \to 0$, where $P \in \mathrm{Cog}(Q)$ with $\mathrm{Ext}^2_\mathcal{C}(P, Q) = 0$. From this we obtain the exact sequence

$$0 = \mathrm{Ext}^1_\mathcal{C}(K,Q) \to \mathrm{Ext}^2_\mathcal{C}(N,Q) \to \mathrm{Ext}^2_\mathcal{C}(P,Q) = 0,$$

proving $\mathrm{Ext}^2_\mathcal{C}(N,Q) = 0$.

(d)$\Rightarrow$(c) By the connecting morphisms of the Ext-functor (see 1.12), (ii) implies that ${}^{\perp c}Q$ is closed under submodules. Hence (i) implies $\mathrm{Cog}(Q) \subset {}^{\perp c}Q$. $\square$

It is easy to see that the above conditions on Q imply that Q is self-pseudo-injective in $\mathcal{C}$ and hence we have by 2.2:

2.5. Corollary. *Let Q be a $\mathrm{Cog}(Q)$-injective weak subgenerator in $\mathcal{C}$. Then $\mathrm{Cog}(Q)$ is closed under extensions in $\mathcal{C}$, and for any $N \in \mathcal{C}$, $\mathrm{Hom}_\mathcal{C}(\mathrm{Re}(N,Q), Q) = 0$.*

Most of the injectivity conditions defined in 2.1 also apply for $\mathcal{C} = \sigma_f[Q]$ (although this is not a Grothendieck category). For example, for a self-pseudo-injective module Q in $\sigma_f[Q]$, $\mathrm{Cog}_f(Q)$ is closed under extensions in $\sigma_f[Q]$. With the proof of 2.3 we obtain:

2.6. w-Π_f-quasi-injective objects.

(1) *For any $Q \in \mathcal{C}$ the following are equivalent:*

(a) Q is w-Π_f-quasi-injective;

(b) $\mathrm{Hom}_{\mathcal{C}}(-,Q)$ respects exact sequences $0 \to K \to L \to N \to 0$, where $N \in \mathrm{Cog}_f(Q)$ and $L \in \mathrm{cop}(Q)$.

(2) If $\mathrm{cog}(Q) = \mathrm{cop}(Q)$ the following are equivalent:

(a) Q is w-Π_f-quasi-injective;

(b) $\mathrm{Hom}_{\mathcal{C}}(-,Q)$ respects exact sequences $0 \to K \to L \to N \to 0$, where $N \in \mathrm{Cog}_f(Q)$ and $L \in \mathrm{cog}(Q)$.

Notice that every module in $\sigma_f[Q]$ is a subfactor of some Q^k. Hence a slight modification of the proof of 2.4 yields:

2.7. $\mathrm{cog}(Q)$-injective objects. *For $Q \in \mathcal{C}$ the following are equivalent:*

(a) Q is $\mathrm{cog}(Q)$-injective;

(b) $\mathrm{Hom}_{\mathcal{C}}(-,Q)$ respects exact sequences $0 \to K \to L \to N \to 0$ in $\sigma_f[Q]$, where $N \in \mathrm{Cog}_f(Q)$;

(c) for each $K \in \mathrm{Cog}_f(Q)$, $\mathrm{Ext}^1_{\sigma_f[Q]}(K,Q) = 0$ (i.e., $\mathrm{Cog}_f(Q) \subset {}^{\perp_{\sigma_f[Q]}}Q$).

2.8. Corollary. *Let $Q \in \mathcal{C}$ be $\mathrm{cog}(Q)$-injective. Then $\mathrm{Cog}_f(Q)$ is closed under extensions in $\sigma_f[Q]$, and for any $N \in \sigma_f[Q]$, $\mathrm{Hom}_{\mathcal{C}}(\mathrm{Re}(N,Q),Q) = 0$.*

3 COTILTING OBJECTS

Dualizing the definitions of tilting and self-tilting modules given in [16] leads to the following notions.

3.1. Definitions. We call $Q \in \mathcal{C}$ a *cotilting object in $\mathcal{C}$* if

(i) Q is $\mathrm{Cog}(Q)$-injective,

(ii) every Q-cogenerated module in $\mathcal{C}$ is Q-presented (i.e., $\mathrm{Cog}(Q) = \mathrm{Cop}(Q)$),

(iii) Q is a weak subgenerator of $\mathcal{C}$ (i.e., $\pi[Q] = \mathcal{C}$).

Q is called *self-cotilting* if it is cotilting in $\pi[Q]$, i.e., if (i) and (ii) hold in the category $\pi[Q]$.

For a cotilting module Q in $\mathcal{C}$ we have $\mathcal{C} = \pi[Q]$ (by (iii)) and so essentially it suffices to investigate self-cotilting objects in detail.

It is obvious that every injective cogenerator is cotilting in $\mathcal{C}$ and (dual to the situation for tilting modules) we have a close connection between cotilting objects and injective cogenerators (see 3.4).

3.2. Q cotilting in $\sigma[Q]$. For any $Q \in \mathcal{C}$ we may ask when Q is cotilting in $\sigma[Q]$. This is clearly the case when Q is an injective cogenerator in $\sigma[Q]$. Notice that in general this does not imply that Q is cotilting in $\pi[Q]$ (i.e., self-cotilting) even when Q is a weak subgenerator in $\mathcal{C}$.

For this consider the Prüfer group $\mathbb{Z}_{p^\infty}$, for any prime number p. This is an injective cogenerator in $\sigma[\mathbb{Z}_{p^\infty}]$ (abelian p-groups) and hence is cotilting in $\sigma[\mathbb{Z}_{p^\infty}]$. Moreover, as a faithful $\mathbb{Z}$-module, $\mathbb{Z}_{p^\infty}$ is a weak subgenerator (i.e., $\pi[\mathbb{Z}_{p^\infty}] = \mathbb{Z}$-Mod). However, $\mathbb{Z}_{p^\infty}$ is not cotilting in $\mathbb{Z}$-Mod by 3.4, since it is injective but not a cogenerator in $\mathbb{Z}$-Mod.

Dualizing the characterizations of self-tilting modules and their proofs (see [16, 4.2]) we obtain characterizations of self-cotilting objects.

3.3. Self-cotilting objects. *For any $Q \in \mathcal{C}$ the following are equivalent:*

(a) Q is self-cotilting;

(b) $\operatorname{Cog}(Q) = \operatorname{Cop}(Q)$ and Q is w-Π-quasi-injective ;

(c) $\operatorname{Cog}(Q) = \operatorname{Cop}(Q)$ and Q is self-Ext-injective in $\pi[Q]$;

(d) $\operatorname{Cog}(Q) = {}^{\perp_Q}Q$;

(e) Q is $\operatorname{Cog}(Q)$-injective, ${}^{\perp_Q}Q$ is closed under submodules, and

　　(i) *for $N \in \pi[Q]$, $\operatorname{Hom}_{\mathcal{C}}(N, Q) = 0 = \operatorname{Ext}^1_{\pi[Q]}(N, Q)$ implies $N = 0$, or*

　　(ii) *for any injective object (some injective cogenerator) $W \in \pi[Q]$, there exists an exact sequence*

$$0 \to Q' \to Q'' \to W \to 0, \quad \text{where } Q', Q'' \in \operatorname{Prod}(Q).$$

Proof. (a)⇔(b) and (b)⇔(c) follow from 2.3 and 2.4, respectively.

(c)⇒(d) and (c)⇒(e)(ii): Consider an exact sequence $0 \to K \to L \to N \to 0$, where L is a submodule of some Q^Λ. With a Q-corepresentation of K (first column, $X \in \operatorname{Cog}(Q)$) and a pushout construction we obtain the commutative exact diagram

$$\begin{array}{ccccccccc}
& & 0 & & 0 & & & & \\
& & \downarrow & & \downarrow & & & & \\
0 & \to & K & \to & L & \to & N & \to & 0 \\
& & \downarrow & & \downarrow & & \parallel & & \\
0 & \to & Q^\Lambda & \to & P & \to & N & \to & 0 \\
& & \downarrow & & \downarrow & & & & \\
& & X & = & X & & & & \\
& & \downarrow & & \downarrow & & & & \\
& & 0 & & 0 & & & &
\end{array}$$

Assume $N \in {}^{\perp_Q}Q$. By 2.5, $\operatorname{Cog}(Q)$ is closed under extensions in $\pi[Q]$ and hence $P \in \operatorname{Cog}(Q)$. Since $\operatorname{Ext}^1_{\pi[Q]}(N, Q^\Lambda) = 0$ the central sequence splits and so $N \in \operatorname{Cog}(Q)$. This proves (d).

Now assume that N is injective in $\pi[Q]$. Then, for any $Y \in \operatorname{Cog}(Q)$, we have the exact sequence

$$0 = \operatorname{Ext}^1_{\pi[Q]}(Y, Q^\Lambda) \to \operatorname{Ext}^1_{\pi[Q]}(Y, P) \to \operatorname{Ext}^1_{\pi[Q]}(Y, N) = 0,$$

and hence $\operatorname{Ext}^1_{\pi[Q]}(Y, P) = 0$. By (c), there exists a copresentation of P,

$$0 \to P \to Q^\Omega \to Y \to 0, \text{ where } Y \in \operatorname{Cog}(Q).$$

Since $\operatorname{Ext}^1_{\pi[Q]}(Y, P) = 0$ this sequence splits and hence $P \in \operatorname{Prod}(Q)$, thus proving (e)(ii).

(e): (ii)⇒(i)　Let W be any injective cogenerator in $\pi[Q]$. The given sequence yields the exact sequence

$$\operatorname{Hom}_{\mathcal{C}}(N, Q'') \to \operatorname{Hom}_{\mathcal{C}}(N, W) \to \operatorname{Ext}^1_{\pi[Q]}(N, Q) = 0,$$

and $\mathrm{Hom}_{\mathcal{C}}(N,Q) = 0$ implies $\mathrm{Hom}_{\mathcal{C}}(N,Q'') = 0$ and hence $\mathrm{Hom}_{\mathcal{C}}(N,W) = 0$, which means $N = 0$.

(d)$\Rightarrow$(c) Let $N \in \mathrm{Cog}(Q)$ and $\Lambda := \mathrm{Hom}_{\mathcal{C}}(N,Q)$. With the canonical sequence on the top and any extension with $X \in \pi[Q]$ on the bottom we have the diagram

$$\begin{array}{ccccccccc} 0 & \to & N & \to & Q^\Lambda & \to & L & \to & 0 \\ & & \beta \downarrow & & \alpha \downarrow & & \| & & \\ 0 & \to & Q & \to & X & \to & L & \to & 0, \end{array}$$

which can be extended by some $\alpha : Q^\Lambda \to X$ (since $\mathrm{Ext}^1_{\pi[Q]}(Q^\Lambda, Q) = 0$) and some $\beta : N \to Q$ commutatively. Since Q is injective with respect to the upper sequence we conclude (by the Homotopy Lemma) that the bottom sequence splits. Hence $\mathrm{Ext}^1_{\pi[Q]}(L, Q) = 0$ implying $L \in {}^{\perp_Q}Q = \mathrm{Cog}(Q)$ and therefore $N \in \mathrm{Cop}(Q)$.

(e)(i)$\Rightarrow$(d) Let $N \in {}^{\perp_Q}Q$. Then $\mathrm{Re}(N,Q) \in {}^{\perp_Q}Q$ and $\mathrm{Hom}_{\mathcal{C}}(\mathrm{Re}(N,Q),Q) = 0$ by 2.5. Now (i) implies $\mathrm{Re}(N,Q) = 0$ which means $N \in \mathrm{Cog}(Q)$. □

3.4. Corollary. *For any self-cotilting module $Q \in \mathcal{C}$, the following are equivalent:*

(a) *Q is a cogenerator in $\pi[Q]$;*

(b) *Q is injective in $\pi[Q]$.*

Proof. (a)$\Rightarrow$(b) is obvious.
(b)$\Rightarrow$(a) follows from 3.3(d), since for Q injective in $\pi[Q]$ clearly ${}^{\perp_Q}Q = \pi[Q]$. □

To make a self-cotilting object $Q \in \mathcal{C}$ cotilting in $\mathcal{C}$ some condition is needed to turn Q into a weak subgenerator in $\mathcal{C}$.

3.5. Cotilting objects. *For any $Q \in \mathcal{C}$ the following are equivalent:*

(a) *Q is cotilting in $\mathcal{C}$;*

(b) *Q is self-cotilting and a weak subgenerator in $\mathcal{C}$;*

If $\mathcal{C}$ has a projective (sub-) generator then (a)-(b) are equivalent to:

(c) $\mathrm{Cog}(Q) = {}^{\perp_c}Q$;

(d) (i) $\mathrm{Ext}^1_{\mathcal{C}}(Q^\Lambda, Q) = 0$, *for any set Λ;*
   (ii) $\mathrm{Ext}^2_{\mathcal{C}}(N, Q) = 0$, *for each $N \in \mathcal{C}$;*
   (iii) *for $N \in \pi[Q]$, $\mathrm{Hom}_{\mathcal{C}}(N,Q) = 0 = \mathrm{Ext}^1_{\mathcal{C}}(N,Q)$ implies $N = 0$;*

(e) (i) *and* (ii) *as in* (e) *and*

   (iv) *for some injective cogenerator $W \in \mathcal{C}$, there exists an exact sequence*

$$0 \to Q' \to Q'' \to W \to 0, \text{ where } Q', Q'' \in \mathrm{Prod}(Q).$$

Proof. The assertions follow by 3.3, 2.4, and the observation that any projective object of $\mathcal{C}$ belongs to $^{\perp_c}Q$. The latter implies that in (c),(d) and (e), Q is a weak subgenerator in $\mathcal{C}$. □

For any R-module Q, $\pi[Q] = R/\mathrm{An}(Q)$-Mod and hence it has a projective generator. So we obtain from 3.3 and 3.5:

3.6. Self-cotilting modules. *For any left R-module Q, put $\overline{R} = R/\mathrm{An}(Q)$. Then the following are equivalent:*

(a) Q *is self-cotilting;*

(b) $\mathrm{Cog}(Q) = \mathrm{Cop}(Q)$ *and Q is w-Π-quasi-injective ;*

(c) Q *is cotilting in $\overline{R}$-Mod;*

(d) $\mathrm{Cog}(Q) = {}^{\perp_{\overline{R}}}Q$;

(e) (i) $\mathrm{Ext}^1_{\overline{R}}(Q^\Lambda, Q) = 0$, *for any set Λ;*

  (ii) $\mathrm{Ext}^2_{\overline{R}}(N, Q) = 0$, *for each $N \in \overline{R}$-Mod;*

  (iii) *for $N \in \overline{R}$-Mod, $\mathrm{Hom}_R(N, Q) = 0 = \mathrm{Ext}^1_{\overline{R}}(N, Q)$ implies $N = 0$;*

(f) (i), (ii) *as in (e) and*

  (iv) *for some injective cogenerator $W \in \overline{R}$-Mod, there exists an exact sequence*
  $$0 \to Q' \to Q'' \to W \to 0, \quad \text{where } Q', Q'' \in \mathrm{Prod}(Q).$$

3.7. Cotilting in R-Mod. By definition, Q is cotilting in R-Mod if and only if it is self-cotilting and faithful. Hence 3.6 yields characterizations of these modules by replacing $\overline{R}$ by R in (c),(d),(e) and (f).

Notice that for this case $(d) \Leftrightarrow (e)$ was proved in Colpi-D'Este-Tonolo [7, Proposition 1.7] and (d) implies $\mathrm{Cog}(Q) = \mathrm{Cop}(Q)$ is shown in [7, Proposition 1.8]. Moreover $(d) \Rightarrow (f)$ corresponds to Angeleri-Hügel-Tonolo-Trlifaj [3, Proposition 2.3].

From 3.5 we obtain the

3.8. Corollary. *For Q cotilting in R-Mod, the following are equivalent:*

(a) Q *is a cogenerator in R-Mod;*

(b) Q *is injective in R-Mod.*

To investigate dualities we introduce a finite version of cotilting objects.

3.9. Definition. We call $Q \in \mathcal{C}$ an *f-cotilting object* if

(i) Q is $\mathrm{cog}(Q)$-injective,

(ii) every finitely Q-cogenerated object in $\mathcal{C}$ is finitely Q-copresented (i.e., $\mathrm{cog}(Q) = \mathrm{cop}(Q)$).

For example, every semisimple R-module is f-cotilting.

Remark. In Angeleri-Hügel-Valenta [2] *finitely cotilting* modules Q are defined as "cotilting" modules (with a slightly different definition) which are finitely generated R-modules such that $\operatorname{Hom}_R(X,Q)$ is a finitely generated $\operatorname{End}_R(Q)$-module, for any finitely generated R-module X. It is easy to see that over noetherian rings such modules are f-cotilting in the sense defined above (compare [2, Corollary 5.2]).

Recall that an object Q is injective in $\sigma_f[Q]$ if and only if it is injective in $\sigma[Q]$ (i.e., Q-injective) and Q is an injective cogenerator in $\sigma_f[Q]$ if and only if it is an injective cogenerator in $\sigma[Q]$. Similar to 3.4 we have now:

3.10. Proposition. *For any f-cotilting object $Q \in \mathcal{C}$, the following are equivalent:*

(a) *Q is a cogenerator in $\sigma_f[Q]$ (in $\sigma[Q]$);*

(b) *Q is injective in $\sigma_f[Q]$ (in $\sigma[Q]$).*

Proof. (a)$\Rightarrow$(b) is obvious.

(b)$\Rightarrow$(a) (compare proof of 3.3, (c)$\Rightarrow$(d)) For any subobject $K \subset Q$ we obtain the commutative exact diagram, where the first column is a Q-copresentation of K ($k \in \mathbb{N}$, $X \in \operatorname{Cog}(Q)$), by a pushout construction

$$\begin{array}{ccccccccc}
& & 0 & & 0 & & & & \\
& & \downarrow & & \downarrow & & & & \\
0 & \to & K & \to & Q & \to & N & \to & 0 \\
& & \downarrow & & \downarrow & & \| & & \\
0 & \to & Q^k & \to & P & \to & N & \to & 0 \\
& & \downarrow & & \downarrow & & & & \\
& & X & = & X & & & & \\
& & \downarrow & & \downarrow & & & & \\
& & 0 & & 0 & & & &
\end{array}.$$

Since P is in $\sigma[Q]$ and Q is Q-injective, the central column splits implying that $P \in \operatorname{Cog}(Q)$. For the same reasons the central row splits and hence $N \in \operatorname{Cog}(Q)$. From this we conclude that Q is a cogenerator in $\sigma[Q]$ (e.g., [14, 16.5]). $\square$

The condition $\operatorname{cog}(Q) = \operatorname{cop}(Q)$ holds trivially provided Q is a cogenerator in $\sigma_f[Q]$ but it need not follow from the condition $\operatorname{Cog}(Q) = \operatorname{Cop}(Q)$. In fact it is related to some finiteness properties.

3.11. Proposition. *Let $Q \in \mathcal{C}$ be w-Π_f-quasi-injective. Then for every $K \in \operatorname{cop}(Q)$, $\operatorname{Hom}_{\mathcal{C}}(K,Q)$ is a finitely generated right $\operatorname{End}_{\mathcal{C}}(Q)$-module.*

Proof. This is obvious. $\square$

Adapting the proofs of 3.3 we obtain characterizations of f-cotilting objects.

3.12. f-cotilting objects. *For any $Q \in \mathcal{C}$ and $S = \operatorname{End}_{\mathcal{C}}(Q)$, the following are equivalent:*

(a) *Q is f-cotilting;*

(b) $\text{cog}(Q) = \text{cop}(Q)$ and Q is w-Π_f-quasi-injective;

(c) $\text{cog}(Q) = \text{cop}(Q)$ and $\text{Cog}_f(Q) \subset {}^{\perp_f} Q$;

(d) $\text{Cog}_f(Q) = {}^{\perp_f} Q$, and for every $K \in \text{cog}(Q)$, $\text{Hom}_\mathcal{C}(K, Q) \in \text{mod-}S$;

(e)   (i) Q is $\text{cog}(Q)$-injective, ${}^{\perp_f} Q$ is closed under submodules,
      (ii) for $N \in \sigma_f[Q]$, $\text{Hom}_\mathcal{C}(N, Q) = 0 = \text{Ext}^1_{\sigma_f[Q]}(N, Q)$ implies $N = 0$, and
      (iii) for every $K \in \text{cog}(Q)$, $\text{Hom}_\mathcal{C}(K, Q) \in \text{mod-}S$.

Proof. (a)⇔(b) and (b)⇔(c) follow from 2.6 and 2.7, respectively.

(c)⇒(d) In the proof of 3.3, (c)⇒(d), take $N \in {}^{\perp_f} Q$ and Λ a finite index set.

(d)⇒(c) Let $N \in \text{cog}(Q)$ and $f_1, \ldots, f_k$ a generating set for the S-module $\text{Hom}_\mathcal{C}(K, Q)$. The f_i's yield a canonical monomorphism $N \to Q^k$. Now procede as in the proof of 3.3, (d)⇒(c).

The remaining implications can also be transferred from 3.3. □

More examples of f-cotilting objects will be given at the end of the paper.

4 REFLEXIVE MODULES AND DUALITIES

To avoid technical complications we restrict our study of dualities to module categories. We investigate dualities induced by any left R-module Q with $S = \text{End}_R(Q)$.

4.1. Canonical functors. Related to ${}_RQ_S$ we have the adjoint pair of functors

$$D : \text{R-Mod} \xrightarrow{\text{Hom}_R(-, Q)} \text{Mod-S}, \quad D' : \text{Mod-S} \xrightarrow{\text{Hom}_S(-, Q)} \text{R-Mod},$$

and for any $N \in \text{R-Mod}$ and $X \in \text{S-Mod}$, the canonical (evaluation) morphisms

$$\Phi_N : N \to D'D(N), \quad n \mapsto [\beta \mapsto (n)\beta],$$
$$\Phi'_X : X \to DD'(X), \quad x \mapsto [\alpha \mapsto \alpha(x)].$$

where

$$\text{Ke}\,\Phi_N = \text{Re}(N, Q), \quad \text{Ke}\,\Phi'_X = \text{Re}(N, Q).$$

4.2. (Semi-)reflexive modules. A module $N \in \text{R-Mod}$ is called *(semi-) Q-reflexive* if $\Phi_N : N \to D'D(N)$ is an isomorphism (epimorphism). Similarly *(semi-) Q-reflexive* objects in Mod-S are defined. It is straightforward to prove:

$N \in \sigma[Q]$ *is semi-Q-reflexive if and only if $N/\text{Re}(N, Q)$ is Q-reflexive.*

The class of all Q-reflexive modules in R-Mod (in Mod-S) is denoted my $\text{Ref}_R(Q)$ (resp., $\text{Ref}_S(Q)$).

Obviously we have the following

4.3. Basic duality. *For any R-module Q, the functor*

$$D = \text{Hom}_R(-, Q) : \text{Ref}_R(Q) \to \text{Ref}_S(Q)$$

defines a duality with inverse $D' = \text{Hom}_S(-, Q)$.

Cotilting Objects and Dualities

For any $N \in$ R-Mod consider an exact sequence $S^{(\Lambda')} \to S^{(\Lambda)} \to D(N) \to 0$ in Mod-S. By left exactness of $\mathrm{Hom}_S(-, Q)$ we obtain the exact sequence

$$0 \to \mathrm{Hom}_S(D(N), Q) \to Q^\Lambda \to Q^{\Lambda'}.$$

Now if $N \in \mathrm{Ref}_R(Q)$, i.e., $N \simeq D'D(N)$, we conclude $N \in \mathrm{Cop}_R(Q)$.

4.4. Classes of modules related to Q. We have the following inclusions:

(1) $D'(\text{Mod-S}) \subset \mathrm{Cop}_R(Q)$, $\quad D(\text{R-Mod}) \subset \mathrm{Cop}_S(Q)$.

(2) $\mathrm{add}\,(Q) \subset \mathrm{Ref}_R(Q) \subset \mathrm{Cop}_R(Q) \subset \mathrm{Cog}_R(Q) \subset$ R-Mod.

(3) $\mathrm{add}\,(Q) \subset \mathrm{cop}_R(Q) \subset \mathrm{cog}_R(Q) \subset \sigma_f[Q]$.

We investigate the properties of the classes considered above in view of certain injectivity conditions.

A submodule $K \subset M$ in called Q-*closed in* M provided $M/K \in \mathrm{Cog}(Q)$.

4.5. w-Π-quasi-injective modules. *Let* $_RQ$ *be w-Π-quasi-injective. Then:*

(1) $\mathrm{Ref}_R(Q)$ *is closed under Q-closed submodules; in particular*, $\mathrm{Ref}_R(Q)$ *is closed under kernels.*

(2) *Every factor module of a Q-reflexive right S-module is semi-Q-reflexive.*

Proof. (1) Let $P \in \mathrm{Ref}_R(Q)$ with a Q-closed submodule $K \subset P$. Then we have an exact commutative diagram,

$$\begin{array}{ccccccccc}
 & & 0 & & 0 & & & & \\
 & & \downarrow & & \downarrow & & & & \\
0 & \to & K & \to & P & \to & Y & \to & 0 \\
 & & \downarrow & & \simeq\downarrow & & \downarrow & & \\
0 & \to & K^{**} & \to & P^{**} & \stackrel{g}{\to} & Y^{**} & & \\
 & & \downarrow & & & & & & \\
 & & 0 & & & & & &
\end{array}$$

where $Y \in \mathrm{Cog}(Q)$. This shows that K is Q-reflexive.

(2) Let $X \in \mathrm{Ref}_S(Q)$. For any epimorphism $X \to L$ we have the exact commutative diagram

$$\begin{array}{ccccc}
X & \to & L & \to & 0 \\
\simeq\downarrow & & \downarrow & & \\
X^{**} & \to & L^{**} & \to & 0,
\end{array}$$

which shows that L is semi-Q-reflexive. $\square$

The question arises to which extent properties of $\mathrm{Ref}_R(Q)$ imply injectivity conditions on Q. Since $\mathrm{Ref}_R(Q)$ is not closed under products we are not able to get a general assertion converse to 4.5. However the situation is different if we restrict our considerations to finite products as we will see below. First we formulate a finite version of 4.5. Since for any $k \in \mathbb{N}$, $Q^k \in \mathrm{Ref}_R(Q)$ and $S^k \in \mathrm{Ref}_S(Q)$, essentially the same proof yields:

4.6. w-Π_f-quasi-injective modules. *Let Q be a w-Π_f-quasi-injective R-module. Then:*

(1) $\text{cop}(Q) \subset \text{Ref}_R(Q)$.

(2) *Every finitely generated right S-module is semi-Q-reflexive.*

(3) *By restriction we have the functor $\text{Hom}_R(-, Q) : \text{cop}(Q) \to \text{mod-}S$.*

4.7. Proposition. *Let $Q \in R$-Mod. Assume $\text{cop}(Q) \subset \text{Ref}_R(Q)$ and that all finitely generated right S-modules are semi-Q-reflexive. Then Q is w-Π_f-quasi-injective.*

Proof. (compare [14, 47.12]) We have to show that $\text{Hom}_R(-, Q) = (-)^*$ is exact on sequences

$$0 \to K \xrightarrow{f} Q^k \to Y \to 0, \text{ where } Y \in \text{Cog}(Q).$$

From the inclusion $\delta : \text{Im} f^* \to K^*$ we obtain the commutative diagram with exact rows

$$\begin{array}{ccccccccc} 0 & \to & K & \xrightarrow{f} & Q^k & \to & Y & \to & 0 \\ & & \Phi_K \downarrow & & \downarrow \Phi_{Q^k} & & \downarrow \Phi_Y & & \\ & & K^{**} & & & & & & \\ & & \delta^* \downarrow & & & & & & \\ 0 & \to & (\text{Im} f^*)^* & \to & (Q^k)^{**} & \to & Y^{**} & & \end{array}$$

Since Φ_{Q^k} is an isomorphism and Φ_Y is a monomorphism, $\Phi_K \delta^*$ is an isomorphism. Now K being Q-reflexive implies that δ^* is also an isomorphism. We have the commutative diagram

$$\begin{array}{ccc} \text{Im} f^* & \xrightarrow{\delta} & K^* \\ \downarrow \Phi_{\text{Im} f^*} & & \downarrow \Phi_{K^*} \\ (\text{Im} f^*)^{**} & \xrightarrow{\delta^{**}} & K^{***}, \end{array}$$

where $\Phi_{\text{Im} f^*}$ is surjective (by our assumptions) and δ^{**} is an isomorphism. This clearly implies that δ is an isomorphism which means that $(-)^*$ is exact on the given sequence. □

Combining the results just derived we obtain:

4.8. Injectivity and duality. *For $Q \in R$-Mod the following are equivalent:*

(a) *Q is w-Π_f-quasi-injective;*

(b) *$\text{cop}(Q) \subset \text{Ref}_R(Q)$ and all finitely generated right S-modules are semi-Q-reflexive;*

(c) *$\text{cop}(Q) \subset \text{Ref}_R(Q)$ and $\text{mod-}S \cap \text{Cog}_S(Q) \subset \text{Ref}_S(Q)$;*

(d) *$\text{Hom}_R(-, Q) : \text{cop}(Q) \to \text{mod-}S \cap \text{Cog}_S(Q)$ is a duality.*

Proof. $(a) \Rightarrow (b)$ follows from 4.6.
$(b) \Leftrightarrow (c)$ is clear by the comments in 4.2.
$(c) \Rightarrow (a)$ holds by Proposition 4.7.
$(c) \Leftrightarrow (d)$ is obvious. □

4.9. Remark. Dualizing the notion of s-$\sum$-quasi-projective used in Sato [10] we may call an R-module Q *s-Π-quasi-injective* if $\mathrm{Hom}_R(-,Q)$ is exact on sequences

$$0 \to K \to Q^\Lambda \to Q^{\Lambda'}, \text{ for any (or finite) sets } \Lambda, \Lambda'.$$

With similar proofs one gets that for such modules, $\mathrm{Hom}_R(-,Q)$ induces an equivalence between the kernels of morphisms $Q^k \to Q^l$, $k,l \in \mathbb{N}$ (a subclass of $\mathrm{cog}(Q)$) and the finitely presented right S-modules (compare [17, 4.6]).

By the definition of f-cotilting modules, 4.8 yields immediately:

4.10. f-cotilting and duality. *For $Q \in R$-Mod the following are equivalent:*

(a) *Q is f-cotilting;*

(b) $\mathrm{cog}(Q) \subset \mathrm{Ref}_R(Q)$ *and all finitely generated right S-modules are semi-Q-reflexive;*

(c) $\mathrm{cog}(Q) \subset \mathrm{Ref}_R(Q)$ *and* $\mathrm{mod}\text{-}S \cap \mathrm{Cog}_S(Q) \subset \mathrm{Ref}_S(Q);$

(d) $\mathrm{Hom}_R(-,Q) : \mathrm{cog}(Q) \to \mathrm{mod}\text{-}S \cap \mathrm{Cog}_S(Q)$ *is a duality.*

Finally we mention some more examples of f-cotilting objects. Clearly an object Q which is an injective cogenerator in $\mathcal{C}$ - or more generally in $\sigma[Q]$ - is f-cotilting. In particular any semisimple object has this property.

Cotilting modules Q are f-cotilting, provided for every $K \in \mathrm{cog}(Q)$, $\mathrm{Hom}_\mathcal{C}(K,Q) \in \mathrm{mod}\text{-}\mathrm{End}_\mathcal{C}(Q)$. Sufficient for the latter condition is that Q is noetherian both as left R- and right $\mathrm{End}_\mathcal{C}(Q)$-module. This situation was considered in Wang-Xu [13, Theorem 2 and 3]. If $_RQ$ is cotilting and artinian the conditions are also satisfied and this is the situation usually considered in representation theory.

It was already mentioned in the remark following 3.9 that over noetherian rings the "finitely cotilting" modules studied in Angeleri-Hügel-Valenta [2] are f-cotilting modules in our sense. Notice that artinian cotilting modules need not be "finitely cotilting" in the sense of [2] since they need not be finitely generated as modules over their endomorphism ring (see Example 2.3 in [1]).

The Morita duality as described in [14, 47.12] is a special case of 4.10. Notice that in [14, 47.12] the class of reflexive R-modules is closed under factor modules and submodules while in 4.10 this class is only closed under submodules.

REFERENCES

[1] L. ANGELERI-HÜGEL, Finitely cotilting modules, *Comm. Algebra* 28(4) (2000), 2147-2172.

[2] L. ANGELERI-HÜGEL, H. VALENTA, A duality result for almost split sequences, *Coll. Math.* 80 (1999), 267-292.

[3] L. ANGELERI-HÜGEL, A. TONOLO, J. TRLIFAJ, Tilting preenvelopes and cotilting precovers, *Algebras and Repres. Theory*, to appear

[4] R.R. COLBY, A generalization of Morita duality and the tilting theorem, *Comm. Algebra* 17(7) (1989), 1709-1722.

[5] R. COLPI, Cotilting bimodules and their dualities, *Interactions Between Ring Theory and Representations of Algebras*, F. van Oystaeyen, M. Saorin (ed), Marcel Dekker (2000), 81-93.

[6] R. COLPI, K.R. FULLER, Cotilting modules and bimodules, *Pac. J. Math.* 192(2) (2000), 275-291.

[7] R. COLPI, G. D'ESTE AND A. TONOLO, Quasi-tilting modules and counter equivalences, *J. Algebra* 191 (1997), 461-494.

[8] R. COLPI, A. TONOLO AND J. TRLIFAJ, Partial cotilting modules and the lattices induced by them, *Comm. Algebra* 25 (1997), 3225-3237.

[9] J. RADA, M. SAORÍN, A. DEL VALLE, Reflective and coreflective subcategories, *Glasgow Math. J.* 42(1) (2000), 97-113.

[10] M. SATO, On equivalences between module categories, *J. Algebra* 59 (1979), 412-420.

[11] B. STENSTRÖM, Rings of Quotients, *Springer-Verlag, Berlin* (1975).

[12] T. WAKAMATSU, Pseudo-projectives and pseudo-injectives in Abelian categories, *Math. Rep. Toyama Univ.* 2 (1979), 133-142.

[13] WANG MINGYI, XU XONGHUA, *-modules, co-*-modules and cotilting modules over Noetherian rings, *Science in China* (Series A) 39(1) (1996), 48-55.

[14] R. WISBAUER, Foundations of Module and Ring Theory, Gordon and Breach, Reading (1991).

[15] R. WISBAUER, On module classes closed under extensions, *Rings and radicals*, Gardner, Liu Shaoxue, Wiegandt (ed.), Pitman RN 346 (1996), 73-97.

[16] R. WISBAUER, Tilting in module categories, *Abelian groups, module theory, and toplogy*, Dikranjan, Salce (ed.), Marcel Dekker LNPAM 201 (1998), 421-444.

[17] R. WISBAUER, Static modules and equivalences, *Interactions Between Ring Theory and Representations of Algebras*, F. van Oystaeyen, M. Saorin (ed), Marcel Dekker (2000), 423-449.

[18] HUANG ZHAOYONG, On a generalization of the Auslander-Bridger transpose, *Comm. Algebra* 27(12) (1999), 5791-5812.

Coherent Components of Auslander-Reiten Quivers Whose DTr-orbits Are Finite

HAILOU YAO Department of Applied Mathematics, Beijing Polytechnic University, 100 Pingleyuan, Chaoyang District, Beijing,100022, P. R. China, E-mail: yaohl@public3.bta.net.cn

Dedicated to Professor Shaoxue Liu for His 70th Birthday

ABSTRACT Let k be an algebraically closed field, A be a finite-dimensional algebra over k, Γ_A be the Auslander-Reiten quiver of A. In this paper we give a necessary and sufficient condition for a coherent component Γ of Γ_A such that the translation quiver obtained from Γ by deleting the DTr-orbits of projectives is connected and has property that its DTr-orbits are finite.

0 INTRODUCTION

Let A be a connected, basic finite-dimensional algebra over an algebraically closed field, mod$A$ be the category of all finitely generated right A-modules, and Γ_A be the Auslander-Reiten quiver of A, then Γ_A is given a combinatorial structure making it a translation quiver. Since Γ_A is an important combinatorial and homological invariant of the category mod$A$, a lot of work has been done on describing all possible shapes of the connected components of Γ_A.

General theorems for the description of shapes of stable components of Γ_A are due to Happel-Preiser-Ringel[3] and Zhang [12]. The former theorem says that if Γ is a stable component containing a τ-periodic module then Γ is a stable tube $\mathbb{Z}A_\infty/(\tau^n)$ with $n \in \mathbb{Z}^+$ if Γ is infinite; otherwise, $\Gamma = \mathbb{Z}\Delta/G$ where the underlying graph $\overline{\Delta}$ of Δ is a Dynkin diagram and G is an automorphism group of $\overline{\Delta}$. The latter theorem says that if Γ is a stable component of an Auslander-Reiten quiver containing no τ-periodic module, then Γ is isomorphic to $\mathbb{Z}\Delta$ where Δ is a valued quiver without cyclic path.

S. Liu has studied the structure of semi-regular components of an Auslander-Reiten quiver[4]. His result is that if Γ is a semi-regular component containing an oriented cycle and no periodic module, then Γ is a ray tube or coray tube; if Γ is a semi-regular component without oriented cycles, then there exists a valued locally finite quiver Δ containing no oriented cycle such that Γ is isomorphic to a full translation sub-quiver of $\mathbb{Z}\Delta$ which is closed for predecessors and successors. In [5] he gives all possible structures of components in the Auslander-Reiten quivers of tilted algebras. Recently, P. Malicki and A. Skowroński have described the shape of an arbitrary coherent and almost cyclic component of Γ_A (see [8]). Their result is that a connected component of Γ_A is coherent and almost cyclic if and only if Γ is a generalized multi-coil, where Γ is said to be almost cyclic if all but at most a finite number of vertices of Γ lie on oriented cycles contained entirely in Γ.

A translation sub-quiver of Γ_A is said to be coherent if the following two conditions are satisfied.

(c1). For each projective module p in Γ there is an infinite sectional path

$$p = x_1 \to x_2 \to \cdots \to x_i \to x_{i+1} \to x_{i+2} \to \cdots.$$

in Γ, called a ray starting at p.

(c2). For each injective module I in Γ there is an infinite sectional path

$$\cdots \to y_{j+2} \to y_{j+1} \to y_j \to \cdots y_2 \to y_1 = I$$

in Γ, called a coray ending in I.

We say that a τ-orbit in Γ is finite if it contains only finitely many vertices.

The aim of this paper is to give the structure of a kind of coherent components of Auslander-Reiten quivers with finite DTr-orbits on the basis of results in [11]. Some classes of algebras have this type of AR-components such as tame hereditary algebras, polynomial growth trivial extension algebras, and some multicoil algebras, etc. One of the motivations of this article is to see which coils have finite DTr-orbits.

We shall denote by $\partial\Gamma$ the translation sub-quiver of Γ obtained by deleting the DTr-orbits of all projective modules in Γ. We have the following results.

THEOREM A. *Let Γ be a coherent component of an Auslander-Reiten quiver. If $\partial\Gamma$ is a connected subquiver of Γ, then Γ can be obtained from a stable tube by finitely many multiple admissible coray-ray insertions (see section 1 for its definition) if and only if each DTr-orbit in Γ is finite.*

REMARK: Such components in Theorem A exist. This is the case (trivially) for tame concealed algebras, but also for polynomial growth self-injective algebras, or for the algebras in the examples in section 4—which are coil enlargements of tame concealed algebras.

COROLLARY B. *If Γ is a component which satisfies the conditions in Theorem A, then Γ is a coil with finite DTr-orbits.*

COROLLARY C. *If Γ is a coil, then each $DTr-$orbit in Γ is finite if and only if Γ can be obtained from a stable tube by finitely many multiple admissible coray-ray insertions.*

The concepts and notations which are not explained in the following can be found in [1,2, 9,10].

1 PRELIMINARIES

From now on we let k be an algebraically closed field, A be a representation-infinite, basic, connected, finite-dimensional algebra over k, mod$A$ be the category of finite-dimensional right A-modules, Γ_A be the Auslander-Reiten quiver of A, and Γ be a connected component of Γ_A. We use τ and τ^- to denote the Auslander-Reiten translations DTr and TrD, respectively. In case there is no possibility of confusion we do not distinguish between an indecomposable module X in mod$A$ and the corresponding vertex $[X]$ in Γ_A.

Assume X to be an indecomposable module in mod$A$, then X is said to be left stable if $\tau^n X \neq 0$ for all positive integers n, and right stable if $\tau^n X \neq 0$ for all negative integers n. we use $_l\Gamma_A$ to denote the full sub-quiver of Γ_A which is generated by all τ-orbits containing no projective module, whereas $_r\Gamma_A$ indicates the one generated by those τ-orbits containing no injective module. The connected components of $_l\Gamma_A$ are called left stable components of Γ_A while those of $_r\Gamma_A$ are called right stable components of Γ_A.

For a translation quiver $\Gamma = (\Gamma_0, \Gamma_1, \tau)$ we can introduce the concepts of left, right stable vertices and stable vertices in Γ. Here we still use the notation Γ and the translation map τ. The context will allow no confusion.

Let $X \longrightarrow Y$ be an arrow in Γ_A, its valuation (d_{XY}, d'_{XY}) in Γ_A is so defined that Y occurs d_{XY} times in the codomain of the source map for X and X occurs d'_{XY} times in the domain of the sink map for Y. If $d_{XY} = d'_{XY} = 1$ we say that the arrow $X \longrightarrow Y$ has trivial valuation. If every arrow in a sub-quiver Γ of Γ_A has trivial valuation we say that Γ has trivial valuation.

Let $\Gamma = (\Gamma_0, \Gamma_1, \tau)$ be a translation quiver, a function $l: \Gamma_0 \longrightarrow \mathbb{N}$ (natural number set) is, by definition, a length function on Γ, if

(i) For any x which is not projective, $l(x) + l(\tau x) = \sum\limits_{y \in x^-} l(y)$,

(ii) For any x which is projective, $l(x) = 1 + \sum\limits_{y \in x^-} l(y)$,

(iii) For any x which is injective, $l(x) = 1 + \sum\limits_{y \in x^+} l(y)$

where, as usual, $x^-(x^+)$ denotes the set of direct predecessors(successors) of x in Γ.

We quote some Lemmas from [11].

Let Γ be an infinite connected component of Γ_A in which each DTr−orbit is finite, then we have

LEMMA 1.1. *Let* $0 \to X \to \bigoplus\limits_{i=1}^{r} Y_i \to Z \to 0$ *be an Auslander-Reiten sequence in* mod$A$ *with the Y_i indecomposable, where X and Z belong to Γ, then $r \leq 4$. In case $r = 4$, one of the Y_i is projective-injective.*

LEMMA 1.2. *Assume that X is injective in Γ and $X \to \bigoplus\limits_{i=1}^{r} Y_i$ is a source map,* then $r \leq 3$ and none of Y_i's is projective.

Dually, assume that Z is projective in Γ and $\bigoplus_{i=1}^{r} Y_i \to Z$ is a sink map, then $r \leq 3$ and none of Y_i's is injective.

LEMMA 1.3. Γ *has trivial valuation.*

Let $\Gamma = (\Gamma_0, \Gamma_1, \tau)$ be a connected translation quiver and let $k(\Gamma)$ be its mesh-category. We are going to define the so-called admissible operations.

DEFINITION 1.1. If $\text{SuppHom}_{k(\Gamma)}(x, -)$ equals an infinite sectional path starting at x,
$$x = x_0 \to x_1 \to x_2 \to \cdots.$$

Let t be an arbitrary positive integer, and Δ denote the following translation quiver, isomorphic to the Auslander-Reiten quiver of the full $t \times t$ lower triangular matrix algebra

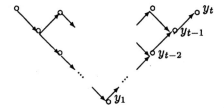

Let Γ' be the translation quiver having as vertices those of Γ, those of Δ, additionally z_{ij} and x'_i (where $i \geq 0$, $j \geq 1$) and having arrows as shown in Figure 1.1.

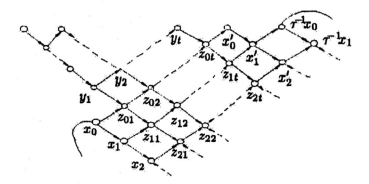

Figure 1.1

The translation τ' is defined as follows: $\tau' z_{ij} = z_{i-1,j-1}$, if $i \geq 2$, $j \geq 2$, $\tau' z_{i1} = x_{i-1}$, if $i \geq 1$, $\tau' z_{0j} = y_{j-1}$ if $j \geq 2$, $p = z_{01}$ is projective, $\tau' x'_0 = y_t, \tau' x'_i = z_{i-1,t}$ if $i \geq 1$, $\tau'(\tau^{-1} x_i) = x'_i$ provided x_i is not injective in Γ, otherwise x'_i is injective in Γ'. For the remaining vertices of Γ or Δ, τ' coincides with the translation of Γ or Δ, respectively. If $t = 0$, the new translation quiver Γ' is obtained from Γ by inserting the ray consisting of the x'_i's.

DEFINITION 1.2. If $\operatorname{SuppHom}_{k(\Gamma)}(x,-)$ consists of two sectional paths starting at x, one infinite and the other finite with at least one arrow ,

$$y_t \longleftarrow \cdots \longleftarrow y_2 \longleftarrow y_1 \longleftarrow x = x_0 \longrightarrow x_1 \longrightarrow x_2 \longrightarrow \cdots.$$

with $t \geq 1$, in particular , x_0 is injective , then Γ' is the translation quiver having as vertices those of Γ, and additional vertices denoted by $p = x'_0$, $z_{ij}(i \geq 1,\ j \geq 1)$ and having arrows as shown in Figure 1.2

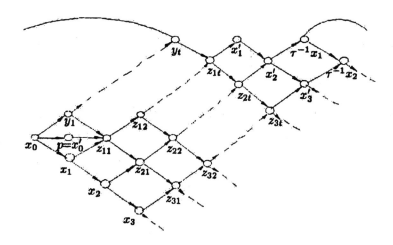

Figure 1.2

The translation τ' is defined as follows: p is projective-injective , $\tau' z_{ij} = z_{i-1,j-1}(i \geq 2,\ j \geq 2)$, $\tau' z_{i1} = x_{i-1}$ $(i \geq 1)$, $\tau' z_{ij} = y_{j-1}(j \geq 2)$, $\tau' x'_i = z_{i-1,t}(i \geq 2)$, $\tau' x'_1 = y_t$, $\tau'(\tau^{-1} x_i) = x'_i$ provided x_i is not injective in Γ, otherwise x'_i is injective in Γ. For the remaining vertices of Γ', τ' coincides with the translation τ of Γ.

DEFINITION 1.3. Assume that $\operatorname{SuppHom}_A(x,-)$ consists of two parallel sectional paths, the first infinite and starting at x, the second finite with at least one arrow and starting at a vertex y_1 such that there is an arrow $x \longrightarrow y_1$, not lying on the first path

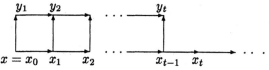

where $t \geq 2$, so that, in particular, x_{t-1} is injective. Moreover, consider the subquiver of Γ obtained by deleting the arrows $y_i \longrightarrow \tau^{-1} y_{i-1}$ assume that its connected component Γ^* containing the vertex x does not contain any of the vertices $\tau^- y_{i-1}$, $2 \leq i \leq t$. Then Γ' is the translation quiver having as vertices those vertices of Γ^* , additional vertices denoted by x'_0, z_{ij}, x'_i(where $i \geq 1$, $1 \leq j \leq t$), and having arrows as in the Figures 1.3 and 1.4 below.

If t is odd

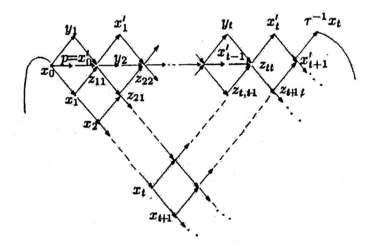

Figure 1.3

If t is even

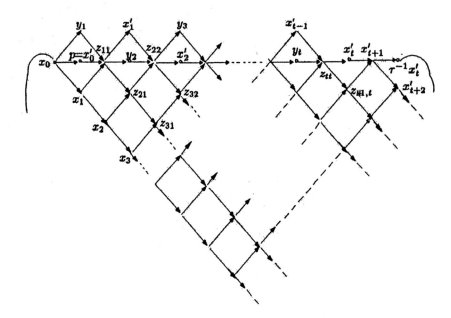

Figure 1.4

The translation τ' of Γ' is defined as follows: x_0' is projective, $\tau' z_{ij} = z_{i-1,j-1}$ if $i \geq 2, 2 \leq j \leq t$, $\tau' z_{i1} = x_{i-1}$ if $i \geq 1$, $\tau' x_i' = y_i$ if $1 \leq i \leq t$, $\tau' x_i' = z_{i-1,t}$ if $i \geq t+1$, $\tau' y_j = x_{j-2}'$ if $i \geq t$ provided x_i is not injective in Γ, otherwise x_i' is injective in Γ'. In both cases, x_{t-1}' is injective. For the remaining vertices of Γ', τ' coincides with the translation τ of Γ.

For the convenience of statements, the ray insertion in definition 1.1 is called $\alpha(t+1)$-insertion. The vertex $x = x_0$ is called an α-insertion vertex. Its dual, i.e. coray insertion in definition 1.2 is called $\alpha^*(t+1)$-insertion. In this case, the vertex x_0 is called an α^*-insertion vertex. The insertion in definition 1.2 is called $\beta(t+1)$-insertion, the vertex x_0 is called a β-insertion vertex. Its dual is called $\beta^*(t+1)$- insertion. The insertion in definition 1.3 is called $\gamma(t+1)$-insertion, and its dual is called $\gamma^*(t+1)$-insertion.

We now introduce a notion as follows.

Assume that Γ is an infinite connected translation quiver and that there is an α-insertion vertex x in Γ_0. Let $m_1,\cdots,m_\mu$ be natural numbers. So, the support of $\text{Hom}_{k(\Gamma)}(x,-)$ equals an infinite sectional path starting at x

$$x = x_0 \longrightarrow x_1 \longrightarrow x_2 \longrightarrow \cdots.$$

We make $\alpha(m_1)$-insertion at the vertex x in Γ to obtain a new infinite connected translation quiver, denoted by Γ_1. Assume that the added projective vertex p_1 (as p in definition 1.1) which is a direct successor of x, is a β^*-insertion vertex in Γ_1. So, the support of $\text{Hom}_{k(\Gamma)}(-,p_1)$ consists of the following sectional paths ending at p_1, one infinite and other finite.

$$\cdots \longrightarrow u_i \longrightarrow u_{i-1} \longrightarrow \cdots \longrightarrow u_2 = x_0 \longrightarrow u_1 = p_1 \longleftarrow p_{11} \longleftarrow \cdots \longleftarrow p_{1,m_1-1}.$$

So, $u_2 = x_0$ is an α^*-insertion vertex. By the definition of α-insertion we know that there is a unique maximal finite sectional path starting at p_1 (denoted by $L_{p_1}^+$). The end vertex z_1 of $L_{p_1}^+$ is still an α-insertion vertex while the predecessor of z_1 (denoted by $x_{1,1}$) on the path $L_{p_1}^+$ is a γ^*-insertion vertex if $m_1 \geq 3$. Moreover, the support of $\text{Hom}_{k(\Gamma)}(-,x_{1,1})$ is of the following form

$$\begin{array}{ccccccccc}
& & y_{1,m_1-1} & \longrightarrow & y_{1,m_1-2} & \longrightarrow & \cdots & \longrightarrow & y_{1,2} & \longrightarrow & y_{1,1} \\
& & \downarrow & & \downarrow & & & & \downarrow & & \downarrow \\
\cdots \longrightarrow x_{1,m_1+1} & \longrightarrow & x_{1,m_1}=x_0 & \longrightarrow & x_{1,m_1-1}=p_1 & \longrightarrow & x_{1,m_1-2} & \longrightarrow & \cdots & \longrightarrow & x_{1,2} & \longrightarrow & x_{1,1}
\end{array}$$

where $y_{1,1} = \tau z$, and $x_{1,m_1-1}(=p_1)$ is projective.

We make $\alpha(m_2)$-insertion at z_1 in Γ_1 to obtain an infinite connected translation quiver, denoted by Γ_2. Obviously the added projective vertex p_2, which is a direct successor of z_1, is still a β^*- insertion vertex. The support of $\text{Hom}_{k(\Gamma)}(-,p_2)$ consists of the following sectional paths ending in p_2, one infinite and the other finite.

$$\cdots \to u_i \to u_{i-1} \to \cdots \to p_1 \to \cdots \to p_2 \leftarrow p_{21} \leftarrow \cdots \leftarrow p_{2,m_2-1}.$$

Let $L_{p_2}^+$ denote the unique maximal finite sectional path starting at p_2 in Γ_2, then the end vertex, denoted by z_2, is still an α- insertion vertex. The direct predecessor of z_2 (denoted by $x_{2,1}$) on the path $L_{p_2}^+$ is still a γ^*-insertion vertex if $m_2 \geq 3$, and the predecessor z_1 (denoted by x_{2,m_2}) of p_2 is still an α^* insertion vertex. Inductively, we obtain an infinite connected translation quiver Γ_μ by $\alpha(m_\mu)$-insertion at the vertex $z_{\mu-1}$. The added projective vertex p_μ, which is a direct successor of $z_{\mu-1}$, is a β^*-insertion vertex. So, the support of $\text{Hom}_{k(\Gamma_\mu)}(-,p_\mu)$ consists of the following sectional paths. One infinite and the other finite,

$$\cdots \longrightarrow u_i \longrightarrow u_{i-1} \longrightarrow \longrightarrow p_1 \longrightarrow \cdots \longrightarrow p_\mu \longleftarrow p_{\mu,1} \longleftarrow \cdots \longleftarrow p_{\mu,m_\mu-1}.$$

Let $L_{p_\mu}^+$ denote the unique maximal finite sectional path starting at p_μ in Γ_μ, the end vertex denoted by z_μ, is both an α-insertion and α^*-insertion vertex.

The direct predecessor of z_μ (denoted by $x_{\mu,1}$) on the path $L^+_{p_\mu}$ is still a γ^*-insertion vertex if $m_\mu \geq 3$, and the predecessor $z_{\mu-1}$ (denoted by x_{μ,m_μ}) of p_μ is an α^*-insertion vertex. The support of $\operatorname{Hom}_{k(\Gamma_\mu)}(-, x_{\mu,1})$ is in the following form if $m_\mu \geq 3$.

$$\begin{array}{ccccccc}
& & y_{\mu,m_\mu-1} & \to & \cdots \to & y_{\mu,2} \to & y_{\mu,1} \\
& & \downarrow & & & \downarrow & \downarrow \\
\cdots \to & x_{\mu,m_\mu}=z_{\mu-1} \to & x_{\mu,m_\mu-1}=p_\mu & \to & \cdots \to & x_{\mu,2} & \to x_{\mu,1}
\end{array}$$

where $y_{\mu,1} = \tau z_\mu$, $x_{\mu,m_\mu-1} = p_\mu$ is injective, the path $y_{\mu,m_\mu-1} \to \cdots \to y_{\mu,2} \to \cdots \to y_{\mu,1}$ and $\cdots \to x_{\mu,2} \to x_{\mu,1}$ are sectional and every sectional path at $x_{\mu,1}$ (corresponding at $y_{\mu,1}$) is a sub-path of one of the paths $y_{\mu,1} \to x_{\mu,1}$ or $\cdots \to x_{\mu,2} \to x_{\mu,1}$ (respectively, of $y_{\mu,m_\mu-1} \to \cdots \to y_{\mu,2} \to y_{\mu,1}$).

We make the following insertions (not necessarily all of them) in Γ_μ.

(1) $\beta^*(m_\mu^*)$-insertion at the vertex p_μ, where $m_\mu^* = m_\mu$.

(2) $\gamma^*(m_\mu^*)$-insertion at the vertex $x_{\mu,1}$, where $m_\mu^* = m_\mu$.

(3) $\alpha^*(m_\mu^*)$-insertion at the vertex z_μ or $x_{\mu,m_\mu} = z_{\mu-1}$ or other coray vertex, where m_μ^* may not be equal to m_μ.

So, we obtain a translation quiver Γ'_μ. From the definition of $\beta^*(m_\mu^*)$-insertion, $\gamma^*(m_\mu^*)$-insertion and $\alpha^*(m_\mu^*)$-insertion we learn that the β^*-insertion vertex p_i's in Γ_μ correspond to the new β^*-insertion vertices (of course, they are still projective), still denoted by p_i's, for $i = 1, \cdots, \mu - 1$, and γ^*-insertion vertices $x_{i,1}$'s in Γ_μ correspond to the new γ^*-insertion vertices ($\operatorname{Hom}(-, x_{i,1})$ is the same as translation sub-quiver as is in Γ_μ) for $i = 1, \cdots, \mu - 1$, and α^*-insertion vertices $x_{i,m_i}(= z_{i-1})$ are still α^*-insertion vertices for $i = 1, 2, \cdots, \mu - 2$. If we do not make $\alpha^*(m_\mu^*)$-insertion at the vertex $x_{\mu,m_\mu} = z_{\mu-1}$ in Γ_μ, then x_{μ,m_μ} also corresponds to a new α^*-insertion vertex in Γ'_μ, still denoted by $x_{\mu,m_\mu}(= z_\mu)$.

We again make the following insertions (not necessarily all of them) in Γ'_μ.

(1) $\beta^*(m_{\mu-1}^*)$-insertion at the vertex $p_{\mu-1}$, where $m_{\mu-1}^* = m_{\mu-1}$.

(2) $\gamma^*(m_{\mu-1}^*)$-insertion at the vertex $x_{\mu-1,1}$, where $m_{\mu-1}^* = m_{\mu-1}$.

(3) $\alpha^*(m_{\mu-1}^*)$-insertion at the vertex $z_{\mu-2} = x_{\mu-1,m_{\mu-1}}$ or the other coray vertex, where $m_{\mu-1}^*$ may not be equal to $m_{\mu-1}$.

So, we obtain a translation quiver $\Gamma'_{\mu-1}$. The β^*-insertion vertices p_i's, γ^*-insertion vertices $x_{i,1}$'s correspond to the new β^*-insertion vertices (still denoted by p_i's) and the new γ^*-insertion vertices (still denoted by $x_{i,1}$'s), respectively, for $i = 1, 2, \cdots, \mu - 2$. The α^*-insertion vertices x_{i,m_i}'s ($=z_{i-1}$) are still α^*-insertion vertices for $i = 1, 2, \cdots, \mu - 3$. $x_{\mu-1,m_{\mu-1}}$ also corresponds to a new α^*-insertion vertex (still denoted by $x_{\mu-1,\mu-1}$) in $\Gamma'_{\mu-1}$ if we do not make $\alpha^*(m_{\mu-1}^*)$-insertion at the vertex $x_{\mu-1,m_{\mu-1}}$ in Γ'_μ.

Inductively, by the μ^*-th step, where μ^* may not be equal to μ, we obtain a translation quiver Γ', which is said to be obtained from Γ by a multiple admissible ray-coray insertion if $\sum_{i=1}^{\mu} m_i = \sum_{j=1}^{\mu^*} m_j^*$. The vertex x is called a multiple admissible ray-coray insertion vertex.

Dually, the definition of multiple admissible coray-ray insertion can be made.

EXAMPLE. Consider a tube $\Gamma_1 = B\mathbb{Z}A_\infty/(1)$, we make an $\alpha(3)$-insertion at x in Γ_1 to obtain a translaton quiver Γ_2.

where the vertical dotted lines have to be identified in order to obtain a stable tube and ray tube (similar in the below). We again make an $\alpha(3)$-insertion at x_0' in Γ_2 to obtain a translation quiver Γ_3. We make an $\beta^*(3)$-insertion at p_2 in Γ_3 to obtain a translation quiver Γ_4.

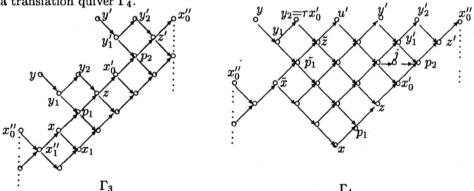

we also make a $\gamma^*(3)$-insertion at $\tilde{p}_1$ in Γ_3 to obtain a translation quiver Γ_5 as follows. Thus, Γ_5 is obtained from the stable tube Γ_1 by one multiple admissible ray-coray insertion.

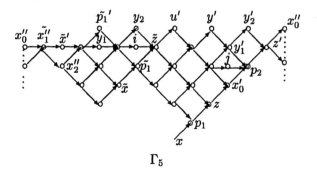

2 SOME LEMMAS

In this section we will establish some preliminary lemmas to prove the theorem A.

Let Γ be a translation quiver without multiple arrows. A mesh with exactly three middle terms will be called exceptional and a projective middle term in an exceptional mesh will be called exceptional projective. Other meshes and projectives will be called ordinary. The set of vertices which are the starting or ending vertex of a mesh in Γ with unique middle term will be called the mouth of Γ.

Let l be a length function in Γ. A ray starting at x will be denoted by $[x, \infty)$. Dually, a coray ending in y will be denoted by $(\infty, y]$.

Firstly, we quote several lemmas from [8].

LEMMA 2.1. *Assume Γ contains a translation sub-quiver of the form*

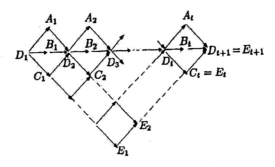

with B_1 projective and $t \geq 2$. then at most one of the modules A_t and B_t can be injective. Moreover, if it is the case, then B_t is injective if t is odd, while A_t is injective if t is even.

Proof. This is Lemma 4.2 in [8]. □

LEMMA 2.2. *Assume Γ contains a translation sub-quiver of the form*

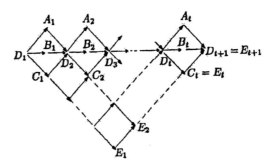

where $t \geq 2$, A_1 or B_1 is projective, and A_t or B_t is injective. Then $A_i^+ = \{D_{i+1}\} = B_i^+$ for $1 \leq i \leq t-1$, and $A_i^- = \{D_i\} = B_i^-$ for $2 \leq i \leq t$. Here A_i^+ and B_i^+ are sets of successors of A_i and B_i, respectively; A_i^- and B_i^- are sets of predecessors of A_i and B_i, respectively.

Proof. This is Lemma 4.5 in [8]. □

LEMMA 2.3. *Γ contains no translation sub-quiver of the form*

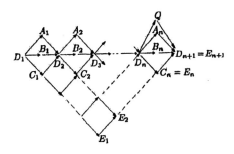

with B_1 projective and Q projective-injective.

Proof. This is Lemma 4.4 in [8]. □

LEMMA 2.4. *Γ contains no translation sub-quiver which is created by identifying the sectional path $A_t \longrightarrow D_{t+1} \longrightarrow \cdots \longrightarrow F_t \longrightarrow E_{t+1} \longrightarrow F_{s+1}$ in the Figures A and B, and $D'_2 \longrightarrow \cdots \longrightarrow M \longrightarrow N$ in the Figures B and C below*

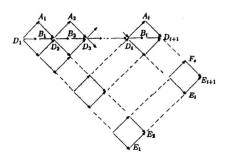

Figure A

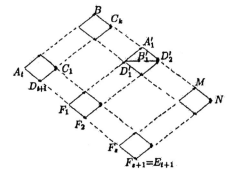

Figure B

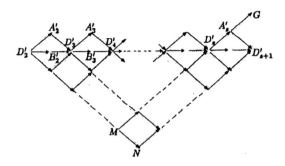

Figure C

where $D_{t+1} \to C_1 \to C_2 \to \cdots \to C_k$ is a sectional path, B_1, B_1' are projective, A_t or B_t is injective, A_s' or B_s' is injective, possibly some (or all) of $C_2, C_3, \cdots, C_k$ are projective, $t \geq 1$, $s \geq 1$, $k \geq 1$ and possibly B or E_1 does not exist.

Proof. This is Lemma 4.7 in [8]. □

LEMMA 2.5. Γ contains no translation sub-quiver of the following form.

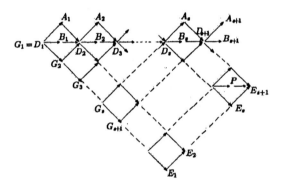

where B_1 is projective and P is projective-injective.

Proof. Assume that Γ contains such a sub-quiver. Using the length function l we obtain the following inequalities

$$l(G_{s+1}) + l(E_{s+1}) \geq l(D_{s+1}) + l(P) + l(E_1) \tag{1}$$

$$l(G_i) + l(D_{i+1}) \geq l(A_i) + l(B_i) + l(G_{i+1}) \quad 1 \leq i \leq s \tag{2}$$

Thus, we have

$$l(G_1) + \sum_{i=1}^{s} l(D_{i+1}) \geq l(G_{s+1}) + \sum_{i=1}^{s}(l(A_i) + l(B_i)). \tag{3}$$

Since $l(A_{2j-1}) + l(A_{2j}) \geq l(D_{2j})$, $(2 \leq 2j \leq s)$ and $l(B_{2j}) + l(B_{2j+1}) \geq l(D_{2j+1})$, $(3 \leq 2j + 1 \leq s)$, we obtain that

$$\sum_{i=1}^{s}(l(A_i) + l(B_i)) \geq \sum_{i=2}^{s} l(D_i) + l(M). \tag{4}$$

where $M = B_s$ if s is even, and $M = A_s$ if s is odd. So we get from (1), (3) and (4) that

$$l(G_1) + l(E_{s+1}) \geq l(P) + l(E_1) + l(B_1) + l(M).$$

This is impossible because $l(P) \geq 1+l(E_{s+1})$ and $l(B_1) \geq 1+l(G_1)$. This completes the proof. □

LEMMA 2.6. Γ *contains no translation sub-quiver in the following form*

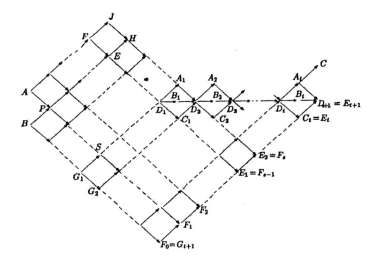

where P is ordinary projective, B_1 is projective and A_t is injective if t is even, and B_t is injective if t is odd.

Proof. Using the length function we have the following inequalities:

$$l(P) + l(D_1) \geq l(S) + l(E), \quad l(B) + l(F_1) \geq l(P) + l(F_0),$$
$$l(S) + l(E_1) \geq l(D_1) + l(F_1), \quad l(F) + l(A_1) \geq l(J) + l(D_1),$$
$$l(A) + l(E) \geq l(P) + l(F).$$

Combining the above inequalities we get that

$$l(B) + l(E_1) + l(A_1) + l(A) \geq l(P) + l(F_0) + l(D_1) + l(J).$$

Since P is projective we have $l(P) \geq 1 + l(A) + l(B)$. So, we obtain

$$l(E_1) + l(A_1) \geq 1 + l(F_0) + l(D_1) + l(J). \tag{$*$}$$

In case $t = 1$, i.e. B_1 is projective-injective, then $E_1 = C_1$. So, we have $l(D_1) = 1 + l(A_1) + l(C_1)$. From $(*)$ we get $0 \geq 2 + l(F_0) + l(J)$. This is a contradiction.

In case $t \geq 2$, we have $l(D_i)+l(E_{i+1}) \geq l(A_i)+l(B_i)+l(E_i)$, $1 \leq i \leq t$. Then we have $\sum_{i=1}^{t+1} l(D_i) \geq \sum_{i=1}^{t}[l(A_i)+l(B_i)]+l(E_1)$ where $D_{t+1}=E_{t+1}$.

If t is odd, let $t=2s+1$, then B_t is injective. We have the following inequalities:
$$l(A_{2j})+l(A_{2j+1}) \geq l(D_{2j+1}), \quad 1 \leq j \leq s$$
$$l(B_{2j-1})+l(B_{2j}) \geq l(D_{2j}), \quad 1 \leq j \leq s$$

Then we get $l(D_1)+l(D_{t+1}) \geq l(A_1)+l(B_t)+l(E_1)$. Since B_t is injective we have $l(B_t) \geq 1+l(D_{t+1})$. Combining these inequalities with (*) we obtain $0 \geq 2+l(J)+l(F_0)$. A contradiction.

Similarly, if t is even, then A_t is injective, let $t=2s$ (with $s \geq 1$), then we have
$$l(A_{2j})+l(A_{2j+1}) \geq l(D_{2j+1}), \quad 1 \leq j \leq s-1$$
$$l(B_{2j-1})+l(B_{2j}) \geq l(D_{2j}), \quad 1 \leq j \leq s$$

We get $l(D_1)+l(D_{t+1}) \geq l(A_1)+l(A_t)+l(E_1)$. Since A_t is injective we have $l(A_t) \geq l(C)+l(D_{t+1})+1$, combining these inequalities with (*) we obtain that $0 \geq 2+l(F_0)+l(C)+l(J)$. A contradiction. This completes the proof. □

LEMMA 2.7. Γ *contains no translation sub-quiver of the following form*

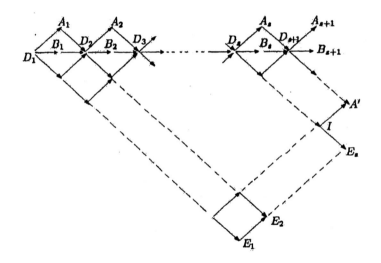

where B_1 is projective and I is injective.

Proof. Using the length function l we have inequalities as follows
$$l(D_i)+l(E_{i+1}) \geq l(A_i)+l(B_i)+l(E_i), \quad 1 \leq i \leq s-1.$$

Thus, we obtain
$$l(E_s)+\sum_{i=1}^{s-1} l(D_i) \geq \sum_{i=1}^{s-1}[l(A_i)+l(B_i)]+l(E_1) \tag{1}$$

Since $l(A_{2j-1}) + l(A_{2j}) \geq l(D_{2j})$ (with $2 \leq 2j \leq s-1$) and $l(B_{2j}) + l(B_{2j+1}) \geq l(D_{2j+1})$ (with $3 \leq 2j+1 \leq s-1$), we then have

$$l(D_1) + \sum_{i=1}^{s-1}(l(A_i) + l(B_i)) \geq \sum_{i=2}^{s-1} l(D_i) + l(B_1) + l(M). \tag{2}$$

where $M = A_{s-1}$ if $(s-1)$ is odd and $M = B_{s-1}$ if $(s-1)$ is even.

So, we get the inequality from (1) and (2) as well as $l(B_1) = 1 + l(D_1)$

$$l(E_s) \geq 1 + l(E_1) + l(M) \tag{3}$$

On the other hand, we have

$$l(D_s) + l(A') \geq l(A_s) + l(B_s) + l(I) \tag{4}$$

$$l(I) \geq 1 + l(A') + l(E_s) \tag{5}$$

So, we have the following inequality from (3), (4) and (5)

$$l(D_s) \geq 2 + l(A_s) + l(B_s) + l(M) \tag{6}$$

As $l(A_{s-1}) + l(A_s) \geq l(D_s)$ and $l(B_{s-1}) + l(B_s) \geq l(D_s)$ as well as $M = A_{s-1}$ or B_{s-1} we get a contradiction from (6). This completes the proof. $\square$

LEMMA 2.8. Γ *contains no translation sub-quiver of the following form.*

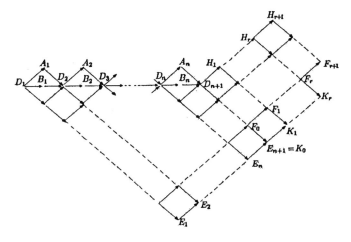

where B_1 is projective and F_r is injective; A_n is injective if n is even, and B_n is injective if n is odd. The path between D_{n+1} and H_{r+1} is sectional.

Proof. In fact, it follows from the proof of Lemma 2.7 that the statement of Lemma 2.8 is true if $r = 0$. So, we may assume that $r \geq 1$.

A similar calculation to the one in the proof of Lemma 2.7 yields

$$l(E_{n+1}) \geq 1 + l(E_1) + l(M) \tag{1}$$

where $M = A_n$ if n is odd, and $M = B_n$ if n is even.

On the other hand , we have the following inequalities

$$l(D_{n+1}) + l(K_r) \geq l(E_{n+1}) + l(H_r)$$
$$l(H_r) + l(F_{r+1}) \geq l(F_r) + l(H_{r+1}).$$

Thus , we obtain that

$$l(D_{n+1}) + l(K_r) + l(F_{r+1}) \geq l(F_r) + l(H_{r+1}) + l(E_{n+1}). \tag{2}$$

We then get from (1) and (2)

$$l(D_{n+1}) + l(K_r) + l(F_{r+1}) \geq 1 + l(E_1) + l(M) + l(H_{r+1}) + l(F_r) \tag{3}$$

By induction on r one can show that $l(M) + l(H_{r+1}) \geq l(D_{n+1})$. So , using the inequality $l(F_r) \geq 1 + l(K_r) + l(F_{r+1})$ we will get a contradiction from (3). The proof is completed. □

LEMMA 2.9. Γ *contains no translation sub-quiver of the following form*

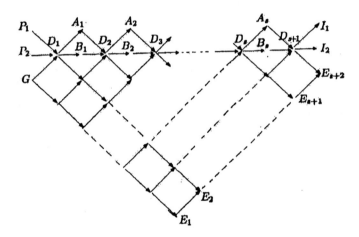

where D_1, P_1 and P_2 are projective , D_{s+1}, I_1 and I_2 are injective. P_1, P_2, I_1 and I_2 may not exist. If $P_1(P_2)$ does not exist , $A_1(B_1)$ is projective. If $I_1(I_2)$ does not exist , $A_s(B_s)$ is injective.

Proof. Assume Γ contains such a translation sub-quiver . Using the length function l we have

$$l(D_i) + l(E_{i+2}) \geq l(A_i) + l(B_i) + l(E_{i+1}), \quad 1 \leq i \leq s.$$

So we obtain

$$l(E_{s+2}) + \sum_{i=1}^{s} l(D_i) \geq \sum_{i=1}^{s}(l(A_i) + l(B_i)) + l(E_2). \tag{1}$$

We also have $l(A_{2j-1}) + l(A_{2j}) \geq l(D_{2j})$ $(1 \leq 2j - 1 \leq s)$ and $l(B_{2j}) + l(B_{2j+1}) \geq l(D_{2j+1})$ $(2 \leq 2j \leq s)$. Thus , we get from (1)

$$l(E_{s+2}) + l(D_1) \geq l(E_2) + l(M) + l(B_1). \tag{2}$$

where $M = A_s$ if s is odd, and $M = B_s$ if s is even.

Now, we have
$$l(G) + l(E_2) \geq l(D_1) + l(E_1). \tag{3}$$

In case s is odd, then $M = A_s$. Thus, we have
$$l(A_s) + l(I_1) \geq l(D_{s+1}). \tag{4}$$
$$l(P_1) + l(B_1) \geq l(D_1). \tag{5}$$

Therefore, we get the following inequalities from (2), (3), (4) and (5)
$$l(E_{s+2}) + l(G) + l(P_1) + l(I_1) \geq l(E_1) + l(D_{s+1}) + l(D_1). \tag{6}$$

Since $l(D_1) \geq l(P_2) + l(P_1) + l(G) + 1$ and $l(D_{s+1}) \geq l(I_1) + l(I_2) + l(E_{s+2}) + 1$, we get from (6) that $0 \geq 2 + l(E_1)$. This is a contradiction.

In case s is even, then $M = B_s$ and we can get a contradiction similarly. This finishes the proof. □

LEMMA 2.10. Γ *contains no translation sub-quiver of the following form*

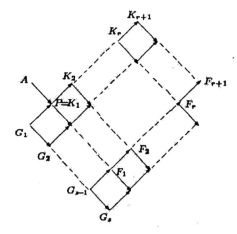

where P is ordinary projective and F_r is injective, the path $G_1 \to P_1 = K_1 \to \cdots \to K_{r+1}$ is sectional.

Proof. Without loss of generality we may assume that there is no projective on the path $K_2 \to \cdots \to K_r$. We consider two cases.

Case I. F_r is not on the path $P = K_1 \to \cdots \to K_{r+1}$ or on the ray starting at P. We then have $l(P) + l(E_r) \geq l(E_1) + l(K_r)$ and $l(K_r) + l(F_{r+1}) \geq l(K_{r+1}) + l(F_r)$, as well as $l(G_1) + l(E_1) \geq l(P) + l(G_s)$.

Thus we obtain that
$$l(G_1) + l(F_{r+1}) + l(E_r) \geq l(G_s) + l(F_r) + l(K_{r+1}). \tag{1}$$

Since P is projective, A is not injective. So, $\tau^- A = K_2$ by Lemma 2.9.

By induction on r one can show that $l(A) + l(K_{r+1}) \geq l(P)$. Thus, we obtain that $0 \geq 2 + l(G_s)$ since $l(F_r) \geq 1 + l(F_{r+1}) + l(E_r)$ and $l(P) \geq 1 + l(G_1) + l(A)$. This is a contradiction.

Case II. $F_r = K_r$ or F_r is on the ray starting at P. Firstly assume that $F_r = K_r$. It is obvious that $r \neq 1$ since P is projective. Without loss of generality we may assume that there is no projective or injective on the path $K_2 \to \cdots \to K_{r-1}$. Since $K_r = F_r$ is injective, K_{r+1} is not projective. Hence we have the following form.

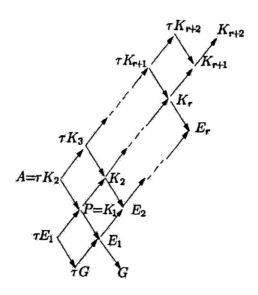

If τG or τK_{r+2} does not exist, we agree $l(\tau G) = 0$ or $l(\tau K_{r+2}) = 0$. Then we have

$$l(\tau K_2) + l(K_{r+1}) \geq l(\tau K_{r+2}) + l(P)$$

$$l(\tau E_1) + l(E_r) = l(\tau G) + l(K_r).$$

So, we obtain

$$l(\tau K_2) + l(\tau E_1) + l(E_r) + l(K_{r+1}) \geq l(P) + l(K_r) + l(\tau K_{r+2}) + l(\tau G).$$

This is impossible since $l(P) \geq 1 + l(\tau K_2) + l(\tau E_1)$ and $l(K_r) \geq 1 + l(K_{r+1}) + l(E_r)$.

If the injective K_r is on the ray starting at P, we can similarly show that it is impossible. This completes the proof. □

PROPOSITION 2.11. *Let Γ be a coherent connected translation quiver with finite τ-orbits. Assume there is a length function l in Γ. If $\partial\Gamma$ is a connected sub-quiver of Γ, then the mesh category $k(\Gamma)$ contains no oriented cycle of projectives.*

Proof. Firstly, by Happel-Preiser-Ringel's theorem [3] we know that $\partial \Gamma$ is a stable tube. Let $p_0 \longrightarrow p_1 \longrightarrow \cdots \longrightarrow p_t = p_0$ be a cycle of projectives in $k(\Gamma)$. We claim that there exists another such a cycle of projectives with each p_s either exceptional or else such that there exist arrows $q \longrightarrow x \longrightarrow p_s$, with q injective.

Let us consider p_0 and define a vertex z as follows. If p_0 is exceptional and not injective, let x be its unique direct predecessor, and $c = \tau^{-t}x$ be such that one of the direct predecessors of c is injective. As Γ is coherent, there exist a ray $[p_0, \infty)$ and a coray $(\infty, \tau^{-t+1}x]$. Set $z' = [p_0, \infty) \cap (\infty, \tau^{-t+1}x]$. Let z denote the direct successor of z' on the ray $[p_0, \infty)$. There is a sectional path between z and c by Lemma 2.7. If p_0 is exceptional and injective, let $z = c$. If p_0 is not exceptional, set $z = p_0$.

Now, let us consider the set of vertices y on the sectional path from z to the mouth such that there exists a ray $[y, \infty)$. This set is not empty since it contains z. Let y be a maximal element in this set (that is, closer to the mouth) and let $[z, y]$ denote the sectional path from z to y. There exists a ray $[v, \infty)$ for each v on $[z, y]$ by Lemma 2.7, 2.8 and 2.10. Let $\mathcal{R}$ denote the mesh-complete translation sub-quiver consisting of all vertices lying on these rays. Then $\mathcal{R} \subseteq \mathrm{SuppHom}_{k(\Gamma)}(p_0, -)$: This is clear if $z = p_0$ and $z = c$. Otherwise it follows from the fact that c and exactly two of its direct predecessors (namely, the one lying on $[z, y]$, and the injective direct predecessor) belong to $\mathrm{SuppHom}_{k(\Gamma)}(p_0, -)$.

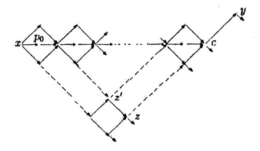

We learn that $\mathcal{R}$ contains no exceptional mesh from Lemmas 2.1- 2.6.

We claim that $[y, \infty)$ contains either an arrow $q \longrightarrow x$, with q injective and x a direct predecessor of a projective p', or else a vertex x being the direct predecessor of an exceptional projective p'. Assume that this is not the case. Since $\mathrm{Hom}_{k(\Gamma)}(p_0, p_1) \neq 0$ and the rays with v on $[z, y]$ exist we must have that p_1 belongs to $[c, y]$ from Lemma 2.7, 2.8 and 2.10. As $\mathrm{Hom}_{k(\Gamma)}(p_i, p_{i+1}) \neq 0$ for all $1 \leq i \leq t - 2$, and $\mathcal{R}$ consists of ordinary meshes. We conclude that $p_2, \cdots, p_{t-1}$ all belong to $[c, y]$. From the definition of y and the assumption that $[y, \infty)$ contains no vertices as required, we have $\mathrm{Hom}_{k(\Gamma)}(p_{t-1}, p_0) = 0$. This is a contradiction.

Note that p', as defined in the previous claim, is the unique projective vertex having a direct predecessor on $[y, \infty)$. For, if this is not the case, the existence of the ray $[p', \infty)$ implies that we have two cases in the following.

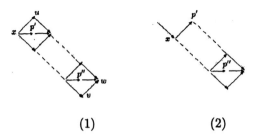

(1)　　　　　　　(2)

Since there exist rays $[y, \infty)$ and $[p', \infty)$, the projective p'' is exceptional. So, we will consider p'' in two cases.

Case I. p'' is projective-injective. By using length function l we have $l(x)+l(w) = l(u)+l(p')+l(p'')+l(v)$ in the diagram(1). Since $l(p') = l(x)+1$, and $l(p'') = l(w)+1$, we obtain that $0 = 2 + l(u) + l(v)$, a contradiction. A similar contradiction can be obtained in the diagram (2).

Case II. p'' is projective but not injective. Then both of cases in the diagram (1) and (2) can be formed into the case in the following sub-quiver.

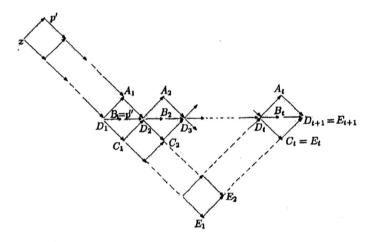

with A_t or B_t injective.

Firstly, we have the inequality $l(x) + l(A_1) \geq l(p') + l(D_1)$. Since p' is projective, we get $l(p') \geq 1 + l(x)$. Hence we obtain

$$l(A_1) \geq 1 + l(D_1). \qquad (*)$$

Secondly, we have the following inequalities

$$l(D_i) + l(E_{i+1}) \geq l(A_i) + l(B_i) + l(E_i), \quad 1 \leq i \leq t.$$

Summing up these inequalities we get

$$\sum_{i=1}^{t+1} l(D_i) \geq \sum_{i=1}^{t}[l(A_i) + l(B_i)] + l(E_1)$$

because $D_{t+1} = E_{t+1}$.

If t is odd, let $t = 2s + 1$ for some $s \geq 1$. We have the following inequalities.

$$l(A_{2j}) + l(A_{2j+1}) \geq l(D_{2j+1}), \quad 1 \leq j \leq s,$$

$$l(B_{2j-1}) + l(B_{2j}) \geq l(D_{2j}), \quad 1 \leq j \leq s.$$

So, we obtain that $l(D_1) + l(D_{t+1}) \geq l(A_1) + l(B_t) + l(E_1)$. From Lemma 2.1 we know that B_t is injective and hence we have $l(B_t) \geq 1 + l(D_{t+1})$. From (*) we get that $0 \geq 2 + l(E_1)$. This is a contradiction.

If t is even, let $t = 2s$ ($s \geq 1$), then A_t is injective by Lemma 2.1. Using the following inequalities

$$l(A_{2j}) + l(A_{2j+1}) \geq l(D_{2j+1}), \quad 1 \leq j \leq s - 1,$$

$$l(B_{2j-1}) + l(B_{2j}) \geq l(D_{2j}), \quad 1 \leq j \leq s.$$

we get $l(D_1) + l(D_{t+1}) \geq l(A_1) + l(A_t) + l(E_1)$. Since $l(A_t) \geq 1 + l(D_{t+1})$, we have that $l(D_1) \geq l(A_1) + l(E_1) + 1$. From (*) we obtain $0 \geq 1 + l(E_1)$, a contradiction. Therefore, p' is the unique projective vertex with a direct predecessor on the ray $[y, \infty)$.

Let i be such that the projectives $p_1, \cdots, p_i$ belong to $\mathcal{R}$ while $p_{i+1} \notin \mathcal{R}$, then either $p_{i+1} = p'$ or the morphism $p_i \longrightarrow p_{i+1}$ factors through p'. If $p_{i+1} = p'$ we replace $p_1, \cdots, p_i, p_{i+1}$ by p', while if $p_i \longrightarrow p_{i+1}$ factors through p', we replace $p_1, \cdots, p_i$ by p'. In this way we replace inductively the given cycle by a cycle of projectives satisfying the required property. Actually we have obtained a cycle of the following form

$$\cdots \longrightarrow p_s \longrightarrow \cdots \longrightarrow z_s \longrightarrow \cdots \longrightarrow y_s \longrightarrow \cdots \longrightarrow p_{s+1} \longrightarrow \cdots,$$

where we have sectional paths $[p_s, z_s]$ pointing to infinity, $[z_s, y_s]$ pointing to the mouth, and $[y_s, x_s]$(where x_s is a direct predecessor of p_{s+1} on this cycle) pointing to infinity.

For each s, let u_s denote the direct predecessor of z_s on the ray $[y_{s-1}, \infty)$. Then we have a cycle

$$\cdots \longrightarrow u_s \longrightarrow \cdots \longrightarrow y_s \longrightarrow \cdots \longrightarrow u_{s+1} \longrightarrow \cdots,$$

where paths correspond to sectional paths, $[u_s, y_s]$ pointing to the mouth, $[y_s, u_{s+1}]$ pointing to infinity.

We will get contradiction by carrying over the corresponding argument verbatim in the proof of Proposition 4.5 in [1]. This completes the proof. □

REMARK. It follows from the proof that all projectives in Γ lie above some cyclical path, and consequently Γ has only finitely many projectives.

3 THE PROOF OF THEOREM A

Firstly, we introduce several kinds of cancellations in a translation quiver $\Gamma = (\Gamma_0, \Gamma_1, \tau)$. Let $k(\Gamma)$ denote its mesh-category.

1). Let p be an ordinary projective vertex in Γ. If $\mathrm{SuppHom}_{k(\Gamma)}(p, -)$ consists of the vertex x'_j and z_{ij} of a mesh-complete translation sub-quiver of Γ of the form

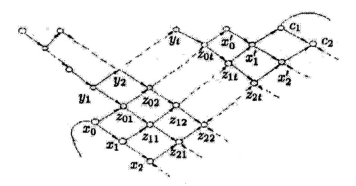

Let $\mathcal{R}$ be the set of the vertices in Γ lying on the sectional paths from the mouth to infinity passing through $z_{01}, z_{02}, \cdots, z_{0t}$ and x'_0. Let Γ' be the translation quiver obtained from Γ by deleting $\mathrm{SuppHom}_{k(\Gamma)}(p, -)$ and replacing the sectional paths $x_i \longrightarrow z_{i1} \longrightarrow \cdots \longrightarrow z_{it} \longrightarrow x'_i \longrightarrow c_i$ (if they exist) by arrows $x_i \longrightarrow c_i$, $i \geq 1$, we define $\tau'c_i = x_{i-1}$ for $i = 1, 2, \cdots$, and for the other vertex x in Γ' we define $\tau'x = \tau x$. We then say that Γ' is obtained from Γ by $\alpha(t+1)-$ cancellation at the vertex p.

2). Let p be a projective-injective vertex in Γ. If $\mathrm{SuppHom}_{k(\Gamma)}(p, -)$ consists of the vertices x'_j and z_{ij} of a mesh-complete translation sub-quiver of the form.

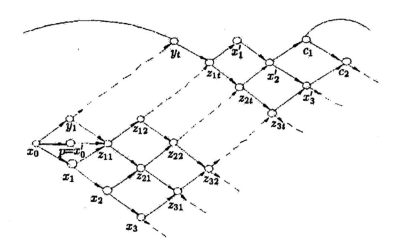

Denote by Γ' the translation quiver obtained by deleting $\mathrm{SuppHom}_{k(\Gamma)}(p, -)$ and replacing the sectional paths $x_i \longrightarrow \cdots \longrightarrow c_{i-1}$ (If they exist) by arrows $x_i \longrightarrow c_{i-1}$ for $i \geq 2$. Define $\tau'c_i = x_i$ for $i \geq 2$, x_0 is injective and $\tau'x = \tau x$ for the other x in Γ'. We then say that Γ' is obtained from Γ by $\beta(t+1)$-cancellation at the vertex p.

3). Let p be exceptional projective but not injective in Γ. If $\mathrm{SuppHom}_{k(\Gamma)}(p, -)$ consists of the vertices x'_i and z_{ij} of a mesh-complete translation sub-quiver of Γ of the form

Coherent Components of Auslander-Reiten Quivers

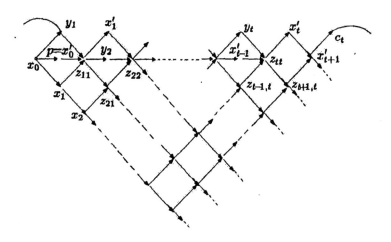

if t is odd, or

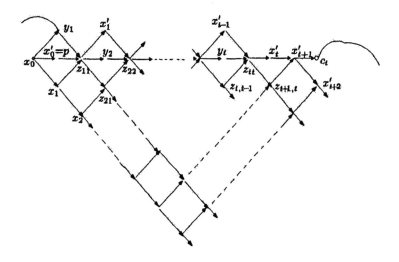

if t is even.

Denote by $\mathcal{R}$ the set of vertices in Γ of the forms $x'_i, i \geq 0, z_{ij}, i \geq 1, 1 \leq j \leq t$ and by Γ' the translation quiver obtained from Γ by deleting $\mathcal{R}$, and replacing the sectional paths $x_i \longrightarrow \cdots \longrightarrow y_{i+1}$ by arrows $x_i \longrightarrow y_{i+1}$ ($0 \leq i \leq t-1$). The sectional paths $y_i \longrightarrow z_{ii} \longrightarrow y_{i+1}$ by arrows $y_i \longrightarrow y_{i+1}$ ($1 \leq i \leq t-1$) and the sectional paths $x_i \longrightarrow \cdots \longrightarrow x'_i \longrightarrow c_{i-1}$ (if they exist) by arrows $x_i \longrightarrow c_{i-1}, i \geq t+1$. Hence Γ' is of the right form. We then say that Γ' is obtained from Γ by $\gamma(t+1)$-cancellation at the vertex p.

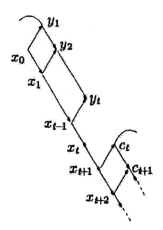

REMARK. i). The dual of $\alpha(t+1)$-cancellation, $\beta(t+1)$-cancellation and $\gamma(t+1)$-cancellation will be called $\alpha^*(t+1)$-cancellation and $\beta^*(t+1)$-cancellation and $\gamma^*(t+1)$-cancellation, respectively.

ii). Assume that there is a length function l in a translation quiver Γ, let Γ' be a translation quiver obtained from Γ by one of the above six kinds of cancellations, then if set $l'(x) = l(x)$ for any x in Γ', l' is a length function in Γ'.

For the convenience of statement we introduce the following notations.

Let Γ be a translation quiver, $\text{SuppHom}_{k(\Gamma)}(p, -)$ consist of vertices $x'_i, i \geq 0$ and $z_{ij}, i \geq 0, 1 \leq j \leq t$ of a mesh-complete translation sub-quiver of Γ, if we have the form of Γ as follows,

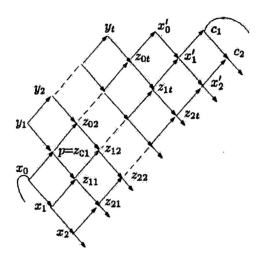

denote by $\mathcal{R}$ the set consisting of points x'_i, $i \geq 0$, $z_{ij}, i \geq 0, 1 \leq j \leq t$. Let Γ' be the translation quiver obtained from Γ by deleting $\mathcal{R}$, and replacing the sectional path $x_i \longrightarrow z_{i1} \longrightarrow \cdots \longrightarrow c_i$ by arrows $x_i \longrightarrow c_i$(if they exist) for $i \geq 1$. We define $\tau' c_i = x_{i-1}$ for $i \geq 1$, and $\tau' x = \tau x$ for any other x in Γ'. We then say that Γ' is obtained from Γ by $\overline{\alpha}(t+1)$-cancellation at p.

In fact, Γ' is obtained from Γ by $(t-1)$ times of $\alpha(1)$-cancellations at $y_t, \cdots, y_2$, successively, and one time of $\alpha(2)$-cancellation at p, finally.

The dual of an $\overline{\alpha}(t+1)$-cancellation is called an $\overline{\alpha}^*(t+1)$-cancellation.

We now set up to prove Theorem A and its corollaries. Corollary B is the immediate consequence of Theorem A.

Proof. of Corollary C. If Γ is a coil, then Γ is coherent by theorem 4.2 in [1]. If each DTr-orbit in Γ is finite, then $\partial\Gamma$ is a stable tube, and so, $\partial\Gamma$ is connected. Therefore, we get corollary C from theorem A. □

In order to prove Theorem A, it is sufficient to prove the following theorem by Lemma 1.1, 1.2 and 1.3.

THEOREM 3.1. *Let Γ be a coherent connected translation quiver with a length function l. If $\partial\Gamma$ is connected, then Γ can be obtained from a stable tube by finitely many times of multiple admissible coray-ray insertions if and only if each τ-orbit in Γ is finite.*

REMARK. The idea of the following proof is similar to 4.6 in [1]. Here we use the method so called 'cancellation'.

Proof. Necessity. From the definition of multiple admissible coray-ray insertions we learn that each τ-orbit in Γ is finite if Γ is obtained from stable tube by only one time of multiple admissible coray-ray insertion. Hence, by induction on the number of multiple admissible coray-ray insertions one can show that each τ-orbit in Γ is finite.

Sufficiency. We have learned that Γ contains only finitely many projective vertices from the remark of Proposition 2.11. So we will show the theorem by induction on the number of projective vertices in Γ.

By Proposition 2.11 there exists a projective vertex $p \in \Gamma_0$ such that $\text{SuppHom}_{k(\Gamma)}(p,-)$ contains no projective vertex. We will consider p in two cases.

Case I. Assume that $\text{SuppHom}_{k(\Gamma)}(-,p)$ contains no projective vertex.

If p is an ordinary projective vertex, consider the sectional path from p pointing to the mouth $p = a_0 \longrightarrow a_1 \longrightarrow \cdots \longrightarrow a_t$ (denoted by L_p^+) with a_t lying on the mouth. Let s be the largest index such that there exists a projective vertex $p' \in \Gamma_0$ and a sectional path $[p', a_s]$ pointing to infinity.

Obviously we can choose p' so that its successors in $[p', a_s]$ are not projective. Observe that $\text{SuppHom}_{k(\Gamma)}(p',-)$ contains no projective vertex. Indeed, by the definition of s, no projective vertex lies on a sectional path pointing to infinity and passing through a_r, $s < r \leq t$. Moreover, by the assumption on p, the sectional path $[a_t, \infty)$ contains no direct predecessor of a projective vertex.

Also, p' is ordinary. In fact, if $p \neq p'$, then $\text{Hom}_{k(\Gamma)}(p, p') = 0$. Thus p' lies above the sectional path $[p, a_t]$ (denoted as L_P^+). On the other hand, there exists a maximal sectional path (denoted by L_p^-) from mouth to the projective $p = a_0$. By Lemma 2.10, any vertex in $\text{SuppHom}_{k(\Gamma)}(p, -)$ is on one of rays starting at a point on $[p, a_t]$. Similar to the proof of Proposition 2.11 one can show that

SuppHom$_{k(\Gamma)}(p,-)$ does not contain an exceptional mesh. Thus, if p' is exceptional, the injective vertex I' in the τ-orbit of $p' = b_0$ must be above the sectional path L_p^+. Since $p = a_0$ is ordinary and $\partial\Gamma$ is a connected sub-quiver of Γ, the vertices (roughly speaking, i.e. vertices between the paths L_p^- and L_p^+) on the sectional paths ending in the vertices a_i and pointing to infinity are not τ-periodic for $i = 0, 1, \cdots, t$. So, the sectional path ending in the injective vertex I' must be finite. This contradicts the condition that Γ is coherent. Thus, p' is ordinary.

Therefore, Γ contains a mesh-complete translation sub-quiver of the form:

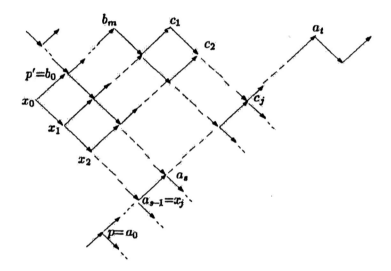

As each τ-orbit in Γ is finite, each τ-orbit of a projective vertex contains an injective vertex and each τ-orbit of an injective vertex contains a projective vertex. Thus, by Lemma 2.9, we have projective vertices $b_0, b_1, \cdots, b_m$ and injective vertices $u_0, u_1, \cdots, u_r$ such that $n(L_{u_0}^+) + \cdots + n(L_{u_r}^+) = n(L_{b_0}^-) + \cdots + n(L_{b_m}^-)$, where $n(L_{u_i}^+)$ and $n(L_{b_j}^-)$ are the numbers of vertices on $L_{u_i}^+$ and $L_{b_j}^-$ for $0 \leq i \leq r$ and $0 \leq j \leq m$, respectively. Here $L_{u_i}^+$ is a maximal sectional path starting at u_i and pointing to the mouth and each vertex in $L_{u_i}^+$ is injective for $i = 0, \cdots, r$ and $L_{b_j}^-$ is a maximal sectional path from the mouth to b_j, and each vertex on $L_{b_j}^-$ is projective.

Without loss of generality we may assume $r = 0$ and $m = 0$ i.e. $n(L_{u_0}^+) = n(L_{b_0}^-)$, denoted by n_0 (if $r \neq 0$ or $m \neq 0$, the proof is similar as in case $r = 0$ and $m = 0$, but the statement will be tediously long). So, we can make an $\alpha(n_0)$-cancellation at $p' = b_0$ to obtain a translation quiver of $\Gamma' = (\Gamma_0', \Gamma_1', \tau')$, which is, by the definition of $\alpha(n_0)$-cancellation, left stable except for finitely many τ'-orbits containing both projectives and injectives.

Then, we make an $\alpha^*(n_0)$-cancellation at u_0 in Γ' to obtain a translation quiver $\Gamma'' = (\Gamma_0'', \Gamma_1'', \tau'')$, which is stable, except for finitely many τ''-orbits containing both projectives and injectives, by the definition of $\alpha^*(n_0)$-insertion and $\alpha(n_0)$-insertion. So, there must exist the following sub-quiver in Γ

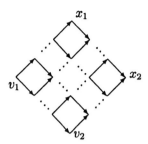

where v_i and x_i belong to Γ_0'' for $i = 1, 2$ and there is a positive integer r such that $\tau^r x_2 = v_1$, and any other vertices in the above sub-quiver lie in $\Gamma_0 \backslash \Gamma_0''$. Moreover, v_1 and x_1 are τ-periodic. So, we have a sub-quiver in Γ'' as follows

and $\tau'' x_2 = v_1$.

Hence, v_1 and x_2 are τ''-periodic in Γ'' since v_1 and x_2 are τ-periodic in Γ.(In fact, let the τ-period of x_2 is n, then $\tau^{n-r} v_1 = x_2$. So, $(\tau'')^{n-r+1} x_2 = (\tau'')^{n-r} v_1 = \tau^{n-r} v_1 = x_2$) Thus, from Happel-Preiser-Ringel's Theorem [3] and the condition that $\partial \Gamma$ is connected we learn that each τ''-orbit in Γ'' is finite. Obviously, Γ'' is coherent by the definition of $\alpha(n_0)$-insertion and $\alpha^*(n_0)$-insertion. It is also evident that Γ'' has at least one projective less than Γ and still there is a length function l'' in Γ''.

If p is exceptional we have two cases to consider.

(i). Suppose that p is injective, then $\text{SuppHom}_{k(\Gamma)}(p, -)$ and $\text{SuppHom}_{k(\Gamma)}(-, p)$ are mesh-complete translation sub-quivers of Γ of the following form.

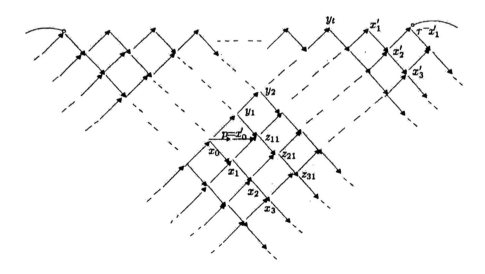

So, we make a $\beta(n_0)$-cancellation at $p = x'_0$ in Γ to obtain a translation quiver $\Gamma' = (\Gamma'_0, \Gamma'_1, \tau')$ where $n_0 = t + 1$. Thus, Γ' is left stable except for finitely many τ'-orbits containing both projectives and injectives.

Again we make an $\alpha^*(n_0)$-cancellation at x_1 in Γ' to obtain a translation quiver $\Gamma'' = (\Gamma''_0, \Gamma''_1, \tau'')$ where $n_0 = t + 1$. Thus, Γ'' is stable except for finitely many τ''- orbits containing both projectives and injectives, by the definition of $\alpha(n_0)$−insertion and $\beta^*(n_0)$−insertion.

Similarly, there must exist the following sub-quiver in Γ

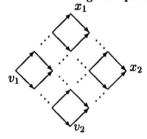

where v_i and x_i belong to Γ''_0 for $i = 1, 2$, and there is a positive integer r such that $\tau^r x_2 = v_1$, and any other vertices in the sub-quiver lie in $\Gamma_0 \backslash \Gamma''_0$. Moreover, v_1 and x_2 are τ-periodic. So, we have sub-quiver in Γ'' as follows

and $\tau'' x_2 = v_1$.

Hence, v_1 and x_2 are τ''-periodic in Γ'' since they are τ-periodic in Γ. Then, by Happel-Preiser-Ringel's Theorem and the condition that $\partial \Gamma$ is connected we learn that each τ''-orbit in Γ'' is finite. Obviously, Γ'' is coherent by the definitions of $\alpha^*(n_0)$- and $\beta(n_0)$-insertions. Γ'' has at least one projective less than Γ, and still there is a length function l'' in Γ''.

(ii) Suppose that p is not injective, then by the Lemmas 2.1–2.4 and 2.7–2.8 , we infer that Γ has the following translation sub-quivers.

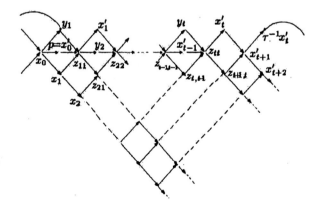

Figure 3.1

if t is odd, or

Coherent Components of Auslander-Reiten Quivers

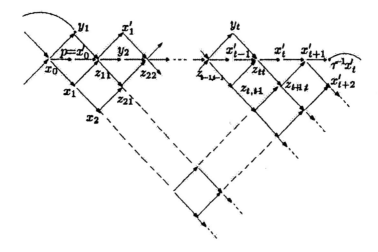

Figure 3.2

if t is even.

We make a $\gamma(n_0)$-cancellation, where $n_0 = t + 1$, at the vertex x_0 in Γ to obtain a translation quiver $\Gamma' = (\Gamma'_0, \Gamma'_1, \tau')$. Hence, Γ' is of the right form. By the definition of $\gamma(n_0)$-insertion we know that Γ' is left stable except for finitely many τ'-orbits containing both projectives and injectives.

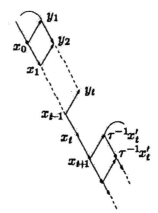

We again make an $\overline{\alpha}^*(n_0)$-cancellation, where $n_0 = t + 1$, at the vertex x_t to obtain a translation quiver $\Gamma'' = (\Gamma''_0, \Gamma''_1, \tau'')$, which is stable except for finitely many τ''-orbits containing both projectives and injectives.

Similar to the above, each τ''-orbit in Γ'' is finite and there is a length function l'' in Γ''. Obviously, Γ'' is coherent, and Γ'' has at least one projective less than Γ.

Case II. There is no projective p such that $\mathrm{SuppHom}_{k(\Gamma)}(-, p)$ contains no projective. So we assume that there are projectives $p_1, p_2, \cdots, p_m$ such that $p_i \in \mathrm{SuppHom}_{k(\Gamma)}(p_{i-1}, -)$ for $i = 2, \cdots, m$ and both $\mathrm{SuppHom}_{k(\Gamma)}(-, p_1)$ and $\mathrm{SuppHom}_{k(\Gamma)}(p_m, -)$ contain no projective.

We make one of the following cancellations at the vertex p_m in Γ.

(1). If p_m is ordinary, we consider the sectional path from p_m to the mouth $p_m = a_0 \longrightarrow a_1 \longrightarrow \cdots \longrightarrow a_t$, with a_t lying on the mouth. Because of the Case I and without loss of generality we may assume that no projective lies on the sectional path starting from the mouth and ending in $a_i (i = 1, 2, \cdots, t)$.

Let $n_m = n(L^-_{p_m})$, where $L^-_{p_m}$ is a sectional path from the mouth to the vertex p_m. Since each τ-orbit in Γ is finite and without loss of generality, we may assume that there exists an injective I_m such that $n_m = n(L^+_{I_m})$ where $L^+_{I_m}$ is a sectional path from I_m to the mouth, and $n(L^+_{I_m})$ (or $n(L^-_{p_m})$) is the number of vertices on the path $L^+_{I_m}$ (or $L^-_{p_m}$). Hence we make an $\alpha(n_m)$–cancellation at the vertex p_m to obtain a translation quiver $\Gamma^{(1)} = (\Gamma^{(1)}_0, \Gamma^{(1)}_1, \tau^{(1)})$, which is left stable except for finitely many $\tau^{(1)}$-orbits containing both projectives and injectives, by the definition of $\alpha(n_m)$-insertion.

(2). If p_m is exceptional and injective, without loss of generality, we may assume that no projective lies on the sectional paths starting from the mouth and passing through the vertices on $L^+_{s(p_m)}$, where $s(p_m)$ is the unique direct successor of p_m.

So, we make a $\beta(n_m)$–cancellation at the vertex p_m, where $n_m = n(L^+_{s(p_m)})$, to obtain a translation quiver $\Gamma^{(1)} = (\Gamma^{(1)}_0, \Gamma^{(1)}_1, \tau^{(1)})$. From the definition of $\beta(n_m)$–insertion we learn that $\Gamma^{(1)}$ is left $\tau^{(1)}$-stable except for finitely many $\tau^{(1)}$-orbits containing both projectives and injectives.

(3). If p_m is exceptional but not injective then by Lemma 2.1-2.4 and 2.7-2.8 we infer that Γ has a translation quiver as in the Figure 3.1 and Figure 3.2 in the Case I. So we make a $\gamma(n_m)$-cancellation at the vertex x_0 in Γ to obtain a translation quiver $\Gamma^{(1)} = (\Gamma^{(1)}_0, \Gamma^{(1)}_1, \tau^{(1)})$, which is left $\tau^{(1)}$-stable, except for finitely many $\tau^{(1)}$-orbits containing both projectives and injectives, by the definition of $\gamma(n_m)$-insertion where $n_m = t + 1$.

From the definition of $\alpha(n_m)$- and $\beta(n_m)$- and $\gamma(n_m)$-cancellation we know that $p_2, \cdots, p_m$ are still projectives in $\Gamma^{(1)}$ and they satisfy the property that $p_i \in \mathrm{SuppHom}_{k(\Gamma^{(1)})}(p_i, -)$ for $i = 2, \cdots, m-1$, and $\mathrm{SuppHom}_{k(\Gamma^{(1)})}(-, p_1)$ contains no projective and $\mathrm{SuppHom}_{k(\Gamma^{(1)})}(p_{m-1}, -)$ contains no projective.

Inductively we make a corresponding cancellation at the vertex p_i in the translation quiver $\Gamma^{(i-1)}$ according as p_i is ordinary or exceptional.

So, at the m-th step we obtain a translation quiver $\Gamma^{(m)} = (\Gamma^{(m)}_0, \Gamma^{(m)}_1, \tau^{(m)})$ which is left $\tau^{(m)}$-stable except for finitely many $\tau^{(m)}$-orbits containing both projectives and injectives. There is a length function $l^{(m)}$ in $\Gamma^{(m)}$ and moreover $\Gamma^{(m)}$ is coherent.

We will make one of the following cancellations in $\Gamma^{(m)}$.

(1). If p_1 is ordinary in Γ then there are injectives $I_1, I_2, \cdots, I_u$ such that $n(L^+_{I_1}) + \cdots + n(L^+_{I_u}) = n(L^-_{p_1}) + \cdots + n(L^-_{p_r})$ since each τ-orbit is finite in Γ. Without loss of generality we may assume $u = 1$ and $r = 1$. So we have $n(L^-_{p_1}) = n(L^+_{I_1})$. Thus, $n_1 = n(L^+_{p_1})$. As we have made an $\alpha(n_1)$-cancellation in $\Gamma^{(m-1)}$ to obtain $\Gamma^{(m)}$, we have a translation sub-quiver in the following form in $\Gamma^{(m)}$

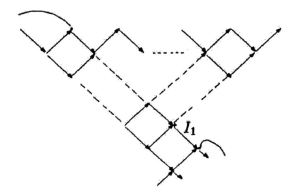

Figure 3.3

Thus we make an $\alpha^*(n_1)$-cancellation at I_1 in $\Gamma^{(m)}$ to obtain a translation quiver $\partial\Gamma^{(m)} = (\partial\Gamma_0^{(m)}, \partial\Gamma_1^{(m)}, \partial\tau^{(m)})$. Obviously $\partial\Gamma^{(m)}$ is coherent and there is a length function $l^{(m)}$ in $\partial\Gamma^{(m)}$.

(2). If p_1 is exceptional and injective, we also have the translation sub-quiver as Figure 3.3. So we also make an $\alpha^*(n_1)$-cancellation at the vertex I_1 to obtain a translation quiver $\partial\Gamma^{(m)} = (\partial\Gamma_0^{(m)}, \partial\Gamma_1^{(m)}, \partial\tau^{(m)})$. Obviously $\partial\Gamma^{(m)}$ is coherent and there is a length function $\partial l^{(m)}$ in $\partial\Gamma^{(m)}$.

(3). If p_1 is exceptional and not injective, there is a translation sub-quiver in the form as in Figure 3.1 and Figure 3.2 in Case I. So we have the following translation sub-quiver in $\Gamma^{(m)}$.

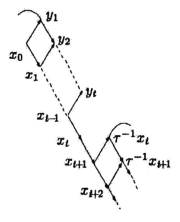

So we make $\overline{\alpha}^*(n_1)$-cancellation in $\Gamma^{(m)} = (\partial\Gamma_0^{(m)}, \partial\Gamma_1^{(m)}, \partial\tau^{(m)})$ to obtain a translation quiver $\partial\Gamma^{(m)}$. Obviously, $\partial\Gamma^{(m)}$ is coherent, and there is a length function $\partial l^{(m)}$ in $\partial\Gamma^{(m)}$. Inductively, we make corresponding cancellation in $\partial^{(i-1)}\Gamma^{(m)}$ according as p_i is ordinary or exceptional in Γ to obtain a translation quiver $\partial^{(i)}\Gamma^{(m)}$, and finally at the m-th step we obtain a translation quiver $\partial^{(m)}\Gamma^{(m)}$, which is coherent and has a length function $\partial^{(m)}l^{(m)}$. Furthermore we know that $\partial^{(m)}\Gamma^{(m)}$ is $\partial^{(m)}$-stable except for finitely many $\tau^{(m)}$- orbits containing both projectives and injectives by the definitions of $\alpha(n_i)$-cancellation and $\alpha^*(n_i)$-cancellation, $\beta(n_i)$-cancellation and $\alpha^*(n_i)$-cancellation as well as $\gamma(n_i)$-cancellation and $\overline{\alpha}^*(n_i)$-cancellation.

It is easy to see that there is at least one $\partial^{(m)}\tau^{(m)}$-periodic vertex in $\partial^{(m)}(\Gamma^{(m)})$. Thus, each $\partial^{(m)}\tau^{(m)}$-orbit is finite by applying Happel-Preiser-Ringel's theorem to the translation quiver $\partial^{(m)}\Gamma^{(m)}$. If we still write $\partial^{(m)}\Gamma^{(m)}$ as Γ', then Γ' has at least one projective less than Γ. By the definitions of cancellations we know that $\partial\Gamma'$ obtained by deleting τ'-orbits of projectives is a connected translation subquiver of Γ'. So, by the assumption of induction Γ' is obtained from a stable tube by finitely many multiple admissible coray-ray insertions. Since Γ can be obtained from Γ' by one time of multiple admissible coray-ray insertion. We complete the proof. □

4 AN EXAMPLE

In this section we will give an example of a finite-dimensional algebra A over an algebraically closed field which has a connected component satisfying the conditions in the main theorem.

Let A be given by the quiver

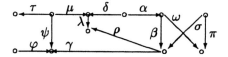

bound by $\alpha\omega = 0$, $\sigma\gamma = 0$, $\mu\lambda = 0$, $\sigma\rho = 0$ and $\delta\lambda = \alpha\beta\rho$. Then Γ_A has a connected component as follows:

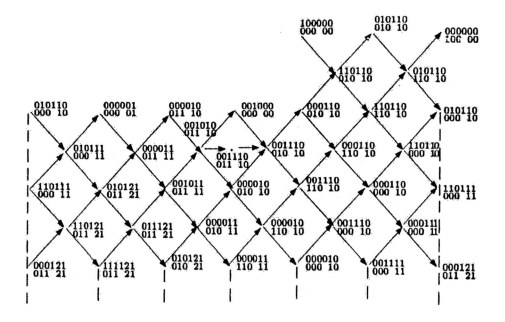

where indecomposable modules are represented by their dimension-vectors and one identifies along the dash lines.

ACKNOWLEDGEMENT

The author would like to express his gratitude to the University of Sherbrooke for her hospitality during his visit. The author would also like to thank professor I.Assem for his many useful suggestions and discussions. Finally,the author would like to thank the referee for the comments. This work is partially supported by Beijing Youth Fund.

REFERENCES

[1] I. Assem and A. Skowroński, Multi-coil algebras, Proceedings of ICRA VI, Canadian Math. Soc. Conference Proceedings, 14(1993), 29–68

[2] G. D'Este and C. M. Ringel, Coherent tubes, J. Algebra 87(1984), 150–201

[3] D. Happel, U. Preiser and C. M. Ringel, Vinberg's characterization of Dynkin diagrams using subadditive functions with application to DTr- periodic modules. Lecture Notes in Mathematics 832 Springer, Berlin, 1980, 280–294.

[4] S. Liu, Semi-stable components of an Auslander-Reiten quiver. J. London Math. Soc. 47(1993), 405–416

[5] S. Liu, The connected components of the Auslander-Reiten quiver of a tilted algebra, J. Algebra, (2)161(1995), 505-523

[6] S. Liu, Shapes of Connected components of the Auslander-Reiten quivers of Artin Algebras. C. M. S. Conference Proceedings, Vol. 19(1996) 109-137

[7] P. Malicki , Generalized coil enlargements of algebras, Colloq. Math. Vol. 76, No. 1(1998) 57-63

[8] P. Malicki and A. Skowroński, Almost cyclic coherent components of an Auslander-Reiten quiver, Preprint

[9] C. M. Ringel, Tame algebras and integral quadratic forms, Lecture Notes in Mathematics 1099, Springer , Berlin, 1984

[10] A. Skowroński, Cycles in module categories, Finite Dimensional Algebras and Related Topics, NATO ASI Series, Series C (Kluwer Academic Publishers), 424(1994), 309-346

[11] H. Yao, Infinite connected components of an Auslander-Reiten quiver in which each DTr- orbit contains only finitely many points, comm. Algebra, 1999, Vol 27 (11),5167-5189

[12] Y. Zhang, The structure of stable components, Can. J. Math. 43(1991),652-672.

Twisted Hopf algebras

PU ZHANG Department of Mathematics, University of Science and Technology of China, Hefei 230026, P R China, E-mail: pzhang@edu.cn

LI-BIN LI Department of Mathematics, University of Science and Technology of China, Hefei 230026, P R China, E-mail: libinli@ustc.edu.cn

ABSTRACT The aim of this paper is to introduce the concept of a twisted Hopf algebra, and then to discuss some properties and to give some constructions of twisted Hopf algebras. Such a twisted Hopf algebra turns out to rise naturally, for example, the positive part U_+ of a quantized enveloping algebra U, and the Ringel-Hall algebras.

INTRODUCTION

Throughout this paper, let K be a field, c a non-zero element in K, and I a set. Denote by $\mathbb{Z}I$ the free abelian group with I as basis, whose element is written as $x = (x_i)_{i \in I}$ with $x_i \in \mathbb{Z}$ and $x_i = 0$ for almost all $i \in I$, and by $\mathbb{N}_0 I$ the subset $\{ x = (x_i)_{i \in I} \in \mathbb{Z}I \mid x_i \in \mathbb{N}_0 \}$. Let (χ_1, χ_2) be a pair of $\mathbb{Z}$-valued bilinear forms on $\mathbb{Z}I$.

The aim of this paper is to introduce the concept of a $(K, c, I, (\chi_1, \chi_2))$-Hopf algebra (or simply, a twisted Hopf algebra), and then to discuss its properties and constructions.

It is well known that a quantized enveloping algebra $U = U_+ \otimes U_0 \otimes U_-$ has a Hopf structure, and that the Hopf operations for U are not closed inside the positive part U_+ (see e.g. [J]), so we naturally hope that inside U_+ there is a "nearly" Hopf algebra structure, this is a motivation of introducing a twisted Hopf algebra. This is particularly natural when we use Ringel-Hall algebras to study U_+ (cf. [R2], [G]).

The authors gratefully acknowledge the support of K. C. Wong Education Foundation, Hong Kong.

The axiomatic difference between a twisted Hopf algebra from a Hopf algebra lies on the axiom of the comultiplication, i.e., the comultiplication $\delta : A \longrightarrow A \otimes A$ of a twisted Hopf algebra A is an algebra homomorphism, not for the componentwise multiplication on $A \otimes A$, but for the multiplication on $A \otimes A$ given by a twisted rule via a pair (χ_1, χ_2) of bilinear forms. This idea of a twisted multiplication should not be considered as a disadvantage, on the contrary, it can be used for any algebra and any coalgebra, as Lusztig ([L]) and Ringel ([R1, R2]) did, see also [G]. Moreover, as we will show in Example 2.6, for any datum k, c, I, χ_1, χ_2, there always exists a $(K, c, I, (\chi_1, \chi_2))$-Hopf algebra. It is proved in Theorem 2.3 that any $(K, c, I, (\chi_1, \chi_2))$-bialgebra is always a $(K, c, I, (\chi_1, \chi_2))$-Hopf algebra. In Theorem 2.4 we point out that the antipode s of a $(K, c, I, (\chi, 0))$-Hopf algebra A gives an $\mathbb{N}_0 I$-graded algebra anti-isomorphism $s : A \longrightarrow A_{\chi^T}$ and an $\mathbb{N}_0 I$-graded coalgebra anti-isomorphism $s : A_\chi \longrightarrow A$. Given a twisted Hopf algebra, by twisting its multiplication or comultiplication, we construct some new twisted Hopf algebras and discuss some relations of their antipodes in §3.

Throughout this paper, let $\chi : \mathbb{Z}I \otimes \mathbb{Z}I \longrightarrow \mathbb{Z}$ be a bilinear form. Such a bilinear form is not symmetric if no otherwise is stated. By χ^T we denote the $\mathbb{Z}$-valued bilinear form on $\mathbb{Z}$ given by $\chi^T(x, y) = \chi(y, x)$.

1 TWISTING AN ALGEBRA AND A COALGEBRA

1.1. Consider an $\mathbb{N}_0 I$-graded K-algebra $A = (A, m, e)$, i.e., $A = \bigoplus_{x \in \mathbb{N}_0 I} A_x$ with $A_0 = K$ is a direct decomposition of K-spaces, such that $A_x A_y \subseteq A_{x+y}$.

If $a \in A_x$, then we denote by x by $|a|$. Define a new multiplication $*$ on A by

$$a * b = c^{\chi(|a|, |b|)} ab \tag{1}$$

for homogeneous elements a, b in A. Denote this multiplication map by m_χ. Then we have

LEMMA. *([R1])* $A_\chi = (A, m_\chi, e)$ *is again an $\mathbb{N}_0 I$-graded K-algebra.*

1.2. By definition (see e.g. [R3], p.206), an $\mathbb{N}_0 I$-graded K-coalgebra $A = (A, \delta, \varepsilon)$ is a direct decomposition of K-spaces $A = \bigoplus_{x \in \mathbb{N}_0 I} A_x$ with $A_0 = K$, such that

(i) there is a K-linear map $\delta : A \longrightarrow A \otimes A$ which is coassociative, i.e. $(id \otimes \delta)\delta = (\delta \otimes id)\delta$;

(ii) the projection ε from A onto $A_0 = K$ is a counit, i.e. $(id \otimes \varepsilon)\delta = id = (\varepsilon \otimes id)\delta$;

(iii) δ respects the grading, i.e. $\delta(A_d) \subseteq \bigoplus_{x+y=d} A_x \otimes A_y$.

We need the following easy fact (see also [LZ])

LEMMA. *Let $A = (A, \delta, \varepsilon)$ be an $\mathbb{N}_0 I$-graded K-coalgebra. Then*

(i) $\delta(1) = 1 \otimes 1$.

(ii) For $a \in A_d$ with $0 \neq d \in \mathbb{N}_0 I$ we have

$$\delta(a) = a \otimes 1 + 1 \otimes a + \sum_{x+y=d; x, y \neq 0} a_x \otimes b_y$$

where $a_x \in A_x, b_y \in A_y$.

In particular, we have $\delta(a) = a \otimes 1 + 1 \otimes a$ for $a \in A_i$, $i \in I$.

1.3. Let $A = (A, \delta, \varepsilon)$ be an $\mathbb{N}_0 I$-graded K-coalgebra. Define a new K-linear map $\delta_\chi : A \longrightarrow A \otimes A$ by

$$\delta_\chi(a) = \sum c^{\chi(|a_1|,|a_2|)}(a_1 \otimes a_2) \qquad (2)$$

where $\delta(a) = \sum a_1 \otimes a_2$ is Sweedler's notation with all factors a_1, a_2 homogeneous. By a direct verification we have the following (see also [LZ])

LEMMA. $A_\chi = (A, \delta_\chi, \varepsilon)$ is again an $\mathbb{N}_0 I$-graded K-coalgebra.

Note that the construction of δ_χ has been introduced by Lusztig for $'\mathbf{f}$ and $\mathbf{f}$ in [L], p.6.

LEMMA 1.4. Let $\chi : \mathbb{Z}I \times \mathbb{Z}I \longrightarrow \mathbb{Z}$ be a bilinear form, and $A = (A, \delta, \varepsilon)$ be an $\mathbb{N}_0 I$-graded coalgebra. Then $_\chi A = (A, {}_\chi\delta, \varepsilon)$ is again an $\mathbb{N}_0 I$-graded coalgebra, if the K-linear map $_\chi\delta: A \longrightarrow A \otimes A$ is defined by

$$_\chi\delta(a) = \sum c^{-\chi(|a_1|,|a_2|)} (a_2 \otimes a_1), \qquad (3)$$

where $\delta(a) = \sum a_1 \otimes a_2$ with all factors homogeneous.

Proof. Note that $_\chi\delta$ is exactly $T\delta_{-\chi}$, where T is the twisted map given by $T(a \otimes b) = b \otimes a$. Thus, the assertion follows from Lemma 1.3 and the fact that if (A, δ, ε) is an $\mathbb{N}_0 I$-graded coalgebra, then so is $(A, T\delta, \varepsilon)$. □

1.5. Let A be an $\mathbb{N}_0 I$-graded algebra. Then $A \otimes A$ is an $(\mathbb{N}_0 I)^2$-graded algebra with componentwise multiplication, where $(A \otimes A)_{(x,y)} = A_x \otimes A_y$ for $x, y \in \mathbb{N}_0 I$.

Note that we can identify $(\mathbb{Z}I)^2 = \mathbb{Z}I \oplus \mathbb{Z}I$ with the free abelian group with I' as basis, and $(\mathbb{N}_0 I)^2 = \mathbb{N}_0 I'$, where I' is a set with $|I'| = 2|I|$. Thus, by Lemma 1.1, in order to twist the tensor algebra $A \otimes A$, one needs to have a $\mathbb{Z}$-valued bilinear form χ on $(\mathbb{Z}I)^2$, i.e., a map $\chi : (\mathbb{Z}I)^4 \longrightarrow \mathbb{Z}$ satisfying

$$\chi(x_1 + x'_1, x_2 + x'_2, y_1, y_2) = \chi(x_1, x_2, y_1, y_2) + \chi(x'_1, x'_2, y_1, y_2) \qquad (4)$$

and

$$\chi(x_1, x_2, y_1 + y'_1, y_2 + y'_2) = \chi(x_1, x_2, y_1, y_2) + \chi(x_1, x_2, y'_1, y'_2). \qquad (5)$$

However, in order to deal with a twisted bialgebra and a twisted Hopf algebra, this is not enough. As observed by Ringel ([R3], p. 227), one should consider a bilinear form on $(\mathbb{Z}I)^2$ given by a pair $\chi = (\chi_1, \chi_2)$ of bilinear forms on $\mathbb{Z}I$, again denoted by χ, i.e.,

$$\chi(x_1, x_2, y_1, y_2) = \chi_1(x_2, y_1) + \chi_2(x_1, y_2). \qquad (6)$$

Such a map $\chi : (\mathbb{Z}I)^4 \longrightarrow \mathbb{Z}$ satisfies not only (4) and (5), but also

$$\chi(x_1 + x'_1, x_2, y_1 + y'_1, y_2) = \chi(x_1, x_2, y_1, y_2) + \chi(x'_1, x_2, y'_1, y_2) \qquad (7)$$

and
$$\chi(x_1, x_2 + x_2', y_1, y_2 + y_2') = \chi(x_1, x_2, y_1, y_2) + \chi(x_1, x_2', y_1, y_2'). \tag{8}$$

Moreover, any map $\chi : (\mathbb{Z}I)^4 \longrightarrow \mathbb{Z}$ with properties (4), (5), (7), (8) is given by (6).

The reason we need not only (4) and (5), but also (7) and (8), is that they are really used in Example 2.6, which guarantees the existence of a twisted Hopf algebra.

1.6. Let A be an $\mathbb{N}_0 I$-graded algebra, and $\chi = (\chi_1, \chi_2)$ be a pair of bilinear forms on $\mathbb{Z}I$. Using (6) and applying Lemma 1.1 to the tensor algebra $A \otimes A$, we obtain the $(\mathbb{N}_0 I)^2$-graded algebra $(A \otimes A)_\chi$ with multiplication $*$ given by

$$(a_1 \otimes a_2) * (b_1 \otimes b_2) = c^{\chi_1(|a_2|,|b_1|) + \chi_2(|a_1|,|b_2|)}(a_1 b_1 \otimes a_2 b_2) \tag{9}$$

for homogeneous elements $a_1, a_2, b_1, b_2 \in A$.

1.7. Dually, let $A = (A, \delta, \varepsilon)$ be an $\mathbb{N}_0 I$-graded coalgebra. Then $A \otimes A$ is an $(\mathbb{N}_0 I)^2$-graded coalgebra with comultiplication $(id \otimes T \otimes id)(\delta \otimes \delta)$, again denoted by δ if no confusion caused, where T is the twisted map. Thus, if $\delta(a) = \sum a_1 \otimes a_2$ and $\delta(b) = \sum b_1 \otimes b_2$, then

$$\delta(a \otimes b) = \sum a_1 \otimes b_1 \otimes a_2 \otimes b_2.$$

Now using (6) and applying Lemma 1.3 to the tensor coalgebra $A \otimes A$, we obtain the $(\mathbb{N}_0 I)^2$-graded coalgebra $(A \otimes A)_\chi = (A \otimes A, \delta_\chi, \varepsilon \otimes \varepsilon)$ with comultiplication

$$\delta_\chi(a \otimes b) = \sum c^{\chi_1(|b_1|,|a_2|) + \chi_2(|a_1|,|b_2|)}(a_1 \otimes b_1 \otimes a_2 \otimes b_2) \tag{10}$$

where $\delta(a) = \sum a_1 \otimes a_2$ and $\delta(b) = \sum b_1 \otimes b_2$ with all a_1, a_2, b_1, b_2 homogeneous.

1.8. In many situations, we shall choose $\chi_2 = 0$ in the pair $\chi = (\chi_1, \chi_2)$. In this case, for convenience, we replace χ_1 by χ, i.e., let χ be a $\mathbb{Z}$-valued bilinear form on $\mathbb{Z}I$. Thus, the algebra structure on $(A \otimes A)_\chi$ is given by Lusztig's rule

$$(a_1 \otimes a_2) * (b_1 \otimes b_2) = c^{\chi(|a_2|,|b_1|)}(a_1 b_1 \otimes a_2 b_2) \tag{11}$$

for homogeneous elements $a_1, a_2, b_1, b_2 \in A$.

Also, the coalgebra structure on $(A \otimes A)_\chi$ is given by

$$\delta_\chi(a \otimes b) = c^{\chi(|b_1|,|a_2|)}(a_1 \otimes b_1 \otimes a_2 \otimes b_2) \tag{12}$$

where $\delta(a) = \sum a_1 \otimes a_2$ and $\delta(b) = \sum b_1 \otimes b_2$ with all a_1, a_2, b_1, b_2 homogeneous.

2 TWISTED HOPF ALGEBRAS

2.1. Let $A = (A, m, e)$ be an $\mathbb{N}_0 I$-graded algebra and $A = (A, \delta, \varepsilon)$ be an $\mathbb{N}_0 I$-graded coalgebra. Then it is clear by definition that $\varepsilon : A \longrightarrow K$ is an algebra homomorphism, and $e : K \longrightarrow A$ is a coalgebra homomorphism. Let $\chi = (\chi_1, \chi_2)$ be a pair of $\mathbb{Z}$-valued bilinear forms on $\mathbb{Z}I$. Then we have the following fact

Twisted Hopf Algebras

LEMMA. $\delta : A \longrightarrow (A \otimes A)_\chi$ is an algebra homomorphism if and only $m : (A \otimes A)_{(\chi_1^T, \chi_2)} \longrightarrow A$ is a coalgebra homomorphism, where the algebra structure on $(A \otimes A)_\chi$ and the coalgebra structure on $(A \otimes A)_{(\chi_1^T, \chi_2)}$ are given via (9) and (10) in 1.6 and 1.7, respectively.

Proof. It is clear that $\varepsilon m = \varepsilon \otimes \varepsilon$, and $\delta(1) = 1 \otimes 1$. For any $a, b \in A$, let $\delta(a) = \sum a_1 \otimes a_2$, $\delta(b) = \sum b_1 \otimes b_2$, with all factors homogeneous. Now, $\delta : A \longrightarrow (A \otimes A)_\chi$ is an algebra homomorphism if and only if $\delta(ab) = \delta(a) \star \delta(b)$, and if and only if

$$\begin{aligned}\delta(ab) &= \sum c^{\chi_1(|a_2|,|b_1|) + \chi_2(|a_1|,|b_2|)} a_1 b_1 \otimes a_2 b_2 \\ &= \sum c^{\chi_1^T(|b_1|,|a_2|) + \chi_2(|a_1|,|b_2|)} a_1 b_1 \otimes a_2 b_2 \\ &= (m \otimes m)\delta_{(\chi_1^T, \chi_2)}(a \otimes b),\end{aligned}$$

which is exactly the claim that $m : (A \otimes A)_{(\chi_1^T, \chi_2)} \longrightarrow A$ is a coalgebra homomorphism. $\square$

2.2. Let $\chi = (\chi_1, \chi_2)$ be a pair of $\mathbb{Z}$-valued bilinear forms on $\mathbb{Z}I$.

DEFINITION: (i) An $\mathbb{N}_0 I$-graded K-algebra $A = (A, m, e)$ together with an $\mathbb{N}_0 I$-graded K-coalgebra $A = (A, \delta, \varepsilon)$ is called a (K, c, I, χ)–bialgebra, or simply, a χ–bialgebra, provided that $\delta : A \longrightarrow (A \otimes A)_\chi$ is an algebra homomorphism, where the algebra structure on $(A \otimes A)_\chi$ is given by (9) in 1.6.

(ii) A (K, c, I, χ)-bialgebra $A = (A, m, e, \delta, \varepsilon)$ is called a (K, c, I, χ)-Hopf algebra, or simply, a χ-Hopf algebra, provided that there is a K-linear map $s : A \longrightarrow A$ satisfying

$$m(id \otimes s)\delta = e\varepsilon = m(s \otimes id)\delta. \qquad (1)$$

The map s is called an antipode of A.

The notion of a χ–bialgebra has been introduced by Ringel in [R3], and a χ-Hopf algebra has been introduced in [LZ] (but in [LZ] we only consider the case $\chi = (\chi_1, 0)$ with χ_1 symmetric). Notice that, to say $A = (A, m, e, \delta, \varepsilon)$ is a (K, c, I, χ)-bialgebra, is equivalent to say it is a $(K, c^{-1}, I, -\chi)$-bialgebra, where $-\chi = (-\chi_1, -\chi_2)$.

Recall that for any K-algebra $A = (A, m, e)$ and any K-coalgebra $C = (C, \delta, \varepsilon)$, the K-space $\mathrm{Hom}_K(C, A)$ become an K-(associative) algebra with identity $e\varepsilon$ via the convolution $\star$ defined by

$$(f \star g)(x) = m(f \otimes g)\delta(x) \qquad (2)$$

for $f, g \in \mathrm{Hom}_K(C, A)$ and $x \in C$, see [S].

Thus, for a (K, c, I, χ)-bialgebra $(A, m, e, \delta, \varepsilon)$, it is easy to see that existence of an antipode above is equivalent to the existence of a K-map s with the following property in the convolution algebra $\mathrm{Hom}_K(A, A)$

$$s \star id = id \star s = e\varepsilon. \qquad (3)$$

It follows that a (K, c, I, χ)-Hopf algebra has a unique antipode.

We have the following basic property of a twisted bialgebra

THEOREM 2.3. *Let $\chi = (\chi_1, \chi_2)$ be an arbitrary pair of $\mathbb{Z}$-valued bilinear forms on $\mathbb{Z}I$, and $A = (A, m, e, \delta, \varepsilon)$ be a (K, c, I, χ)-bialgebra. Then there is a graded K-map $s : A \longrightarrow A$ such that $A = (A, m, e, \delta, \varepsilon, s)$ is a (K, c, I, χ)-Hopf algebra.*

Proof. Let $A = \bigoplus_{x \in \mathbb{N}_0 I} A_x$. Define a K-map $s_r : A \longrightarrow A$ inductively. Define $s_r(1) = 1$. For $a \in A_d$, $0 \neq d \in \mathbb{N}_0 I$, by Lemma 1.2 we have

$$\delta(a) = a \otimes 1 + 1 \otimes a + \sum_{x+y=d; x, y \neq 0} a_x \otimes b_y$$

where $a_x \in A_x, b_y \in A_y$. By induction $s_r(b_y)$ has been well defined, it follows that we define

$$s_r(a) = -a - \sum_{x+y=d; x, y \neq 0} a_x s_r(b_y).$$

Thus, we have $m(id \otimes s_r)\delta = e\varepsilon$, i.e., s_r is a right inverse of id in the convolution algebra $\mathrm{Hom}_K(A, A)$. Similarly, one has a left inverse s_l of id. It follows that $s_r = s_l = s$, i.e., s is the antipode of A. $\square$

2.4. Now consider an important special case, i.e., $\chi = (\chi_1, 0)$. In this case, we shall use (11) and (12) in 1.8, and we have the following basic property

THEOREM. *Let $\chi : \mathbb{Z}I \times \mathbb{Z}I \longrightarrow \mathbb{Z}$ be a bilinear form. Let $A = (A, m, e, \delta, \varepsilon, s)$ be a (K, c, I, χ)-Hopf algebra. Then we have*

(i) $s : A \longrightarrow A_{\chi^T}$ is an $\mathbb{N}_0 I$-graded algebra anti-isomorphism, where $\chi^T(x, y) = \chi(y, x)$ for $x, y \in \mathbb{N}_0 I$, and the algebra structure on A_{χ^T} is given via (1) in 1.1.

(ii) $s : A_\chi \longrightarrow A$ is an $\mathbb{N}_0 I$-graded coalgebra anti-isomorphism, where the coalgebra structure on A_χ is given by (2) in 1.3.

(iii) $_{\chi^T} A = (A, m, e, {}_{\chi^T}\delta, \varepsilon)$ is a $(K, c, I, -\chi^T)$-Hopf algebra with antipode s^{-1}, where the coalgebra structure on $_{\chi^T} A$ is given by (3) in 1.4.

Proof. By Theorem 2.3 we know that s is graded.

In order to prove (i), we first prove that $s : A \longrightarrow A_{\chi^T}$ is an algebra anti-homomorphism. Since $s(1) = 1$, it suffices to prove $s(ab) = s(b) \star s(a)$ for $a, b \in A$, or, $sm = m_{\chi^T}(s \otimes s)T$, where T is the twisted map and m_{χ^T} is the multiplication in A_{χ^T}.

Since $e(\varepsilon \otimes \varepsilon)$ is the unit of the convolution algebra $\mathrm{Hom}_K((A \otimes A)_{\chi^T}, A)$, it follows that it suffices to prove that there holds $(sm) \star m = m \star (m_{\chi^T}(s \otimes s)T) = e(\varepsilon \otimes \varepsilon)$ in the convolution algebra $\mathrm{Hom}_K((A \otimes A)_{\chi^T}, A)$.

Let $a, b \in A$ be homogeneous, $\delta(a) = \sum a_1 \otimes a_2$, and $\delta(b) = \sum b_1 \otimes b_2$ with all factors homogeneous. Since $\delta : A \longrightarrow (A \otimes A)_\chi$ is an algebra homomorphism, it follows that $\delta(ab) = \sum c^{\chi(|a_2|, |b_1|)}(a_1 b_1 \otimes a_2 b_2)$, and hence by applying (12) in 1.8 we have

$$\begin{aligned}((sm) \star m)(a \otimes b) &= m(sm \otimes m)\delta_{\chi^T}(a \otimes b) \\ &= \sum c^{\chi^T(|b_1|, |a_2|)} s(a_1 b_1) a_2 b_2 \\ &= \sum c^{\chi(|a_2|, |b_1|)} s(a_1 b_1) a_2 b_2 \\ &= m(s \otimes id)\delta(ab) \\ &= \varepsilon(ab) = (e(\varepsilon \otimes \varepsilon))(a \otimes b).\end{aligned}$$

Twisted Hopf Algebras

On the other hand, we have

$$\begin{aligned}(m \star (m_{\chi^T}(s \otimes s)T))(a \otimes b) &= m(m \otimes m_{\chi^T}(s \otimes s)T)\delta_{\chi^T}(a \otimes b) \\ &= \sum c^{\chi^T(|b_1|,|a_2|)} m(m \otimes m_{\chi^T}(s \otimes s)T)(a_1 \otimes b_1 \otimes a_2 \otimes b_2) \\ &= \sum c^{\chi(|a_2|,|b_1|)} a_1 b_1 m_{\chi^T}((s(b_2) \otimes s(a_2))) \\ &= \sum c^{\chi(|a_2|,|b_1|)+\chi^T(|b_2|,|a_2|)} a_1 b_1 s(b_2) s(a_2) \\ &= \sum c^{\chi(|a_2|,|b|)} a_1 b_1 s(b_2) s(a_2) \\ &= \sum c^{\chi(|a_2|,|b|)} a_1 s(a_2) \varepsilon(b),\end{aligned}$$

it follows that if $b \notin A_0$, then $\varepsilon(b) = 0$, and hence

$$(m \star (m_{\chi^T}(s \otimes s)T))(a \otimes b) = 0 = (e(\varepsilon \otimes \varepsilon))(a \otimes b);$$

and if $b \in A_0$, then $\chi(|a_2|,|b|) = 0$, and hence

$$\begin{aligned}(m \star (m_{\chi^T}(s \otimes s)T))(a \otimes b) &= \sum a_1 s(a_2) \varepsilon(b) \\ &= m(id \otimes s)\delta(a)\varepsilon(b) \\ &= \varepsilon(a)\varepsilon(b) \\ &= (e(\varepsilon \otimes \varepsilon))(a \otimes b).\end{aligned}$$

This proves that $s : A \longrightarrow A_{\chi^T}$ is an algebra anti-homomorphism.

In order to prove (ii), we first prove that $s : A_\chi \longrightarrow A$ is a coalgebra anti-homomorphism. It is clear that $\varepsilon s = \varepsilon$. It remains to prove that $\delta s = T(s \otimes s)\delta_\chi$. Since $(e \otimes e)\varepsilon$ is the unit of the convolution algebra $\mathrm{Hom}_K(A, (A \otimes A)_\chi)$, it follows that it suffices to prove $(\delta s) \star \delta = \delta \star (T(s \otimes s)\delta_\chi) = (e \otimes e)\varepsilon$ in the convolution algebra $\mathrm{Hom}_K(A, (A \otimes A)_\chi)$. Let a be a homogeneous element in A. Since δ is an algebra homomorphism from A to $(A \otimes A)_\chi$, it follows that

$$\begin{aligned}((\delta s) \star \delta)(a) &= \sum \delta s(a_1) \star \delta(a_2) \\ &= \sum \delta(s(a_1)a_2) = \delta m(s \otimes id)\delta(a) \\ &= \delta \varepsilon(a) = \varepsilon(a)(1 \otimes 1) \\ &= (e \otimes e)\varepsilon(a).\end{aligned}$$

Let $\delta(a_1) = \sum a_{11} \otimes a_{12}$ and $\delta(a_2) = \sum a_{21} \otimes a_{22}$ with all factors homogeneous. Then

$$|a_{11}| + |a_{12}| = |a_1|, \quad |a_{21}| + |a_{22}| = |a_2|, \quad |a_1| + |a_2| = |a|;$$

and we have

$$\begin{aligned}(\delta \star (T(s \otimes s)\delta_\chi))(a) &= \sum c^{\chi(|a_{21}|,|a_{22}|)}(a_{11} \otimes a_{12}) \star (s(a_{22}) \otimes s(a_{21})) \\ &= \sum c^{\chi(|a_{21}|,|a_{22}|)+\chi(|a_{12}|,|a_{22}|)} a_{11} s(a_{22}) \otimes a_{12} s(a_{21}).\end{aligned}$$

On the other hand, by using the coassociativity of δ, we have

$$\begin{aligned}(\delta \otimes \delta)\delta &= (\delta \otimes id \otimes id)(id \otimes \delta)\delta = (\delta \otimes id \otimes id)(\delta \otimes id)\delta \\ &= (id \otimes \delta \otimes id)(\delta \otimes id)\delta = (id \otimes \delta \otimes id)(id \otimes \delta)\delta,\end{aligned}$$

and hence we have

$$\sum a_{11} \otimes a_{12} \otimes a_{21} \otimes a_{22} = \sum a_1 \otimes a_{211} \otimes a_{212} \otimes a_{22} \tag{4}$$

where $\delta(a_{21}) = \sum a_{211} \otimes a_{212}$ with all factors homogeneous.

Now define a K-linear map $L: A \otimes A \otimes A \otimes A \longrightarrow A \otimes A \otimes A \otimes A$ by

$$L(a_1 \otimes a_2 \otimes a_3 \otimes a_4) = c^{\chi(|a_2|+|a_3|,|a_4|)}(a_1 \otimes a_2 \otimes a_3 \otimes a_4)$$

for all a_1, a_2, a_3, a_4 homogeneous.

By applying the K-linear map $O = (m \otimes id)(id \otimes T)(id \otimes m \otimes id)(id \otimes id \otimes s \otimes s)L$ to the both sides of (4), and applying the definition of antipode, we get

$$\sum c^{\chi(|a_{12}|+|a_{21}|,|a_{22}|)} a_{11} s(a_{22}) \otimes a_{12} s(a_{21})$$
$$= \sum c^{\chi(|a_{211}|+|a_{212}|,|a_{22}|)} a_1 s(a_{22}) \otimes a_{211} s(a_{212})$$
$$= \sum c^{\chi(|a_{21}|,|a_{22}|)} a_1 s(a_{22}) \otimes a_{211} s(a_{212})$$
$$= \sum c^{\chi(|a_2|-|a_{22}|,|a_{22}|)} a_1 s(a_{22}) \otimes a_{211} s(a_{212})$$
$$= \sum c^{\chi(|a_2|-|a_{22}|,|a_{22}|)} a_1 s(a_{22}) \otimes \varepsilon(a_{21})$$
$$= \sum c^{\chi(|a_2|-|a_{22}|,|a_{22}|)} a_1 s(\varepsilon(a_{21}) a_{22}) \otimes 1.$$

We claim that

$$c^{\chi(|a_2|-|a_{22}|,|a_{22}|)} a_1 s(\varepsilon(a_{21}) a_{22}) \otimes 1 = a_1 s(\varepsilon(a_{21}) a_{22}) \otimes 1$$

In fact, if $|a_{21}| \neq 0$, then $\varepsilon(a_{21}) = 0$, and hence the both sides are 0; if $|a_{21}| = 0$, then $|a_2| - |a_{22}| = 0$, and hence the claim follows.

In this way, we see that

$$\sum c^{\chi(|a_{12}|+|a_{21}|,|a_{22}|)} a_{11} s(a_{22}) \otimes a_{12} s(a_{21})$$
$$= \sum c^{\chi(|a_2|-|a_{22}|,|a_{22}|)} a_1 s(\varepsilon(a_{21}) a_{22}) \otimes 1$$
$$= \sum a_1 s(\varepsilon(a_{21}) a_{22}) \otimes 1$$
$$= \sum a_1 s(a_2) \otimes 1$$
$$= \varepsilon(a) \otimes 1$$
$$= (e \otimes e)\varepsilon(a),$$

where we have used the property of the antipode and the counit property.

Altogether, we have proved that $(\delta s) \star \delta = \delta \star (T(s \otimes s)\delta_\chi)$. This proves that $s: A_\chi \longrightarrow A$ is a coalgebra anti-homomorphism.

Now it remains to prove that s is invertible, and the assertion (iii).

By Theorem 2.3, in order to prove that $_{\chi^T}A = (A, m, e, {}_{\chi^T}\delta, \varepsilon)$ is a $(K, c, I, -\chi^T)$-Hopf algebra, it suffices to prove that it is a $(K, c, I, -\chi^T)$-bialgebra.

By Lemma 1.4 we know that $_{\chi^T}A = (A, {}_{\chi^T}\delta, \varepsilon)$ is an $\mathbb{N}_0 I$-graded coalgebra. We only need to prove that $_{\chi^T}\delta : A \longrightarrow (A \otimes A)_{-\chi^T}$ is an algebra homomorphism, i.e., $_{\chi^T}\delta(ab) = {}_{\chi^T}\delta(a) \ast {}_{\chi^T}\delta(b)$ for homogeneous elements $a, b \in A$, where $\ast$ denotes the multiplication in $(A \otimes A)_{-\chi^T}$.

In fact, since $\delta : A \longrightarrow (A \otimes A)_\chi$ is an algebra homomorphism, it follows that

$$\delta(ab) = \sum c^{\chi(|a_2|,|b_1|)} a_1 b_1 \otimes a_2 b_2,$$

and hence

$$\begin{aligned}&{}_{\chi^T}\delta(ab)\\&= \sum c^{\chi(|a_2|,|b_1|) - \chi^T(|a_1|+|b_1|,|a_2|+|b_2|)} a_2 b_2 \otimes a_1 b_1 \\&= \sum c^{\chi(|a_2|,|b_1|) - \chi(|a_2|+|b_2|,|a_1|+|b_1|)} a_2 b_2 \otimes a_1 b_1.\end{aligned} \tag{5}$$

On the other hand,
$$_{\chi^T}\delta(a) = \sum c^{-\chi^T(|a_1|,|a_2|)} a_2 \otimes a_1 = \sum c^{-\chi(|a_2|,|a_1|)} a_2 \otimes a_1,$$
$$_{\chi^T}\delta(b) = \sum c^{-\chi^T(|b_1|,|b_2|)} b_2 \otimes b_1 = \sum c^{-\chi(|b_2|,|b_1|)} b_2 \otimes b_1,$$

and hence in $(A \otimes A)_{-\chi^T}$ we have

$$\begin{aligned}
& _{\chi^T}\delta(a) * {}_{\chi^T}\delta(b) \\
& = \sum c^{-\chi(|a_2|,|a_1|) - \chi(|b_2|,|b_1|) - \chi^T(|a_1|,|b_2|)} a_2 b_2 \otimes a_1 b_1 \\
& = \sum c^{-\chi(|a_2|,|a_1|) - \chi(|b_2|,|b_1|) - \chi(|b_2|,|a_1|)} a_2 b_2 \otimes a_1 b_1.
\end{aligned} \tag{6}$$

By comparing (5) and (6) we see that $_{\chi^T}\delta(ab) = {}_{\chi^T}\delta(a) * {}_{\chi^T}\delta(b)$.

Finally, let s' be the antipode of $(K, c, I, -\chi^T)$-Hopf algebra $_{\chi^T}A = (A, m, e, {}_{\chi^T}\delta, \varepsilon)$. We prove that $s's(a) = ss'(a) = a$ for any $a \in A_d$ by using induction on d. This is clear for $a \in A_0$. By Lemma 1.2 we have for $d \neq 0$

$$\delta(a) = a \otimes 1 + 1 \otimes a + \sum_{x+y=d; x,y \neq 0} a_x \otimes b_y$$

with $a_x \in A_x, b_y \in A_y$. Then by (3) in 1.4 we have

$$_{\chi^T}\delta(a) = a \otimes 1 + 1 \otimes a + \sum_{x+y=d; x,y \neq 0} c^{-\chi(y,x)} c_y \otimes c_x.$$

By using $m(id \otimes s)\delta(a) = e\varepsilon(a) = 0$ and $m(id \otimes s') {}_{\chi^T}\delta(a) = e\varepsilon(a) = 0$ we get

$$s(a) = -a - \sum_{x+y=d; x,y \neq 0} a_x s(b_y); \quad s'(a) = -a - \sum_{x+y=d; x,y \neq 0} c^{-\chi(y,x)} b_y s'(a_x).$$

By (i) we have known that $s' : A \longrightarrow A_{-\chi}$ is an algebra anti-homomorphism, it follows from induction that

$$\begin{aligned}
s's(a) & = -s'(a) - \sum_{x+y=d; x,y \neq 0} s'(b_y) * s'(a_x) \\
& = a + \sum_{x+y=d; x,y \neq 0} c^{-\chi(y,x)} b_y s'(a_x) - \sum_{x+y=d; x,y \neq 0} c^{-\chi(y,x)} b_y s'(a_x) \\
& = a.
\end{aligned}$$

Dually, we have $ss'(a) = a$.

This completes the proof. $\square$

REMARK: We do not know what is the corresponding result for a pair $\chi = (\chi_1, \chi_2)$ of $\mathbb{Z}$-valued bilinear forms on $\mathbb{Z}I$, with $\chi_2 \neq 0$.

The following lemma is useful in verifying a given map $s : A \longrightarrow A$ being the antipode of a $(K, c, I, (\chi, 0))$-bialgebra A.

LEMMA 2.5. *Let χ be a $\mathbb{Z}$-valued bilinear form on $\mathbb{Z}I$, $A = (A, m, e, \delta, \varepsilon)$ be a (K, c, I, χ)-bialgebra, and $s : A \longrightarrow A_{\chi^T}$ an $\mathbb{N}_0 I$-graded algebra anti-homomorphism, where $\chi^T(x, y) = \chi(y, x)$ for $x, y \in \mathbb{N}_0 I$. Assume that A is generated as algebra by a subset X consisting of homogeneous elements of A, such that (1) in 2.2 holds for all $a \in X$. Then s is the antipode of χ-Hopf algebra A.*

Proof. By assumption, it suffices to prove that if $m(id \otimes s)\delta(a) = \varepsilon(a)$ and $m(id \otimes s)\delta(b) = \varepsilon(b)$, then $m(id \otimes s)\delta(ab) = \varepsilon(ab)$; and if $m(s \otimes id)\delta(a) = \varepsilon(a)$ and $m(s \otimes id)\delta(b) = \varepsilon(b)$, then $m(s \otimes id)\delta(ab) = \varepsilon(ab)$, where a, b are homogeneous elements in A.

Let $\delta(a) = \sum a_1 \otimes a_2$, $\delta(b) = \sum b_1 \otimes b_2$, with all factors homogeneous. Then

$$\delta(ab) = \sum c^{\chi(|a_2|,|b_1|)} a_1 b_1 \otimes a_2 b_2,$$

and hence

$$\begin{aligned} m(id \otimes s)\delta(ab) &= \sum c^{\chi(|a_2|,|b_1|)} a_1 b_1 s(a_2 b_2) \\ &= \sum c^{\chi(|a_2|,|b_1|)+\chi^T(|b_2|,|a_2|)} a_1 b_1 s(b_2) s(a_2) \\ &= \sum c^{\chi(|a_2|,|b|)} a_1 b_1 s(b_2) s(a_2) \\ &= \sum c^{\chi(|a_2|,|b|)} a_1 s(a_2) \varepsilon(b), \end{aligned}$$

it follows that if $b \notin A_0$, then $\varepsilon(b) = 0$ and $m(id \otimes s)\delta(ab) = 0 = \varepsilon(ab)$; and if $b \in A_0$, then

$$m(id \otimes s)\delta(ab) = \sum a_1 s(a_2)\varepsilon(b) = m(id \otimes s)\delta(a)\varepsilon(b) = \varepsilon(a)\varepsilon(b) = \varepsilon(ab).$$

Also we have

$$\begin{aligned} m(s \otimes id)\delta(ab) &= \sum c^{\chi(|a_2|,|b_1|)} s(a_1 b_1) a_2 b_2 \\ &= \sum c^{\chi(|a_2|,|b_1|)+\chi^T(|b_1|,|a_1|)} s(b_1) s(a_1) a_2 b_2 \\ &= \sum c^{\chi(|a|,|b_1|)} s(b_1) s(a_1) a_2 b_2 \\ &= \sum c^{\chi(|a|,|b_1|)} s(b_1) b_2 \varepsilon(a), \end{aligned}$$

and by the same argument we know that

$$m(s \otimes id)\delta(ab) = \varepsilon(a)\varepsilon(b) = \varepsilon(ab).$$

$\square$

Let $\chi = (\chi_1, \chi_2)$ be a pair of bilinear forms on $\mathbb{Z}I$. We will show the existence of a $(K, c, I, (\chi_1, \chi_2))$-Hopf algebra.

EXAMPLE 2.6. Consider the free K-algebra $'\mathbf{F}$ with 1 with generators θ_i, $i \in I$. For each $x = (x_i)_{i \in I} \in \mathbb{N}_0 I$ with $l = \sum_{i \in I} x_i$, let $'\mathbf{F}_x$ be the K-space with basis all words $\theta_{i_1} \cdots \theta_{i_l}$ such that for any $i \in I$ the number of occurances of i in the sequence $i_1, \cdots, i_l$ is equal to x_i. This is just the grading induced by the weight function $w : k\langle \theta_i, i \in I \rangle \longrightarrow \mathbb{Z}I$ where $w(\theta_i)$ is i-th coordinate vector in $\mathbb{Z}I$. Then $'\mathbf{F} = \bigoplus_{x \in \mathbb{N}_0 I} '\mathbf{F}_x$ is an $\mathbb{N}_0 I$-graded algebra.

Let $\delta : '\mathbf{F} \longrightarrow ('\mathbf{F} \otimes '\mathbf{F})_\chi$ be the unique algebra homomorphism such that $\delta(\theta_i) = \theta_i \otimes 1 + 1 \otimes \theta_i$, $i \in I$, and $\varepsilon : '\mathbf{F} \longrightarrow '\mathbf{F}_0 = K$ be the projection. Then $'\mathbf{F}$ is a (K, c, I, χ)-bialgebra. For a proof see [R3], p.228, in particular, we emphasize that (7) and (8) in 1.5 is needed. It follows that it is a (K, c, I, χ)-Hopf algebra by Theorem 2.3.

If in addition $\chi_2 = 0$, then we can determine the antipode s of $'\mathbf{F}$.

In fact, let $s : {}'\mathbf{F} \longrightarrow {}'\mathbf{F}_{\chi^T}$ be the unique algebra anti-homomorphism such that $s(1) = 1$ and $s(\theta_i) = -\theta_i$ for all $i \in I$, where $\chi^T(x,y) = \chi(y,x)$ for $x,y \in \mathbb{Z}I$. Then s is exactly the antipode of $'\mathbf{F}$ by Lemma 2.5, since we have

$$m(id \otimes s)\delta(\theta_i) = 0 = m(s \otimes id)\delta(\theta_i), \quad i \in I.$$

However, we do not know what is the antipode of $'\mathbf{F}$ for $\chi_2 \neq 0$.

EXAMPLE 2.7. Let $\Delta = (a_{ij})$ be an $n \times n$ generalized Cartan matrix with symmetrization $(d_i)_{1 \leq i \leq n}$, in the sense of [K]. Let U be Drinfeld-Jimbo's quantized enveloping algebra of the Kac-Moody algebra of type Δ. Then we have $U = U_+ \otimes U_0 \otimes U_-$.

Let v be an indeterminate, $(I, \cdot)$ be the Cartan datum determined by Δ. Denote by $'\mathbf{f}$ the free $(\mathbb{Q}(v), v, I, \cdot)$-Hopf algebra defined above, with antipode s, and by J the ideal of $'\mathbf{f}$ generated by the quantum Serre relations:

$$\theta_{ij}^+ = \underset{0}{\leq} t \leq 1 - a_{ij} \to \sum (-1)^t \begin{bmatrix} 1 - a_{ij} \\ t \end{bmatrix}_{d_i} \theta_i^{1-a_{ij}-t} \theta_j \theta_i^t, \quad i \neq j, \qquad (7)$$

where $\begin{bmatrix} 1-a_{ij} \\ t \end{bmatrix}_{d_i}$ is Gaussian binomal coefficients. Then the positive part U_+ and the negative part U_- are isomorphic to $\mathbf{f} = {}'\mathbf{f}/J$ as algebra, see [L].

Since $s : {}'\mathbf{f} \longrightarrow {}'\mathbf{f}_\chi$ with $\chi = \cdot$ is an algebra anti-isomorphism, it is easy to verify that $s(J) \subseteq J$, it follows that $U_+ \simeq \mathbf{f}$ is a $(\mathbb{Q}(v), v, I, \cdot)$-Hopf algebra.

3 SOME CONSTRUCTIONS OF TWISTED HOPF ALGEBRAS

The aim of this section is to construct some new twisted Hopf algebras from a given twisted Hopf algebra.

PROPOSITION 3.1. Let $\psi, \phi, \chi_1, \chi_2 : \mathbb{Z}I \times \mathbb{Z}I \longrightarrow \mathbb{Z}$ be bilinear forms. Let $A = (A, m, e, \delta, \varepsilon)$ be a (K, c, I, χ)– bialgebra, where $\chi = (\chi_1, \chi_2)$. Then $A = (A, m_\psi, e, {}_\phi\delta, \varepsilon)$ is a $(K, c, I, \tilde{\chi})$-Hopf algebra, where $\tilde{\chi} = (\tilde{\chi}_1, \tilde{\chi}_2)$ with

$$\tilde{\chi}_1 = \psi - \phi + \chi_2, \quad \tilde{\chi}_2 = \psi - \phi^T + \chi_1.$$

Proof. First, by Lemma 1.1 we know that $A = (A, m_\psi, e)$ is an $\mathbb{N}_0 I$-graded K-algebra. Second, by Lemma 1.4 we know that $(A, {}_\phi\delta, \varepsilon)$ is an $\mathbb{N}_0 I$-graded coalgebra.

Now, we need to prove that ${}_\phi\delta : A_\psi \longrightarrow (A_\psi \otimes A_\psi)_{\tilde{\chi}}$ is an algebra homomorphism, i.e., ${}_\phi\delta(a *_1 b) = {}_\phi\delta(a) *_2 {}_\phi\delta(b)$ for homogeneous elements $a, b \in A$, where $*_1$ denotes the multiplication in A_ψ, and $*_2$ denotes the multiplication in $(A_\psi \otimes A_\psi)_{\tilde{\chi}}$.

In fact, let $\delta(a) = \sum a_1 \otimes a_2$ and $\delta(b) = \sum b_1 \otimes b_2$ with all factors homogeneous. Since $\delta : A \longrightarrow (A \otimes A)_\chi$ is an algebra homomorphism, it follows that

$$\delta(ab) = \sum c^{\chi(|a_1|, |a_2|, |b_1|, |b_2|)} a_1 b_1 \otimes a_2 b_2,$$

and hence

$${}_\phi\delta(a *_1 b) = c^{\psi(|a|, |b|)} {}_\phi\delta(ab) = \sum c^{t(a_1, a_2, b_1, b_2)} a_2 b_2 \otimes a_1 b_1,$$

where

$$t(a_1, a_2, b_1, b_2) = \psi(|a|, |b|) + \chi(|a_1|, |a_2|, |b_1|, |b_2|) - \phi(|a_1| + |b_1|, |a_2| + |b_2|). \quad (1)$$

On the other hand, $_\phi\delta(a) = \sum c^{-\phi(|a_1|,|a_2|)} a_2 \otimes a_1$, $_\phi\delta(b) = \sum c^{-\phi(|b_1|,|b_2|)} b_2 \otimes b_1$, and hence in $(A_\psi \otimes A_\psi)_{\tilde{\chi}}$ we have

$$_\phi\delta(a) *_2 \, _\phi\delta(b) = \sum c^{t'(a_1,a_2,b_1,b_2)} a_2 b_2 \otimes a_1 b_1,$$

where $t'(a_1, a_2, b_1, b_2)$ is

$$-\phi(|a_1|, |a_2|) - \phi(|b_1|, |b_2|) + \psi(|a_2|, |b_2|) + \psi(|a_1|, |b_1|) + \tilde{\chi}(|a_2|, |a_1|, |b_2|, |b_1|). \quad (2)$$

Note that

$$\psi(|a|, |b|) = \psi(|a_1|, |b_1|) + \psi(|a_2|, |b_1|) + \psi(|a_1|, |b_2|) + \psi(|a_2|, |b_2|),$$

$$\chi(|a_1|, |a_2|, |b_1|, |b_2|) = \chi_1(|a_2|, |b_1|) + \chi_2(|a_1|, |b_2|),$$

and

$$\tilde{\chi}(|a_2|, |a_1|, |b_2|, |b_1|) = \tilde{\chi}_1(|a_1|, |b_2|) + \tilde{\chi}_2(|a_2|, |b_1|),$$

and that

$$\tilde{\chi}_1 = \psi - \phi + \chi_2, \quad \tilde{\chi}_2 = \psi - \phi^T + \chi_1,$$

by comparing (1) and (2) we see that $t(a_1, a_2, b_1, b_2) = t'(a_1, a_2, b_1, b_2)$, and hence $_\phi\delta(a *_1 b) = \, _\phi\delta(a) *_2 \, _\phi\delta(b)$.

Thus, $A = (A, m_\psi, e, \, _\chi\delta, \varepsilon)$ is a $(K, c, I, \tilde{\chi})$-bialgebra algebra, and hence by Theorem 2.3 it is a $(K, c, I, \tilde{\chi})$-Hopf algebra. □

REMARK: If $\chi_2 = 0$, $\psi = 0$, $\phi = \chi_1^T$ in Proposition 3.1, then Theorem 2.4(iii) gives the relation between the antipodes of the two twisted Hopf algebras in Proposition 3.1. But we do not know the general situation.

COROLLARY 3.2. *Let* $\chi : \mathbb{Z}I \times \mathbb{Z}I \longrightarrow \mathbb{Z}$ *be a bilinear form,* $A = (A, m, e, \delta, \varepsilon)$ *be a* $(K, c, I, (\chi^T, \chi))$- *bialgebra. Then* $A = (A, m, e, \, _\chi\delta, \varepsilon)$ *is a Hopf algebra.*

Proof. Note that a $(K, c, I, 0)$-Hopf algebra is exactly a K-Hopf algebra. Now, the assertion follows directly from Proposition 3.1. □

Since $T\delta = \, _0\delta$, by Proposition 3.1, we have

COROLLARY 3.3. *Let* $A = (A, m, e, \delta, \varepsilon)$ *be a* $(K, c, I, (\chi_1, \chi_2))$- *bialgebra. Then* $A = (A, m, e, T\delta, \varepsilon)$ *is a* $(K, c, I, (\chi_2, \chi_1))$-*Hopf algebra.*

PROPOSITION 3.4. *Let* $\psi, \phi, \chi_1, \chi_2 : \mathbb{Z}I \times \mathbb{Z}I \longrightarrow \mathbb{Z}$ *be bilinear forms. Let* $A = (A, m, e, \delta, \varepsilon)$ *be a* $(K, c, I, (\chi_1, \chi_2))$- *bialgebra. Then* $A = (A, m_\psi, e, \delta_\phi, \varepsilon)$ *is a* $(K, c, I, \tilde{\chi})$-*Hopf algebra, where* $\tilde{\chi} = (\tilde{\chi}_1, \tilde{\chi}_2)$ *with*

$$\tilde{\chi}_1 = \psi + \phi^T + \chi_1, \quad \tilde{\chi}_2 = \psi + \phi + \chi_2.$$

Twisted Hopf Algebras

Proof. By Corollary 3.3, $A = (A, m, e, T\delta, \varepsilon)$ is a $(K, c, I, (\chi_2, \chi_1))$-Hopf algebra, now applying Proposition 3.1 to it, and note that $_{-\phi^T}(T\delta) = \delta_\phi$, we then get the desired result. $\square$

COROLLARY 3.5. *Let $\chi : \mathbb{Z}I \times \mathbb{Z}I \longrightarrow \mathbb{Z}$ be a bilinear form. Let $A = (A, m, e, \delta, \varepsilon)$ be a $(K, c, I, (\chi, \chi^T))$- bialgebra. Then $A = (A, m, e, \delta_{-\chi}, \varepsilon)$ is a Hopf algebra.*

In general, we do not know a relation between the antipode s of $A = (A, m, e, \delta, \varepsilon)$ and the antipode s' of $A = (A, m_\psi, e, \delta_\phi, \varepsilon)$. However, if $\psi = -\phi$ in Proposition 3.4, then we have the following

PROPOSITION 3.6. *Let $\phi, \chi_1, \chi_2 : \mathbb{Z}I \times \mathbb{Z}I \longrightarrow \mathbb{Z}$ be bilinear forms. Let $A = (A, m, e, \delta, \varepsilon, s)$ be a $(K, c, I, (\chi_1, \chi_2))$- Hopf algebra. Then the antipode of $A = (A, m_{-\phi}, e, \delta_\phi, \varepsilon, s)$ is also s.*

Proof. By Proposition 3.4, $A = (A, m_{-\phi}, e, \delta_\phi, \varepsilon)$ is a $(K, c, I, (\chi_1, \phi-\phi^T+\chi_2))$-Hopf algebra, say, with antipode s'. Let $a \in A_d, 0 \neq d \in \mathbb{N}_0 I$. Then by Lemma 1.2 we have
$$\delta(a) = a \otimes 1 + 1 \otimes a + \sum_{x+y=d; x, y \neq 0} a_x \otimes b_y$$
and
$$\delta_\phi(a) = a \otimes 1 + 1 \otimes a + \sum_{x+y=d; x, y \neq 0} c^{\phi(x,y)} a_x \otimes b_y,$$
where $a_x \in A_x, b_y \in A_y$. By using $m(id \otimes s)\delta(a) = e\varepsilon(a) = 0$ and $m_{-\phi}(id \otimes s') \delta_\phi(a) = e\varepsilon(a) = 0$ we get
$$s(a) = -a - \sum_{x+y=d; x, y \neq 0} a_x s(b_y); \quad s'(a) = -a - \sum_{x+y=d; x, y \neq 0} a_x s'(b_y),$$
and hence $s' = s$ by induction. $\square$

PROPOSITION 3.7. *Let $\psi : \mathbb{Z}I \times \mathbb{Z}I \longrightarrow \mathbb{Z}$ be a bilinear form. Then $A = (A, m, e, \delta, \varepsilon)$ is a $(K, c, I, (0, -\psi))$- Hopf algebra if and only if $A = (A, m_\psi, e, \delta_\psi, \varepsilon)$ is a $(K, \sqrt{c}, I, (\chi, 0))$-Hopf algebra, where χ is the symmetrization of ψ, i.e., $\chi(x, y) = \psi(x, y) + \psi^T(x, y) = \psi(x, y) + \psi(y, x)$.*

Proof. We only need to prove that $\delta : A \longrightarrow (A \otimes A)_{(0, -\psi)}$ is an algebra homomorphism (with respect to c) if and only if $\delta_\psi : A_\psi \longrightarrow (A_\psi \otimes A_\psi)_{(\chi, 0)}$ is an algebra homomorphism (with respect to $\sqrt{c}$).

Let $*_1, *_2$, and $*_3$ denote the multiplications in $(A \otimes A)_{(0,-\psi)}$, in A_ψ, and in $(A_\psi \otimes A_\psi)_{(\chi,0)}$, respectively. For homogeneous elements $a, b \in A$, let $\delta(a) = \sum a_1 \otimes a_2$, $\delta(b) = \sum b_1 \otimes b_2$ with all factors homogeneous. Then $\delta(ab) = \delta(a) *_1 \delta(b)$ (using c) means that
$$\delta(ab) = \sum c^{-\psi(|a_1|,|b_2|)} a_1 b_1 \otimes a_2 b_2. \tag{3}$$
While $\delta_\psi(a *_2 b) = \delta_\psi(a) *_3 \delta_\psi(b)$ (using $\sqrt{c}$) means that
$$\sqrt{c}^{\psi(|a|,|b|)} \delta_\psi(ab) = \sum \sqrt{c}^{t(a_1, a_2, b_1, b_2)} a_1 b_1 \otimes a_2 b_2,$$

where

$$t(a_1, a_2, b_1, b_2) = \psi(|a_1|, |a_2|) + \psi(|b_1|, |b_2|) + \chi(|a_2|, |b_1|) + \psi(|a_1|, |b_1|) + \psi(|a_2|, |b_2|);$$

or simply, means that

$$\delta_\psi(ab) = \sum \sqrt{c}^{\psi(|a_1|,|a_2|)+\psi(|b_1|,|b_2|)+\psi(|b_1|,|a_2|)-\psi(|a_1|,|b_2|)} a_1 b_1 \otimes a_2 b_2. \quad (4)$$

It remains to prove the equivalence of (3) and (4). By definition of δ_ψ (remember that we should use $\sqrt{c}$ to weight $a_1 b_1 \otimes a_2 b_2$), this is clear. $\square$

ACKNOWLEDGEMENT

The authors thank the referee for the helpful comments.

REFERENCES

[G] J. A. Green: Hall algebras, hereditary algebras and quantum groups, Invent. Math. 120(1995), 361-377.

[J] J. C. Jantzen: Lectures on quantum groups, Graduate Studies in Math., vol. 6, Amer. Math. Soc. 1995.

[K] V. G. Kac: Infinite dimensional Lie algebras, Progress in Math. 44, Birkhäuser, Boston·Basel·Berlin 1983.

[L] G. Lusztig: Introduction to quantum groups, Progress in Math. 110, Birkhäuser, Boston·Basel·Berlin 1993.

[LZ] L. B. Li and P. Zhang: Twisted Hopf algebras, Ringel-Hall algebras and Green's categories, preprint.

[R1] C. M. Ringel: Hall algebras revisited, Israel Mathematica Conference Proceedings. Vol.7 (1993), 171-176.

[R2] C. M. Ringel: Hall algebras and quantum groups, Invent. Math. 101(1990), 583-592.

[R3] C. M. Ringel: Green's Theorem on Hall algebras, Canad. Math. Soc. Conf. Proc. Vol.19(1996), 185-245; AMS Providence RI 1996.

[S] M. Sweedler: Hopf Algebra, Benjamin, New York, 1969.